IET CIRCUITS, DEVICES AND SYSTEMS SERIES 23

Understandable Electric Circuits

Other volumes in this series:

Volume 2	**Analogue IC design: the current-mode approach** C. Toumazou, F.J. Lidgey and D.G. Haigh (Editors)
Volume 3	**Analogue-digital ASICs: circuit techniques, design tools and applications** R.S. Soin, F. Maloberti and J. France (Editors)
Volume 4	**Algorithmic and knowledge-based CAD for VLSI** G.E. Taylor and G. Russell (Editors)
Volume 5	**Switched currents: an analogue technique for digital technology** C. Toumazou, J.B.C. Hughes and N.C. Battersby (Editors)
Volume 6	**High-frequency circuit engineering** F. Nibler *et al.*
Volume 8	**Low-power high-frequency microelectronics: a unified approach** G. Machado (Editor)
Volume 9	**VLSI testing: digital and mixed analogue/digital techniques** S.L. Hurst
Volume 10	**Distributed feedback semiconductor** lasers J.E. Carroll, J.E.A. Whiteaway and R.G.S. Plumb
Volume 11	**Selected topics in advanced solid state and fibre optic sensors** S.M. Vaezi-Nejad (Editor)
Volume 12	**Strained silicon heterostructures: materials and devices** C.K. Maiti, N.B. Chakrabarti and S.K. Ray
Volume 13	**RFIC and MMIC design and technology** I.D. Robertson and S. Lucyzyn (Editors)
Volume 14	**Design of high frequency integrated analogue filters** Y. Sun (Editor)
Volume 15	**Foundations of digital signal processing: theory, algorithms and hardware design** P. Gaydecki
Volume 16	**Wireless communications circuits and systems** Y. Sun (Editor)
Volume 17	**The switching function: analysis of power electronic circuits** C. Marouchos
Volume 18	**System on chip: next generation electronics** B. Al-Hashimi (Editor)
Volume 19	**Test and diagnosis of analogue, mixed-signal and RF integrated circuits: the system on chip approach** Y. Sun (Editor)
Volume 20	**Low power and low voltage circuit design with the FGMOS transistor** E. Rodriguez-Villegas
Volume 21	**Technology computer aided design for Si, SiGe and GaAs integrated circuits** C.K. Maiti and G.A. Armstrong
Volume 22	**Nanotechnologies** M. Wautelet *et al.*

Understandable Electric Circuits

Meizhong Wang

The Institution of Engineering and Technology

Published by The Institution of Engineering and Technology, London, United Kingdom

First published 2005
Reprinted 2009
English translation 2010

The Institution of Engineering and Technology
Michael Faraday House
Six Hills Way, Stevenage
Herts, SG1 2AY, United Kingdom

www.theiet.org

British Library Cataloguing in Publication Data
A catalogue record for this product is available from the British Library

ISBN 978-0-86341-952-2 (paperback)
ISBN 978-1-84919-114-2 (PDF)

Typeset in India by MPS Ltd, A Macmillan Company
Printed in the UK by CPI Antony Rowe, Chippenham

Contents

Preface

The book *Understandable Electric Circuits* is based on my teaching notes for the circuit analysis course that I have taught for many years at Canadian and Chinese institutions. The English version of this book continues in the spirit of its successful Chinese version, which was published by the 'Higher Education Press', the largest and the most prominent publisher of educational books in China, in 2005 and reprinted in 2009.

This unique and well-structured book provides *understandable* and effective introduction to the fundamentals of DC/AC circuits, including current, voltage, power, resistor, capacitor, inductor, impedance, admittance, dependent/independent sources, basic circuit laws/rules (Ohm's law, KVL/KCL, voltage/current divider rules), series/parallel and wye/delta circuits, methods of DC/AC analysis (branch current and mesh/node analysis), the network theorems (superposition, Thevenin's/Norton's theorems, maximum power transfer, Millman's and substitution theorems), transient analysis, *RLC* circuits and resonance, mutual inductance and transformers and more.

Key features

As an aid to readers, the book provides some noteworthy features:

- Clear and easy-to-understand written style, procedures and examples.
- Outlining (boxing) of all important principles, concepts, laws/rules and formulas to emphasize and locate important facts and points.
- Objectives at the beginning of each chapter to highlight to readers the knowledge that is expected to be obtained in the chapter.
- Summary at the end of each chapter to emphasize the key points and formulas in the chapter, which is convenient for students reviewing before exams.
- Laboratory experiments at the end of each chapter are convenient for hands-on practice. They also include how to use basic electrical instruments such as the multimeter and oscilloscope.
- Tables organizing and summarizing variables, values and formulas, which clearly present the important information.

Suitable readers

This book is intended for college and university students, technicians, technologists, engineers or any other professionals who require a solid foundation in the basics of electric circuits.

It targets an audience from all sectors in the fields of electrical, electronic and computer engineering such as electrical, electronics, computers, communications, control and automation, embedded systems, signal processing, power electronics, industrial instrumentation, power systems (including renewable energy), electrical apparatus and machines, nanotechnology, biomedical imaging and more. It is also suitable for non-electrical or electronics readers. It provides readers with the necessary foundation for DC/AC circuits in related fields.

To make this book more reader friendly, the concepts, new terms, laws/rules and theorems are explained in an easy-to-understand style. Clear step-by-step procedures for applying methods of DC/AC analysis and network theorems make this book easy for readers to learn electric circuits themselves.

Acknowledgements

Special thanks to Lisa Reading, the commissioning editor for books at the Institution of Engineering and Technology. I really appreciate her belief in my ability to write this book, and her help and support in publishing it. I also appreciate the support from Bianca Campbell, books and journals sales manager, Suzanne Bishop, marketing manager, Felicity Hull, marketing executive, and Jo Hughes, production controller.

In addition, I would like to express my sincere gratitude to Ramya Srinivasan (project manager of my production process from MPS Ltd) for her highly efficient work and good guidance/suggestions that have helped to refine the writing of this book.

I would also like to express my gratitude to Ying Nan, an electrical and computer engineer, for taking the time to edit some chapters of this book.

In addition, it is my good fortune to have help and support from my family members: My husband, Li Wang (an electronics and physics instructor), for translating several chapters of this book from Chinese to English even though he is very weak after having several operations. My daughter Alice Wang (a busy PhD student), who deserves a special acknowledgement for her patience and dedication to editing some chapters and all the experiments in the book, and also proofreading the entire book. And finally, my son Evan Wang has given a hand in editing several chapters. They deserve sincere acknowledgement for their time and energy. My special thanks to all of them.

Chapter 1

Basic concepts of electric circuits

Objectives

After completing this chapter, you will be able to:

- understand the purpose of studying electric circuits
- know the requirements of a basic electric circuit
- become familiar with circuit symbols
- become familiar with the schematics of electric circuits
- understand the concepts of current and voltage
- understand resistance and its characteristics
- become familiar with the ammeter, voltmeter and ohmmeter
- know the difference between the electron flow and the conventional current flow
- know the concept of reference directions of voltage and current
- know how to apply Ohm's law

1.1 Introduction

1.1.1 Why study electric circuits?

Electrical energy is the great driving force and the supporting pillar for modern industry and civilization. Our everyday life would be unthinkable without electricity or the use of electronic products.

Any complex electrical and electronic device or control system is founded from the basic theory of electric circuits. Only when you have grasped and understood the basic concepts and principles of electric circuits can you further study electrical, electronic and computer engineering and other related areas.

When you start reading this book, perhaps you have already chosen the electrical or the electronic fields as your professional goal – a wise choice! Electrical, electronic and computer engineering has made and continues to make incredible contributions to most aspects of human society – a truth that cannot be neglected. Moreover, it may have a bigger impact on human civilization in the future. Therefore, experts forecast that demand for professionals in this field will grow continuously. This is good news for people who have chosen these areas of study.

Reading this book or other electric circuit book is a first step into the electrical, electronic and computer world that will introduce you to the foundation of the professions in these areas.

1.1.2 Careers in electrical, electronic and computer engineering

Nowadays, electrical, electronic and computer technology is developing so rapidly that many career options exist for those who have chosen this field. As long as you have gained a solid foundation in electric circuits and electronics, the training that most employers provide in their branches will lead you into a brand new professional career very quickly.

There are many types of jobs for electrical and electronic engineering technology. Only a partial list is as follows:

- Electrical engineer
- Electronics engineer
- Electrical design engineer
- Control and automation engineer
- Process and system engineer
- Instrument engineer
- Robotics engineer
- Product engineer
- Field engineer
- Reliability engineer
- Integrated circuits (IC) design engineer
- Computer engineer
- Power electronics engineer
- Electrical and electronics engineering professor/lecturer
- Designer and technologist
- Biomedical engineering technologist
- Electrical and electronics technician
- Hydro technician
- Electrician
- Equipment maintenance technician
- Electronic test technician
- Calibration/lab technician
- Technical writer for electronic products
- Electronic repair

Electrical and electronic technicians, technologists, engineers and experts will be in demand in the future, so you definitely do not want to miss this good opportunity.

1.1.3 Milestones of electric circuit theory

Many early scientists have made great contributions in developing the theorems of electrical circuits. The laws and physical quantities that they discovered are named after them, and all are important milestones in the field of electric engineering. We list here only the ones that are described in this book.

- *Coulomb* is the unit of electric charge; it was named in the honour of Charles Augustin de Coulomb (1736–1806), a French physicist. Coulomb developed Coulomb's law, which is the definition of the electrostatic force of attraction and repulsion, and the principle of charge interactions (attraction or repulsion of positive and negative electric charges).
- *Faraday* is the unit of capacitance; it was named in the honour of Michael Faraday (1791–1867), an English physicist and chemist. He discovered that relative motion of the magnetic field and conductor can produce electric current, which we know today as the Faraday's law of electromagnetic induction. Faraday also discovered that the electric current originates from the chemical reaction that occurs between two metallic conductors.
- *Ampere* is the unit of electric current; it was named in the honour of André-Marie Ampère (1775–1836), a French physicist. He was one of the main discoverers of electromagnetism and is best known for defining a method to measure the flow of current.
- *Ohm* is the unit of resistance; it was named in the honour of Georg Simon Ohm (1789–1854), a German physicist. He established the relationship between voltage, current and resistance, and formulated the most famous electric circuit law – Ohm's law.
- *Volt* is the unit of voltage; it was named in the honour of Alessandro Volta (1745–1827), an Italian physicist. He constructed the first electric battery that could produce a reliable, steady current.
- *Watt* is the unit of power; it was named in the honour of James Watt (1736–1819), a Scottish engineer and inventor. He made great improvements in the steam engine and made important contributions in the area of magnetic fields.
- *Lenz's law* was named in the honour of Heinrich Friedrich Emil Lenz (1804–1865), a Baltic German physicist. He discovered that the polarity of the induced current that is produced in the conductor of the magnetic field always resists the change of its induced voltage; this is known as Lenz's law.
- *Maxwell* is the unit of magnetic flux; it was named in the honour of James Clerk Maxwell (1831–1879), a Scottish physicist and mathematician. The German physicist Wilhelm Eduard Weber (1804–1891) shares the honour with Maxwell (1 Wb $= 10^8$ Mx). Maxwell had established the Maxwell's equations that represent perfect ways to state the fundamentals of electricity and magnetism.
- *Hertz* is the unit of frequency; it was named in the honour of Heinrich Rudolf Hertz (1857–1894), a German physicist and mathematician. He

was the first person to broadcast and receive radio waves. Through the low-frequency microwave experiment, Hertz confirmed Maxwell's electromagnetic theory.

- *Henry* is the unit of inductance; it was named in the honour of Joseph Henry (1797–1878), a Scottish-American scientist. He discovered self-induction and mutual inductance.
- *Joule* is the unit of energy; it was named in the honour of James Prescott Joule (1818–1889), an English physicist. He made great contributions in discovering the law of the conservation of energy. This law states that energy may transform from one form into another, but is never lost. Joule's law was named after him and states that heat will be produced in an electrical conductor.

The majority of the laws and units of measurement stated above will be used in the later chapters of this book. Being familiar with them will be beneficial for further study of electric circuits.

1.2 Electric circuits and schematic diagrams

1.2.1 Basic electric circuits

An electric circuit is a closed loop of pathway with electric charges flowing through it. More specifically, an electric circuit can be defined as a sum of all electric components in the closed loop of pathway with flowing electric charges, such as an electric circuit that includes resistors, capacitors, inductors, power sources, switches, wires, etc. (these electric components will be explained later).

Electric circuit

A closed loop of pathway with electric charges or current flowing through it.

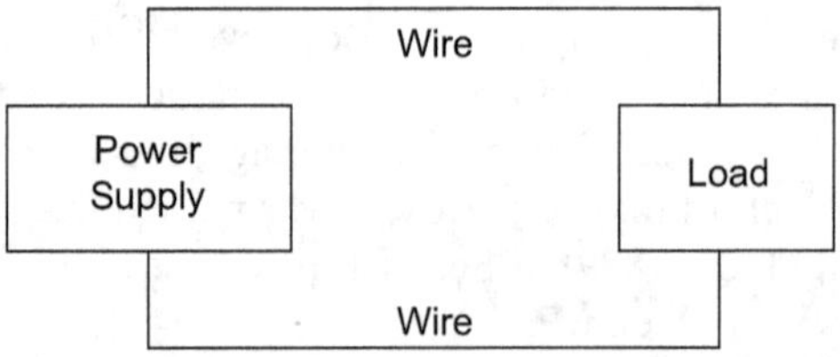

Figure 1.1 Requirements of a basic circuit

A basic electric circuit contains three components: the power supply, the load and the wires (conductors) (Figure 1.1). *Wires* connect the power supply and the load, and carry electric charges through the circuit. A *power supply* is a device that supplies electrical energy to the load of the circuit; it can convert other forms of energy to electrical energy. The electric battery and generator are examples of power supply. For example:

- the battery converts chemical energy into electrical energy.
- the hydroelectric generator converts hydroenergy (the energy of moving water) into electrical energy.
- the thermo generator converts heat energy into electrical energy.
- the nuclear power generator converts nuclear energy into electrical energy.
- the wind generator converts wind energy into electrical energy.
- the solar generator converts solar energy into electrical energy.

Load is a device that is usually connected to the output terminal of an electric circuit. It consumes or absorbs electrical energy from the source. The load may be any device that can receive electrical energy and convert it into other forms of energy. For example:

- electric lamp converts electrical energy into light energy
- electric stove converts electrical energy into heat energy
- electric motor converts electrical energy into mechanical energy
- electric fan converts electrical energy into wind energy
- speaker converts electrical energy into sound energy

Therefore, light bulb, electric stove, electric motor, electric fan and speaker are all electric loads.

Requirements of a basic circuit

- Power supply (power source) is a device that supplies electrical energy to a load; it can convert the other energy forms into electrical energy.
- Load is a device that is connected to the output terminal of an electric circuit, and consumes electrical energy.
- Wires connect the components in a circuit together, and carry electric charges through the circuit.

Figure 1.2 is an example of a simple electric circuit – a flashlight (or electric torch) circuit. In this circuit the battery is the power supply and the small light bulb is the load, and they are connected together by wires.

1.2.2 Circuit schematics (diagrams) and symbols

Studying electric circuits usually requires drawing or recognizing circuit diagrams. Circuit diagrams can make electric circuits easier to understand,

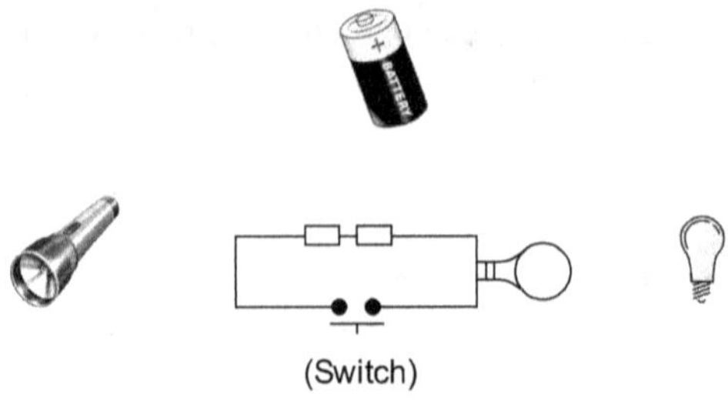

Figure 1.2 The flashlight circuit

analyse and calculate. It is not very difficult to draw a realistic pictorial representation of the flashlight circuit as shown in Figure 1.2, but when studying more theories of electric circuits, circuits can be more and more complex and drawing the pictorial representation of the circuits will not be very realistic.

The more common electric circuits are usually represented by schematics. A *schematic* is a simplified circuit diagram that shows the interconnection of circuit components. It uses standard graphic circuit symbols according to the layout of the actual circuit connection. This is a way to draw circuit diagrams far more quickly and easily.

The *circuit symbols* are the idealization and approximation of the actual circuit components. For example, both the battery and the direct current (DC) generator can convert other energy forms into electrical energy and produce DC voltage. Therefore, they are represented by the same circuit symbol – the DC power supply E.

The electric lamp, electric stove, electric motor and other loads can be represented by a circuit symbol – the resistor R, since all of them have the same characteristic of converting electrical energy into other forms of energies and consuming electrical energy.

The different circuit components are represented by different circuit symbols. Table 1.1 lists some commonly used electric circuit symbols in this book. The most commonly used circuit symbols are the resistor, capacitor, inductor, power supply, ground, switch, etc.

Schematics are represented by circuit symbols according to the layout of the actual circuit connection. The schematic of the flashlight circuit (Figure 1.2) is shown in Figure 1.3.

Further study of this book will help you understand all the circuit elements in Table 1.1.

Table 1.1 The commonly used circuit schematic symbols

Component	**Circuit symbol**
DC power supply	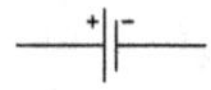 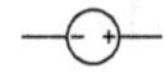
AC power supply	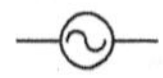
Current source	
Lamp	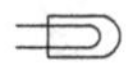
Connected wires	
Unconnected wires	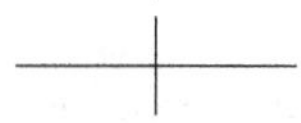
Fixed resistor	
Variable resistor	 or 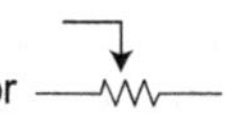
Capacitor	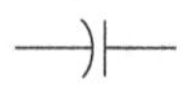
Inductor	
Switch	
Speaker	
Ground	
Fuse	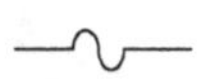
Ohmmeter	
Ammeter	
Voltmeter	
Transformer	

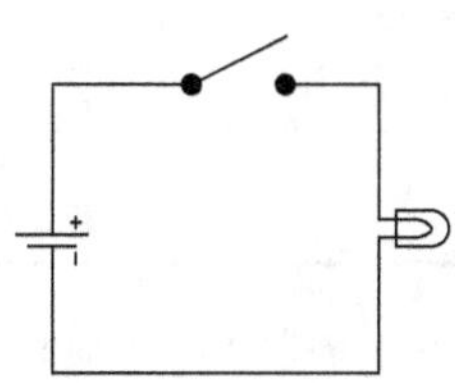

Figure 1.3 Schematic of the flashlight circuit

1.3 Electric current

There are several key circuit quantities in electric circuit theory: electric current, voltage, power, etc. These circuit quantities are very important to study in electric circuits, and they will be used throughout this book. This section will discuss one of them – the electric current.

1.3.1 Current

Although we cannot see electric charges or electric current in the electric circuits, they are analogous to the flow of water in a water hose or pipe. Water current is a flow of water through a water circuit (faucet, pipe or hose, etc.); electric current is a flow of electric charges through an electric circuit (wires, power supply and load).

Water is measured in litres or gallons, so you can measure the amount of water that flows out of the tap at certain time intervals, i.e. litres or gallons per minute or hour. Electric current is measured by the amount of electric charges that flows past a given point at a certain time interval in an electric circuit. If Q represents the amount of charges that is moving past a point at time t, then the current I is:

$$\text{Current} = \frac{\text{Charge}}{\text{Time}} \quad \text{or} \quad I = \frac{Q}{t}$$

If you have learned calculus, current also can be expressed by the derivative: $i = \mathrm{d}q/\mathrm{d}t$.

Electric current *I*

- Current is the flow of electric charges through an electric circuit.
- Current I is measured by the amount of charges Q that flows past a given point at a certain time t: $I = Q/t$

Quantity	Quantity symbol	Unit	Unit symbol
Charge	Q	Coulombs	C
Time	t	Seconds	s
Current	I	Amperes	A

Note: Italic letters have been used to represent the quantity symbols and non-italic letters to represent unit symbols.

A current of 1 A means that there is 1 C of electric charge passing through a given cross-sectional area of wire in 1 s:

$$1\,\mathrm{A} = \frac{1\,\mathrm{C}}{1\,\mathrm{s}}$$

More precisely, 1 A of current actually means there are about 6.25×10^{18} charges passing through a given cross-sectional area of wire in 1 s, since 1 C is approximately equal to 6.25×10^{18} charges ($1\ \mathrm{C} \approx 6.25 \times 10^{18}$ charges), as shown in Figure 1.4.

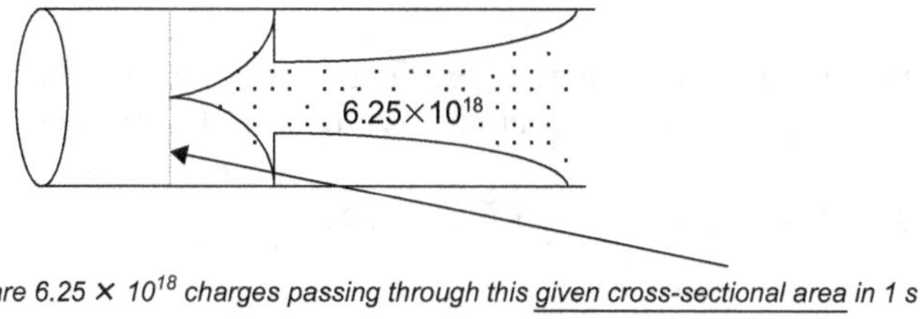

Figure 1.4 1 A of current

Example 1.1: If a charge of 100 C passes through a given cross-sectional area of wire in 50 s, what is the current?

Solution:

Since $Q = 100$ C and $t = 50$ s

$$I = \frac{Q}{t} = \frac{100\,\mathrm{C}}{50\,\mathrm{s}} = 2\,\mathrm{A}$$

1.3.2 Ammeter

Ammeter is an instrument that can be used to measure current, and its symbol is Ⓐ. It must be connected in series with the circuit to measure current, as shown in Figure 1.5.

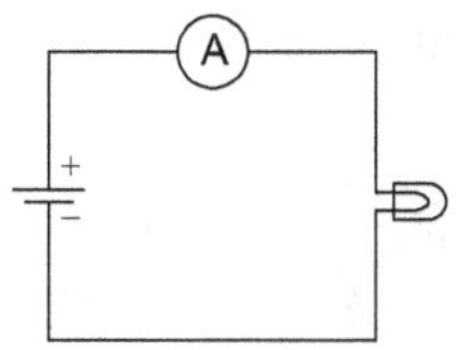

Figure 1.5 Measuring current with an ammeter

1.3.3 The direction of electric current

When early scientists started to work with electricity, the structure of atoms was not very clear, and they assumed at that time the current was a flow of positive charges (protons) from the positive terminal of a power supply (such as a battery) to its negative terminal.

Which way does electric charge really flow? Later on, scientists discovered that electric current is in fact a flow of negative charges (electrons) from the negative terminal of a power supply to its positive terminal. But by the time the real direction of current flow was discovered, a flow of positive charges (protons) from the positive terminal of a power supply to its negative terminal had already been well established and used commonly in electrical circuitry.

Currently, there are two methods to express the direction of electric current. One is known as the *conventional current flow* version, in which the current is defined as a flow of positive charges (protons) from the positive terminal of a power supply to its negative terminal. The other is called *electrons flow* version, in which the current is defined as a flow of negative charges (electrons) from the negative terminal of a power supply unit to its positive terminal. These two methods are shown in Figure 1.6.

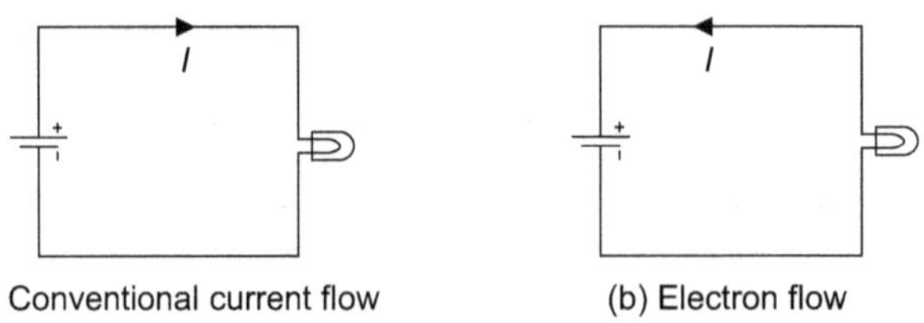

Figure 1.6 The direction of electric current

Because the charge or current cannot be seen in electric circuits, it will make no difference as to which method is used, and it will not affect the analysis, design, calculation, measurement and applications of the electric circuits as long as one method is used consistently. In this book, the conventional current flow version is used.

Conventional and electron current flow version

- Conventional current flow is defined as a flow of *positive* charges (protons) from the positive terminal of a power supply to its *negative* terminal.
- Electron flow version is defined as a flow of *negative* charges (electrons) from the negative terminal of a power supply to its *positive* terminal.

1.4 Electric voltage

1.4.1 Voltage/electromotive force

We have analysed the flow of water in the water circuit to the flow of electric current in the electric circuit. The concept of a water circuit can help develop an understanding of another important circuit quantity – voltage.

The concept of voltage works on the principle of a water gun. The trigger of a water gun is attached to a pump that squirts water out of a tiny hole at the muzzle. If there is no pressure from the gun (the trigger is not pressed), there will be no water out of the muzzle. Low-pressure squirting produces thin streams of water over a short distance, while high pressure produces a very powerful stream over a longer distance.

Just as water pressure is required for a water gun or water circuit, electric pressure or *voltage* is required for an electric circuit. Voltage is responsible for the pushing and pulling of electrons or current through an electric circuit. The higher the voltage, the greater the current will be.

Let us further analyse the voltage by using the previous flashlight or torch circuit in Figure 1.2. If only a small lamp is connected with wires without a battery in this circuit, the flashlight will not work. Since electric charges in the wire (conductor) randomly drift in different directions, a current cannot form in a specific direction. Once the battery is connected to the load (lamp) by wires, the positive electrode of the battery attracts the negative charges (electrons), and the negative electrode of the battery repels the electrons. This causes the electrons to flow in one direction and produce electric current in the circuit.

The battery is one example of a *voltage source* that produces *electromotive force* (EMF) between its two terminals. When EMF is exerted on a circuit, it moves electrons around the circuit or causes current to flow through the circuit since EMF is actually 'the electron-moving force'. It is the electric pressure or force that is supplied by a voltage source, which causes current to flow in a circuit. EMF produced by a voltage source is analogous to water pressure produced by a pump in a water circuit.

Voltage is symbolized by V (italic letter), and its unit is volts (non-italic letter V). EMF is symbolized by E, and its unit is also volts (V).

Electromotive force (EMF)
EMF is an electric pressure or force that is supplied by a voltage source, which causes electric current to flow in a circuit.

1.4.2 Potential difference/voltage

Assuming there are two water tanks A and B, water will flow from tank A to B only when tank A has a higher water level than tank B, as shown in Figure 1.7.

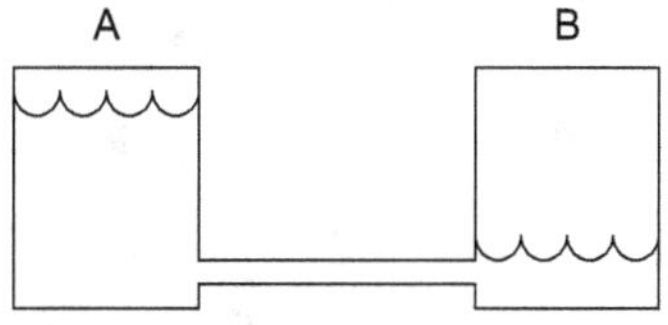

Figure 1.7 Water-level difference

Common sense tells us that 'water flows to the lower end', so water will only flow when there is a water-level difference. It is the water-level difference that produces the potential energy for tank A, and work is done when water flows from tank A to B.

This concept can also be used in the electric circuit. As water will flow between two places in a water circuit only when there is a water-level difference, current will flow between two points in an electric circuit only when there is an electrical potential difference.

For instance, if a light bulb is continuously kept on, i.e. to maintain continuous movement of electrons in the circuit, the two terminals of the lamp need to have an electrical potential difference. This potential difference or voltage is produced by the EMF of the voltage source, and it is the amount of energy or work that would be required to move electrons between two points. Work is represented by W and measured in joules (J). The formula may be expressed as:

$$\text{Voltage} = \frac{\text{Work}}{\text{Charge}} \quad \text{or} \quad V = \frac{W}{Q}$$

If you have learned calculus, voltage can also be expressed by the derivative $v = \mathrm{d}w/\mathrm{d}q$.

Voltage V (or potential difference)

V is the amount of energy or work required to move electrons between two points: $V = W/Q$.

Quantity	Quantity symbol	Unit	Unit symbol
Voltage	V	Volt	V
Work (energy)	W	Joule	J
Charge	Q	Coulomb	C

For example, if 1 J of energy is used to move a 1 C charge from point a to b, it will have a 1 V potential difference or voltage across two points, as shown in Figure 1.8.

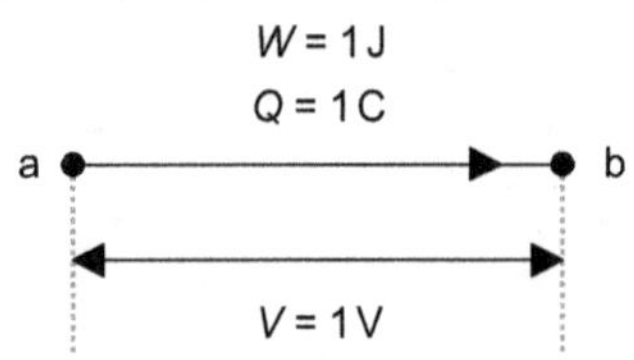

Figure 1.8 Potential difference or voltage

Although voltage and potential difference are not exactly same, the two are used interchangeably. Current will flow between two points in a circuit only when there is a potential difference. The voltage or the potential difference always exists between two points.

There are different names representing voltage or potential difference in electric circuits, such as the source voltage, applied voltage, load voltage, voltage drop, voltage rise, etc. What are the differences between them?

The EMF can be called *source voltage* or *applied voltage* since it is supplied by a voltage source and applied to the load in a circuit. Voltage across the two terminals of the load is called the *load voltage*. Voltage across a component in a circuit is sometimes called *voltage drop* when current flows from a higher potential point to a lower potential point in the circuit, or *voltage rise* when current flows from a lower potential point to a higher potential point in the circuit.

Source voltage or applied voltage (E or V_S):
EMF can be called source voltage or applied voltage (the EMF is supplied by a voltage source and applied to the load in a circuit).
Load voltage (V): Voltage across the two terminals of the load.

Voltage drop:
Voltage across a component when current flows from a higher potential point to a lower potential point.

Voltage rise:
Voltage across a component when current flows from a lower potential point to a higher potential point.

1.4.3 Voltmeter

Voltmeter is an instrument that can be used to measure voltage. Its symbol is (V).

The voltmeter should be connected in parallel with the circuit component to measure voltage, as shown in Figure 1.9.

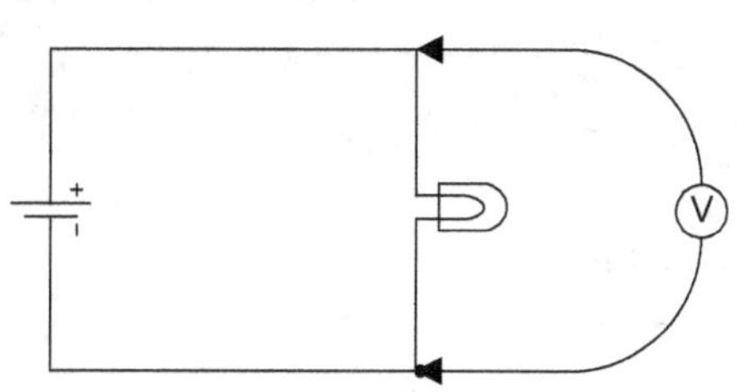

Figure 1.9 Measuring voltage with a voltmeter

1.5 Resistance and Ohm's law

1.5.1 Resistor

Let us use the water current as an example again to explain the resistor. What will happen when we throw some rocks into a small creek? The speed of the water current will slow down in the creek. This is because the rocks (water resistance) 'resist' the flow of water. A similar concept may also be used in an electric circuit. The resistor (current resistance) 'resists' the flow of electrical current. The higher the value of resistance, the smaller the current will be. The resistance of a conductor is a measure of how difficult it is to resist the current flow.

As mentioned in section 1.2, the lamp, electric stove, motor and other such loads may be represented by resistor R because once this kind of load is connected to an electric circuit, it will consume electrical energy, cause resistance and reduce current in the circuit.

Sometimes resistor R will need to be adjusted to a different level for different applications. For example, the intensity of light of an adjustable lamp can be adjusted by using resistors. A resistor can also be used to maintain a safe current level in a circuit. A resistor is a two-terminal component of a circuit that is designed to resist or limit the flow of current. There are a variety of resistors with different resistance values for different applications.

The resistor and resistance of a circuit have different meanings. A *resistor* is a component of a circuit. The *resistance* is a measure of a material's opposition to the flow of current, and its unit is ohms (Ω).

Resistor (R): A two-terminal component of a circuit that limits the flow of current.

Resistance (*R*): The measure of a material's opposition to the flow of current.

Resistors are of many different types, materials, shapes and sizes, but all of them belong to one of the two categories, either fixed or variable. A fixed resistor has a 'fixed' resistance value and cannot be changed. A variable resistor has a resistance value that can be easily changed or adjusted manually or automatically.

Symbols of the resistor

- Fixed resistor 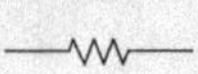
- Variable resistor 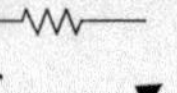 or 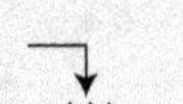

1.5.2 Factors affecting resistance

There is no 'perfect' electrical conductor; every conductor that makes up the wires has some level of resistance no matter what kind of material it is made from. There are four main factors affecting the resistance in a conductor: the cross-sectional area of the wire (A), length of the conductor (ℓ), temperature (T) and resistivity of the material (ρ) (Figure 1.10).

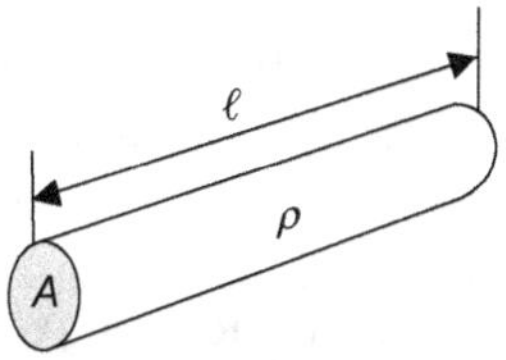

Figure 1.10 Factors affecting resistance

- Cross-sectional area of the wire A: More water will flow through a wider pipe than that through a narrow pipe. Similarly, the larger the diameter of the wire, the greater the cross-sectional area, the less the resistance in the wire and the more the flow of current.
- Length ℓ: The longer the wire, the more the resistance and the more the time taken for the current to flow.
- Resistivity ρ: It is a measure for the opposition to flowing current through a material of wire, or how difficult it is for current to flow through a material. The different materials have different resistivity, i.e. more or less resistance in the materials.
- Temperature T: Resistivity of a material is dependent upon the temperature surrounding the material. Resistivity increases with an increase in temperature for most materials. Table 1.2 lists resistivity of some materials at 20 °C.

Table 1.2 Table of resistivities (ρ)

Material	**Resistivity ρ ($\Omega \cdot m$)**
Copper	1.68×10^{-8}
Gold	2.44×10^{-8}
Aluminium	2.82×10^{-8}
Silver	1.59×10^{-8}
Iron	1.0×10^{-7}
Brass	0.8×10^{-7}
Nichrome	1.1×10^{-6}
Tin	1.09×10^{-7}
Lead	2.2×10^{-7}

Factors affecting resistance can be mathematically expressed with the following formula:

$$R = \rho \frac{\ell}{A}$$

Factors affecting resistance
$R = \rho\left(\frac{\ell}{A}\right)$ where A is the cross-sectional area, ℓ the length, T the temperature and ρ the resistivity (conducting ability of a material for a wire).

Note: ρ is a Greek letter pronounced 'rho' (see Appendix A for a list of Greek letters).

Example 1.3: There is a copper wire 50 m in length with a cross-sectional area of 0.13 cm^2. What is the resistance of the wire?

Solution:

$\ell = 50\ \text{m} = 5\,000\ \text{cm};\ A = 0.13\ \text{cm}^2;$
$\rho = 1.68 \times 10^{-8}\ \Omega \cdot \text{m} = 1.68 \times 10^{-6}\ \Omega \cdot \text{cm}$ (copper)

$$R = \rho \frac{\ell}{A} = \frac{(1.68 \times 10^{-6}\ \Omega \cdot \text{cm})(5\,000\ \text{cm})}{0.13\ \text{cm}^2} \approx 0.0646\ \Omega$$

The resistance of this copper wire is 0.0646 Ω. Although there is resistance in the copper wire, it is very small. A 50-m-long wire only has 0.0646 Ω resistance; thus we can say that copper is a good conducting material. Copper and aluminium are commonly used conducting materials with reasonable price and better conductivity.

1.5.3 Ohmmeter

Ohmmeter is an instrument that can be used to measure resistance. Its symbol is (Ω). The resistor must be removed from the circuit to measure resistance as shown in Figure 1.11.

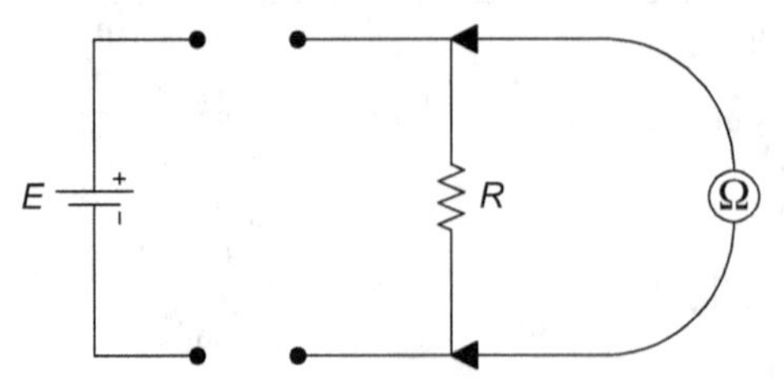

Figure 1.11 Measuring resistance with an ohmmeter

Voltmeter (V): An instrument that is used to measure voltage; it should be connected in parallel with the component.

Ammeter (A): An instrument that is used to measure current; it should be connected in series in the circuit.

Ohmmeter (Ω): An instrument that is used to measure resistance, and the resistor must be removed from the circuit to measure the resistance.

1.5.4 Conductance

Conductance (G) is a term that is opposite of the term resistance. It is the ability of a material to pass current rather than resist it, or how easy rather than how difficult it is for current to flow through a circuit. Conductance is the conductivity of the material; the less the resistance R of the material, the greater the conductance G, the better the conductivity of the material, and vice versa.

The factors that affect resistance are the same for conductance, but in the opposite way. Mathematically, conductance is the reciprocal of resistance, i.e.

$$G = \frac{1}{R} \quad \text{or} \quad G = \frac{A}{\rho \ell} \quad \left(\because R = \rho \frac{\ell}{A}\right)$$

Increasing the cross-sectional area (A) of the wire or reducing the wire length (ℓ) can get better conductivity. This can be seen from the equation of conductance. It is often preferable and more convenient to use conductance in parallel circuits. This will be discussed in later chapters.

Conductance *G*

G is the reciprocal of resistance: $G = 1/R$

The SI unit of conductance is the siemens (S). Some books use a unit mho (℧) for conductance, which was derived from spelling *ohm* backwards and with an upside-down Greek letter omega ℧. Mho actually is the reciprocal of ohm, just as conductance G is the reciprocal of resistance R.

Example 1.4: What is the conductance if the resistance R is 22 Ω?

Solution: $G = 1/R = 1/22\ \Omega \approx 0.0455$ S or 0.0455 ℧

1.5.5 Ohm's law

Ohm's law is a very important and useful equation in electric circuit theory. It precisely expresses the relationship between current, voltage and resistance with a simple mathematical equation. Ohm's law states that current through a conductor in a circuit is directly proportional to the voltage across it and inversely proportional to the resistance in it, i.e.

$$I = \frac{V}{R} \quad \text{or} \quad I = \frac{E}{R}$$

Any form of energy conversion from one type to another can be expressed as the following equation:

$$\text{Effect} = \frac{\text{Cause}}{\text{Opposition}}$$

In an electric circuit, it is the voltage that *causes* current to flow, so current flow is the result or effect of voltage, and resistance is the *opposition* to the current flow. Replacing voltage, current and resistance into the above expression will obtain Ohm's law:

$$\text{Current} = \frac{\text{Voltage}}{\text{Resistance}}$$

Ohm's law

- Ohm's law expresses the relationship between I, V and R.
- I through a conductor is directly proportional to V, and inversely proportional to R: $I = V/R$ or $I = E/R$.

1.5.6 *Memory aid for Ohm's law*

Using mathematics to manipulate Ohm's law, and solving for V and R respectively, we can write Ohm's law in several different forms:

$$V = IR; \quad I = V/R; \quad R = V/I$$

These three equations can be illustrated in Figure 1.12 as a memory aid for Ohm's law. By covering one of the three variables from Ohm's law in the diagram, we can get the right form of Ohm's law to calculate the unknown.

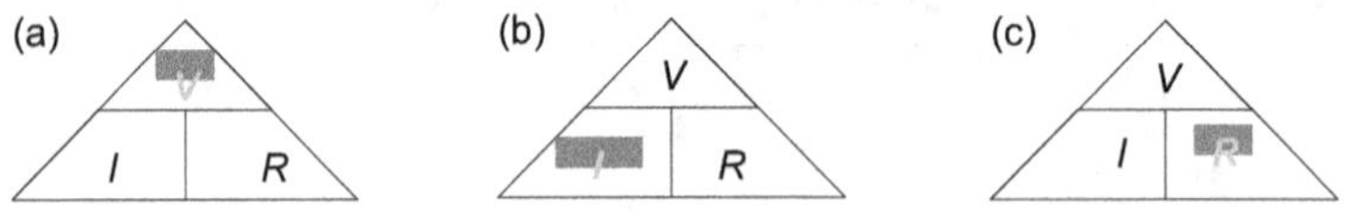

Figure 1.12 Memory aid for Ohm's law. (a) $V = IR$. (b) $I = V/R$. (c) $R = V/I$

1.5.7 *The experimental circuit of Ohm's law*

The experimental circuit with a resistor of 125 Ω in Figure 1.13 may prove Ohm's law. If a voltmeter is connected in the circuit and the source voltage is measured, $E = 2.5$ V. Also, connecting an ammeter and measuring the current in the circuit will result in $I = 0.02$ A. With Ohm's law we can confirm that current in the circuit is indeed 0.02 A:

$$I = E/R = 2.5\ V/125\ \Omega = 0.02\ \mathrm{A}$$

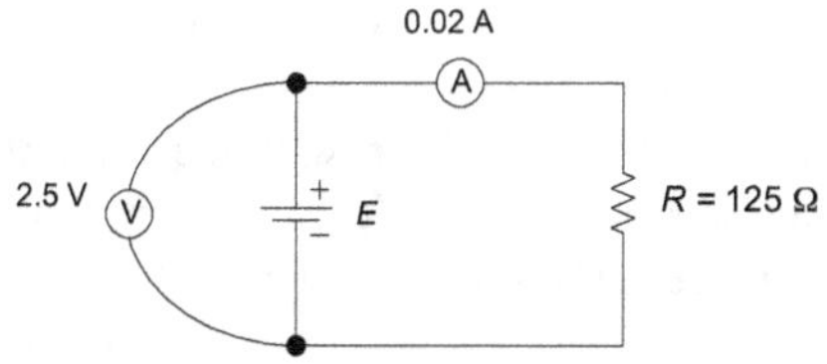

Figure 1.13 The experimental circuit of Ohm's law

1.5.8 *I–V characteristic of Ohm's law*

Using a Cartesian coordinate system, voltage V (x-axis) is plotted against current I (y-axis); this graph of current versus voltage will be a straight line, as shown in Figure 1.14.

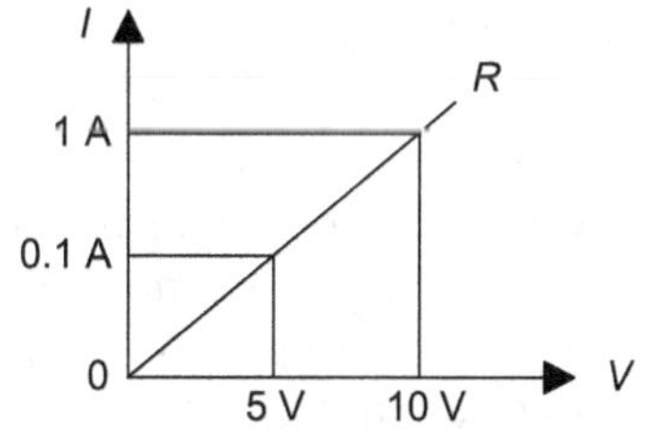

Figure 1.14 I–V characteristics (R = 10 Ω)

When voltage V is 10 V and current is 1 A, $R = V/I = 10$ V/1 A = 10 Ω.
When voltage V is 5 V and current is 0.5 A, $R = V/I = 5$ V/0.1 A = 10 Ω.

So the straight line in Figure 1.14 describes the current–voltage relationship of a 10-Ω resistor. The different lines with different slopes on the I–V characteristic can represent the different values of resistors. For example, a 20-Ω resistor can be illustrated as in Figure 1.15.

Since I–V characteristic shows the relationship between current I and voltage V for a resistor, it is called the I–V characteristic of Ohm's law.

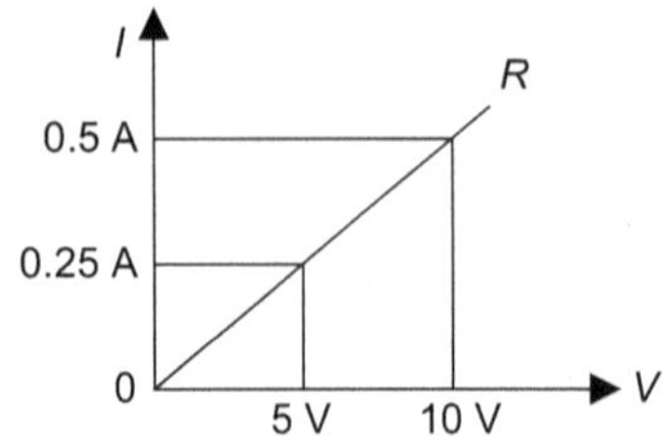

Figure 1.15 I–V characteristics (R = 20 Ω)

The *I–V* characteristic of the straight line illustrates the behaviour of a *linear* resistor, i.e. the resistance does not change with the voltage or current. If the voltage decreases from 10 to 5 V, the resistance still equals 20 Ω as shown in Figure 1.15. When the relationship of voltage and current is not a straight line, the resultant resistor will be a *non-linear* resistor.

1.5.9 Conductance form of Ohm's law

Ohm's law can be written in terms of conductance as follows:

$$I = GV \text{ (since } G = 1/R, \text{ and } I = V/R)$$

1.6 Reference direction of voltage and current

1.6.1 Reference direction of current

When performing circuit analysis and calculations in many situations, the actual current direction through a specific component or branch may change sometimes, and it may be difficult to determine the actual current direction for a component or branch. Therefore, it is convenient to assume an arbitrarily chosen current direction (with an arrow), which is the concept of reference direction of current. If the resultant mathematical calculation for current through that component or branch is positive ($I > 0$), the actual current direction is consistent with the assumed or reference direction. If the resultant mathematical calculation for the current of that component is negative ($I < 0$), the actual current direction is opposite to the assumed or reference direction. As shown in Figure 1.16, the solid line arrows indicate the reference

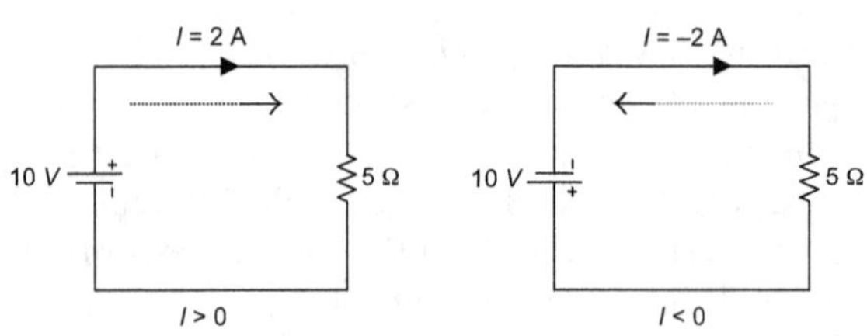

Figure 1.16 Reference direction of current

current directions and the dashed line arrows indicate the actual current directions.

Reference direction of current

Assuming an arbitrarily chosen direction as the reference direction of current I:

- If $I > 0$ the actual current direction is consistent with the reference current direction.
- If $I < 0$ the actual current direction is opposite to the reference current direction.

Figure 1.17 shows two methods to represent the reference direction of current:

- Expressed with an arrow, the direction of the arrow indicates the reference direction of current.
- Expressed with a double subscription, for instance I_{ab}, indicates the reference direction of current is from point a to b.

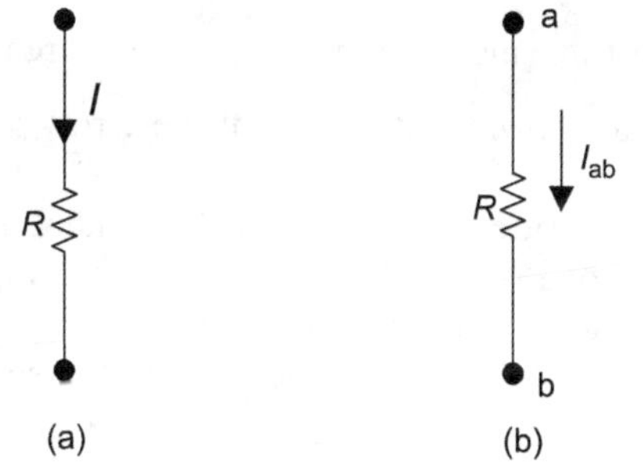

Figure 1.17 Reference direction of current I. (a) Arrow indicates the reference I direction. (b) Double subscription indicates the reference I direction

1.6.2 Reference polarity of voltage

Similar to the current reference direction, the voltage reference polarity is also an assumption of arbitrarily chosen polarity. If the resultant calculation for voltage across a component is positive ($V > 0$), the actual voltage polarity is consistent with the assumed reference polarity. If the resultant calculation is negative ($V < 0$), the actual voltage polarity is opposite to the assumed reference polarity. As shown in Figure 1.18, the positive (+) and negative (−) polarities represent the reference voltage polarities, and arrows represent the actual voltage polarities.

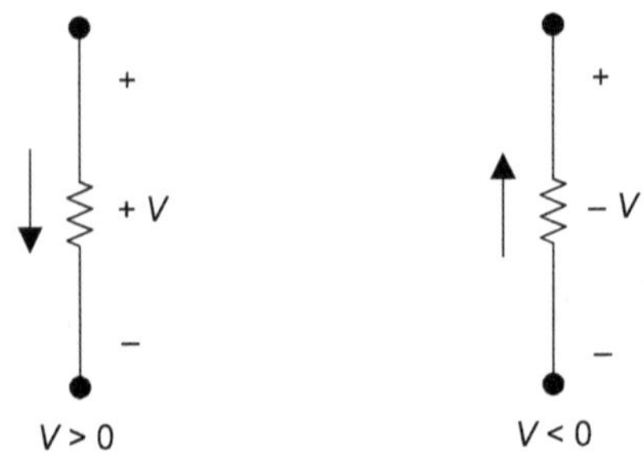

Figure 1.18 Reference polarity of voltage

Reference polarity of voltage

Assuming an arbitrarily chosen voltage polarity as the reference polarity of voltage:

- If $V > 0$ the actual voltage polarity is consistent with the reference voltage polarity.
- If $V < 0$ the actual voltage polarity is opposite to the reference voltage polarity.

Figure 1.19 shows three methods to indicate the reference polarity of voltage:

- Expressed with an arrow, the direction of the arrow points from positive to negative.
- Expressed with polarities, positive sign (+) indicates a higher potential position, and negative sign (−) indicates a lower potential position.
- Expressed with a double subscription, for instance V_{ab}, indicates that the potential position a is higher than the potential position b.

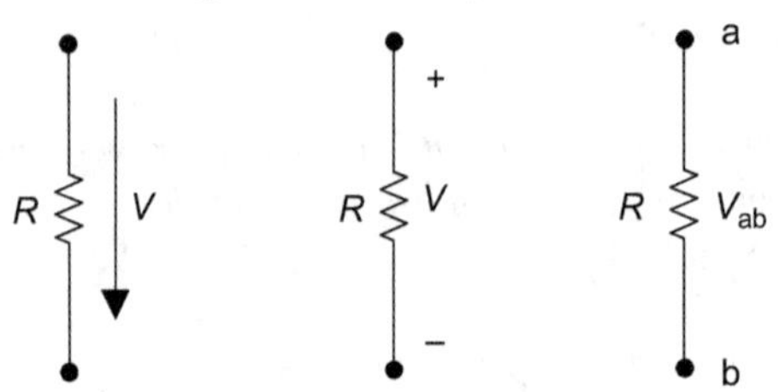

Figure 1.19 Methods indicating the reference polarity of voltage

1.6.3 Mutually related reference polarity of current/voltage

If the reference direction of current is assigned by flow from the positive side to the negative side of voltage across a component (the reference arrow pointing from + to −), then the reference current direction and reference voltage polarity is consistent. In other words, along with the current reference direction is the voltage from positive to negative polarity. This is called the mutually related reference direction or polarity of current/voltage. In this case, if we only know one reference direction or polarity, it is also possible to determine the other, and this is shown in Figure 1.20.

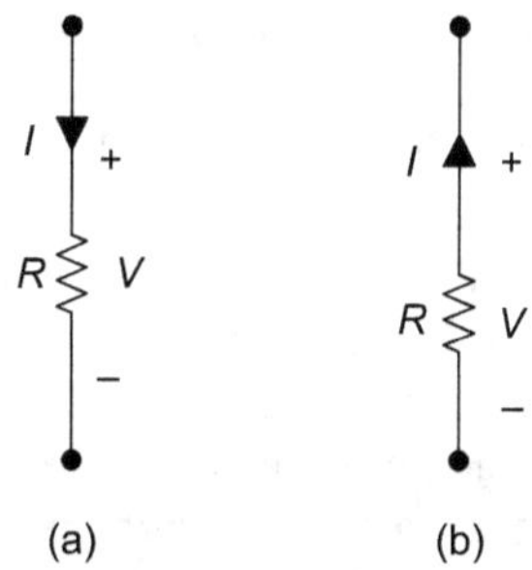

Figure 1.20 (a) Mutually related reference polarity of I and V. (b) Non-mutually related reference polarity of I and V

Mutually related reference polarity of *V* and *I*

If the reference *I* direction is assigned by an arrow pointing from + to − of voltage across a component, then the reference *I* direction and reference *V* polarity is consistent.

Summary

Milestones of the electric circuits

Name of scientist	Nationality	Name of unit/law	Named for
Charles Augustin de Coulomb	French	Coulomb	Unit of charge (C)
Alessandro Volta	Italian	Volt	Unit of voltage (V)
André-Marie Ampère	French	Ampere	Unit of current (A)
Georg Simon Ohm	German	Ohm	Unit of resistance (Ω)
James Watt	Scottish	Watt	Unit of power (W)
Friedrich Emil Lenz	German	Lenz	Lenz's law
James Clerk Maxwell	Scottish	Maxwell	Unit of flux (maxwell), Maxwell's magnetic field equation
Wilhelm Eduard Weber	German	Webber	Unit of flux (weber) 1 Wb = 10^8 Mx
Heinrich Rudolf Hertz	German	Hertz	Unit of frequency (Hz)
Kirchhoff	German	Kirchhoff	Kirchhoff's current and voltage laws
Joseph Henry	Scottish-American	Henry	Unit of inductance (H)
James Prescott Joule	British	Joule	Unit of energy (J)
Michael Faraday	British	Faraday	Unit of capacitance (F)

Basic concepts

- Electric circuit: A closed loop of pathway with electric current flowing through it.
- Requirements of a basic circuit:
    - Power supply (power source): A device that supplies electrical energy to a load.
    - Load: A device that is connected to the output terminal of a circuit, and consumes electrical energy.
    - Wires: Wires connect the power supply unit and load together, and carry current flowing through the circuit.
- Schematic: A simplified circuit diagram that shows the interconnection of circuit components, and is represented by circuit symbols.
- Circuit symbols: The idealization and approximation of the actual circuit components.
- Electric current (I): A flow of electric charges through an electric circuit: $I = Q/t$ (or $I = \mathrm{d}q/\mathrm{d}t$).
- Current direction:
    - Conventional current flow version: A flow of positive charge (proton) from the positive terminal of a power supply to its negative terminal.
    - Electron flow version: A flow of negative charge (electron) from the negative terminal of a power supply unit to its positive terminal.
- Ammeter: An instrument used for measuring current, represented by the symbol Ⓐ. It should be connected in series in the circuit.
- Electromotive force (EMF): An electric pressure or force supplied by a voltage source causing current to flow in a circuit.
- Voltage (V) or potential difference: The amount of energy or work that would be required to move electrons between two points: $V = W/Q$ (or $v = \mathrm{d}w/\mathrm{d}t$).
- Source voltage or applied voltage (E or V_S): EMF can be called source voltage or applied voltage. The EMF is supplied by a voltage source and applied to the load in a circuit.
- Load voltage (V): Voltage across two terminals of the load.
- Voltage drop: Voltage across a component when current flows from a higher potential point to a lower potential point in a circuit.
- Voltage rise: Voltage across a component when current flows from a lower point to a higher point in a circuit.
- Voltmeter: An instrument used for measuring voltage. Its symbol is Ⓥ and it should be connected in parallel with the component.
- Resistor (R): A two-terminal component of a circuit that limits the flow of current.
- Resistance (R): Measure of a material's opposition to the flow of current.
- Factors affecting resistance: $R = \rho(\ell/A)$, where cross-sectional area (A), length (ℓ), temperature (T) and resistivity (ρ).
- Ohmmeter: An instrument used for measuring resistance. Its symbol is (Ω) and the resistor must be removed from the circuit to measure the resistance.

- Conductance (G): It is the reciprocal of resistance: $G = 1/R$.
- Ohm's law: It expresses the relationship between current I, voltage V and resistance R.

$$I = \frac{V}{R} \quad \text{or} \quad I = \frac{E}{R}$$

- Conductance form of Ohm's law: $I = GV$.
- Reference direction of current: Assuming an arbitrarily chosen current direction as the reference direction of current:
    - If $I > 0$ actual current direction is consistent with the reference current direction.
    - If $I < 0$ actual current direction is opposite to the reference current direction.
- Reference polarity of voltage: Assuming an arbitrarily chosen voltage polarity as the reference polarity of voltage:
    - If $V > 0$ actual voltage polarity is consistent with the reference voltage polarity.
    - If $V < 0$ actual voltage polarity is opposite to the reference voltage polarity.
- Mutually related polarity of voltage and current: If the reference current direction is assigned by an arrow pointing from + to − voltage of the component, then the reference current direction and reference voltage polarity is consistent.
- Symbols and units of electrical quantities:

Quantity	**Quantity symbol**	**Unit**	**Unit symbol**
Charge	Q	Coulomb	C
EMF	E	Volt	V
Work (energy)	W	Joule	J
Resistance	R	Ohm	Ω
Resistivity	ρ	Ohm · metres	Ω · m
Conductance	G	Siemens or mho	S or ℧
Current	I	Ampere	A
Voltage	V or E	Volt	V

Experiment 1: Resistor colour code

Objectives

- Become familiar with the breadboard
- Interpret the colour code for resistors
- Measure resistors with a multimeter (ohmmeter function)

Equipment and components

- Breadboard
- Resistors: 12 Ω (2), 100 Ω (2), 2.7 kΩ (2), 3.9 kΩ, 8.2 kΩ, 1.1 MΩ, 15 kΩ, 470 Ω, 18 Ω, 56 kΩ, 4.7 kΩ
- Digital multimeter

Background information

Breadboard guide

- The Universal Solderless Breadboard, or usually known as the breadboard, is one type of circuit board. It offers an easy way to change components or wire connections on the breadboard without soldering.
- The breadboard is a good training tool and is usually used in the lab to perform experiments on electric or electronic circuits, or for professionals to build temporary electrical or electronic circuits to try out ideas for circuit designs.
- Figure L1.1 is a photograph of a small breadboard, and Figure L1.2 is what the underneath of the breadboard looks like.

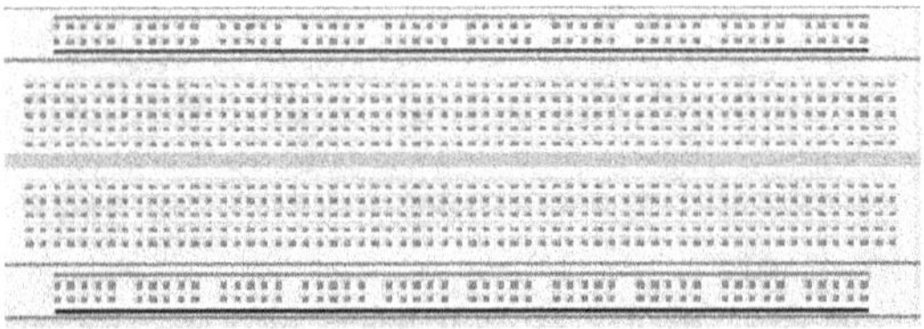

Figure L1.1 A breadboard

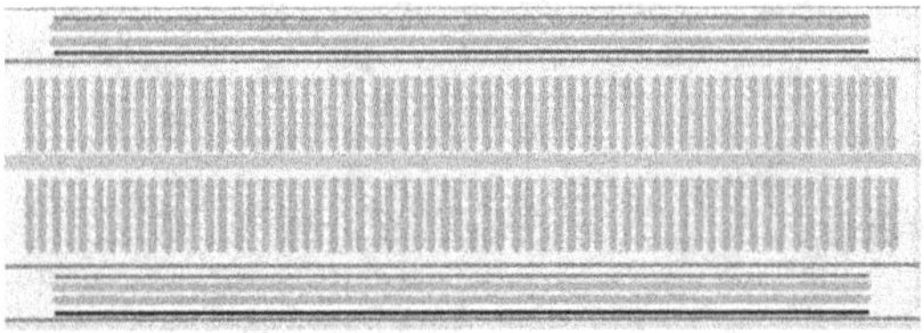

Figure L1.2 The underneath of the breadboard

- The breadboard contains an array of holes where the leads of components and jumper wires can be inserted. The bottom of the board has many strips of metal, which is laid out as shown in Figure L1.2. These strips connect the holes on top of the board. The top and bottom rows will be used to connect the power supply. Figure L1.3(a) is a simple circuit, and Figure L1.3(b) shows how to build this circuit on the breadboard.

Resistor colour code guide (four band)

- Most resistors are very small and it is hard to print the values on them. Usually the small resistors have different colour bands on them, and the

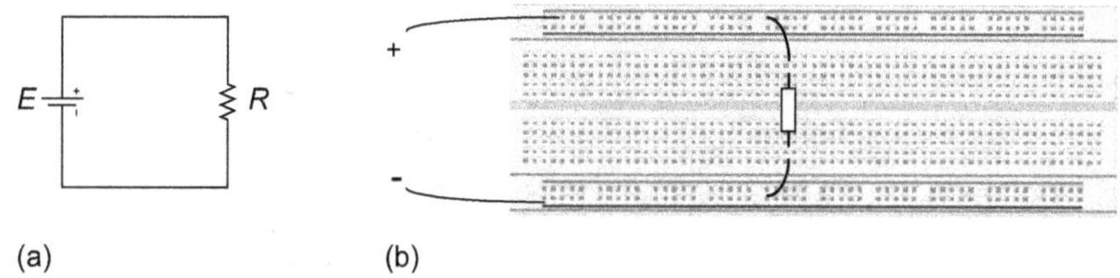

Figure L1.3 Building a simple circuit on the breadboard

standard resistor colour code can be used to interpret the values of different resistors.

- To determine the value of a resistor from the colour band markings, hold the resistor so that the colour bands are closest to the left end as shown in Figure L1.4.

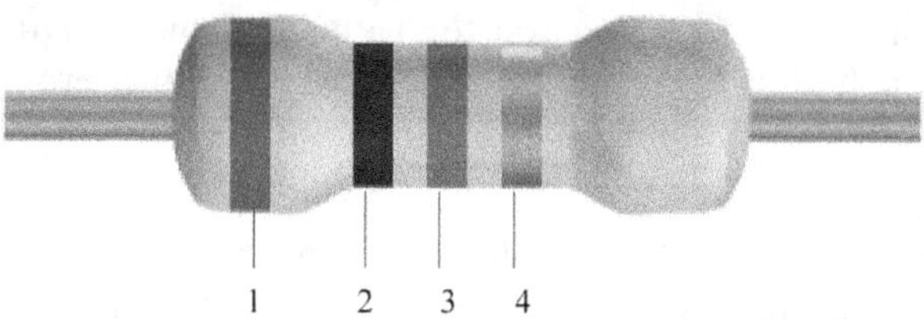

Figure L1.4 Resistor colour bands

- The first two colour bands on the left side of the resistor represent two digits (0–9), the third band represents the number of zeros to add to the integers (multiplier), and the fourth band represents the tolerance of the resistance.
- The resistor colour code is shown in Table L1.1, and the tolerance of the resistance is shown in Table L1.2.

Table L1.1 Resistor colour code

Colour	**Digit**
Black	0
Brown	1
Red	2
Orange	3
Yellow	4
Green	5
Blue	6
Violet	7
Grey	8
White	9

Note: Memory aid: **B**etter **Be** **R**ight **O**r **Y**our **G**reat **B**ig **V**enture **G**oes **W**est
Source: http://www.wikihow.com/Discussion:Remember-Electrical-Resistor-Color-Codes

Table L1.2 Tolerance of resistance

Colour	Tolerance (%)
Gold	±5
Silver	±10

Example: If a resistor has colour bands of brown (1), black (0), red (add two zeros) and silver (tolerance is 10%) from left to right side respectively, it indicates that its resistance lies between 900 and 1100 Ω:

$$R = 1\,000\,\Omega \pm 10\% = 1\,000 \pm 100 = 900\,\text{to}\,1100\,\Omega$$

Example: If one needs to find a resistor with the value of 470 Ω, its colour bands should be yellow (4), violet (7) and brown (add one zero).

Multimeter guide

- A multimeter or VOM (volt–ohm–millimetre) is an electrical and electronic measuring instrument that combines functions of voltmeter, ammeter, ohmmeter, etc. There are two types of multimeters: digital multimeter (DMM) and analog multimeter. DMM is a very commonly used instrument, since it is easier to use and has a higher level of accuracy.
- Method for measuring resistance with a DMM (ohmmeter function):
  - Turn off the power supply if the resistor has been connected in the circuit.
  - Insert the multimeter's leads into the sockets labelled COM and V/Ω as shown in Figure L1.5, and turn on the multimeter.
  - Turn the central selector switch pointing to the ohms range (with Ω sign), and to where the maximum range of the estimated resistor value is closed.
  - Make the measurement by connecting the resistor in parallel with the two leads of the multimeter (connect or touch one lead from the multimeter to one end of the resistor, and connect or touch the other lead of the multimeter to the other end of the resistor).
  - If measuring an unknown resistor, adjust the multimeter range from the maximum to the lower range until suitable resistance is read.
  - Turn off the multimeter.

Note: To get more accurate measurement result, be sure not to grab the resistor's leads with your hands when you are measuring resistors, since you will add your own resistance (in parallel) to the resistor. Better insert the resistors into the holes of the breadboard to measure them.

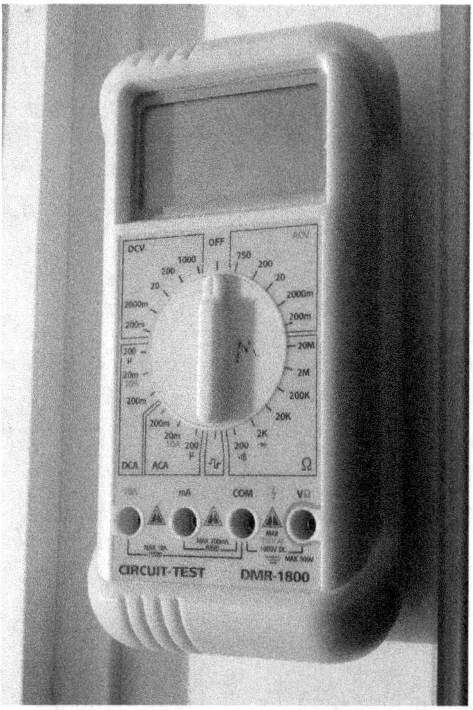

Figure L1.5 Multimeter

Procedure

1. Familiarize with the resistor colour code. Use the resistor colour code chart in Table L1.1 to find the six resistors listed in Table L1.3 from the lab. List the colour band identification for these six resistors, and fill in the 'Colour code' and the 'Resistance range' columns in Table L1.3.

Table L1.3

Resistor	Colour code	Resistance range	Measured value
Example: 470 Ω	Yellow, violet, brown, gold	446.5 Ω – 493.5 Ω	470.5 Ω
12 Ω			
100 Ω			
2.7 kΩ			
3.9 kΩ			
82 kΩ			
1.1 MΩ			

2. Get the multimeter to function as an ohmmeter, and measure the six resistors in Table L1.3 with multimeter and fill in the 'Measured value' column in Table L1.3.
3. Construct the circuits shown in Figure L1.6 on the breadboard with the right resistors. Show your circuits to the instructor to get check-up and signature.

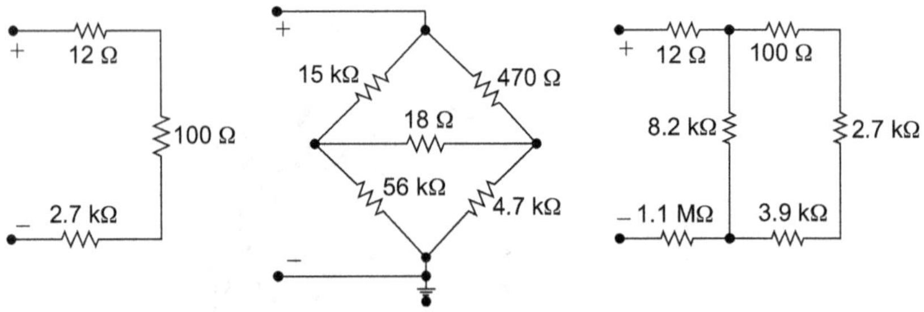

Figure L1.6 Construct circuits on the breadboard

Conclusion

The conclusion may include the following information:

- lab objectivities accomplished
- results, errors and error analysis
- problems encountered during the experiment and their solutions
- knowledge and skills obtained from the lab

Chapter 2

Basic laws of electric circuits

Objectives

After completing this chapter, you will be able to:

- define energy and power
- calculate power
- know the reference directions of power
- analyse and calculate circuits with Kirchhoff's voltage law (KVL)
- analyse and calculate circuits with Kirchhoff's current law (KCL)
- define the branch, node, network and loop
- understand the concepts of the ideal voltage source and the actual voltage source.

2.1 Power and energy

2.1.1 Work

You may have learned in physics that work is the result when a force acts on an object and causes it to move a certain distance. Work (W) is the product of the force (F) and the displacement (S) in the direction of the motion.

Work

$W = F \times S$

where W is, if using for example a force of 1 N to lift an object to 1 m, the 1 J of work done in overcoming the downward force of gravity as shown in Figure 2.1.

Quantity	Quantity Symbol	Unit	Unit symbol
Work	W	Joule	J
Force	F	Newton	N
Displacement	S	Meter	m

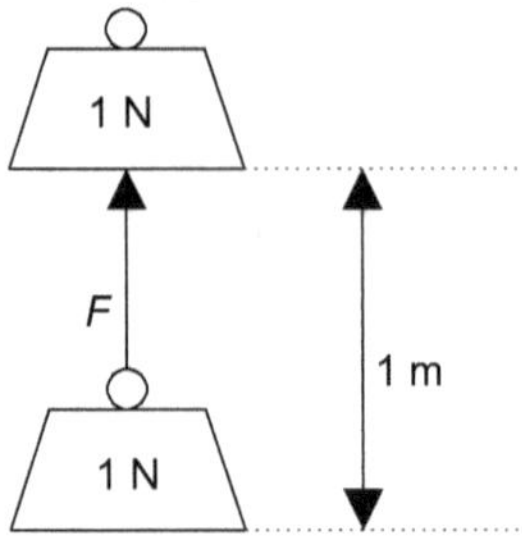

Figure 2.1 Work

Note: When the force (F) and displacement (S) do not point in the same direction, the formula to calculate work will be:

$W = (F \cos \theta)S$

- where the angle θ is the angle between force (F) and displacement (S),
- when θ is 0 degree, $\cos 0° = 1$, $W = (F\cos\theta)S = FS$.

It is the same in an electric circuit. Work is done after the electrons or charges are moved to a certain distance in a circuit as a result of applying an electric field force from the power supply.

2.1.2 Energy

Energy is the ability to do work; it is not work itself, but a transfer of energy. Even though you can't ever really see it, you use energy to do work every day. For example, after you eat and sleep, your body converts the stored energy to keep you doing daily work, such as walking, running, reading, writing, etc.

The law of conservation of energy is one of most important rules in natural science. It states that energy can neither be created nor destroyed, but can only be converted from one form to another. 'Converted' means 'never disappeared' in physics terms. For example:

Electrical generator: mechanical energy → electrical energy.
Lamp: electrical energy → light energy.
Battery: chemical energy → electrical energy.

2.1.3 Power

Power refers to the speed of energy conversion or consumption; it is a measure of how fast energy is transforming or being used. For example in Figure 2.1, 1 N object lifted to 1 m may have different time rates depending on the amount of power applied. If a higher power is applied to the object (an adult is lifting it), it will take a shorter period of time to lift it; and if a lower power is applied to the object (a kid is lifting it), it will take a longer period of time to lift it. So power is defined as the rate of doing work, or the amount of work done per unit of time.

Our daily consumption of electricity is electrical *energy*, and not electrical *power*. The hydro bill that you receive is for electrical power – the amount of electrical energy consumed in 1 or 2 months.

Energy and power

- Energy is the ability to do work.
- Power is the speed of energy conversion, or work done per unit of time: Power = Work/time or $P = W/t$

Quantity	Quantity symbol		Unit	Unit symbol
Electrical Power:				
Work or Energy	W		Joule	J
Time	t		Second	s
Power	P		Watt	W
		Or:	Kilowatt-hour	kWh
			Hour	h
			Watt	W

Electrical power is the speed of electrical energy conversion or consumption in an electric circuit, and it is a measure of how fast electrons or charges are moving in a circuit.

Since current is the amount of charge (Q) that flows past a given point at the certain time: $I = Q/t$ and voltage is the amount of work that is required to move electrons between two points: $V = W/Q$ or $Wc = QV$.

Substituting work W into the power equation gives $P = W/t = QV/t = IV$. It can also be expressed as the form of a derivative:

$$p = (\mathrm{d}w/\mathrm{d}t) = (\mathrm{d}w/\mathrm{d}q)(\mathrm{d}q/\mathrm{d}t) = vi$$

Substituting Ohm's law into the power equation $P = IV$ obtains the other two different power equations:

$$P = VI = (IR)I = I^2R \qquad (\text{Ohm's law}: V = IR)$$

$$P = VI = V\frac{V}{R} = \frac{V^2}{R} \qquad \left(\text{Ohm's law}: I = \frac{V}{R}\right)$$

(Electrical) Power P

$P = IV = I^2R = V^2/R \qquad (\text{or } P = IE = E^2/R)$

Quantity	Quantity symbol	Unit	Unit symbol
Power	P	Watt	W

The above three power equations can be illustrated in Figure 2.2 as the memory aid for power equations. By covering power in any diagram, the correct equation will be obtained to calculate the unknown power.

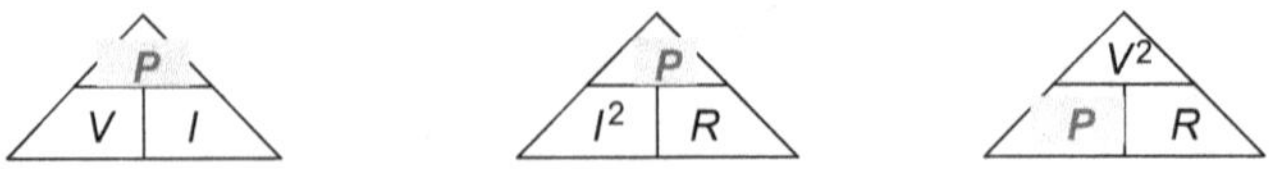

Figure 2.2 Memory aid for power equations

Example 2.1: In a circuit, voltage $V = 10$ V, current $I = 1$ A and resistance $R = 10\ \Omega$, calculate the power in this circuit by using three power equations, respectively.

Solution:

$$P = IV = (1\,\text{A})(10\,\text{V}) = 10\,\text{W}$$

$$P = I^2R = (1\,\text{A})^2(10\,\Omega) = 10\,\text{W}$$

$$P = V^2/R = (10\,\text{V})^2/10\,\Omega = 10\,\text{W}$$

Example 2.1 proved that the three power equations are equivalent since each equation leads to the same value of power at 10 W.

If power is given in a circuit, using mathematical skill to manipulate the power equations and solving for current I and voltage V, respectively, we can express current I and voltage V as follows:

since $P = I^2R$ or $I^2 = P/R$, so $I = \sqrt{P/R}$
since $P = V^2/R$ or $V^2 = PR$, so $V = \sqrt{PR}$

Example 2.2: If power consumed on a 2.5 Ω resistor is 10 W in a circuit, calculate the current flowing through this resistor.

Solution:

$$I = \sqrt{P/R} = \sqrt{10\,\text{W}/2.5\,\Omega} = 2\,\text{A}$$

2.1.4 The reference direction of power

When a component in a circuit has mutually related reference polarity of current and voltage (refer to chapter 1, section 1.6.3), power is positive, i.e. $P > 0$, meaning the component absorption (or consumption) of energy. When a component in a circuit has non-mutually related reference polarity of current and voltage, power is negative, i.e. $P < 0$, meaning the component releasing (or providing) of energy. The concept of the reference direction of power can be illustrated in Figure 2.3.

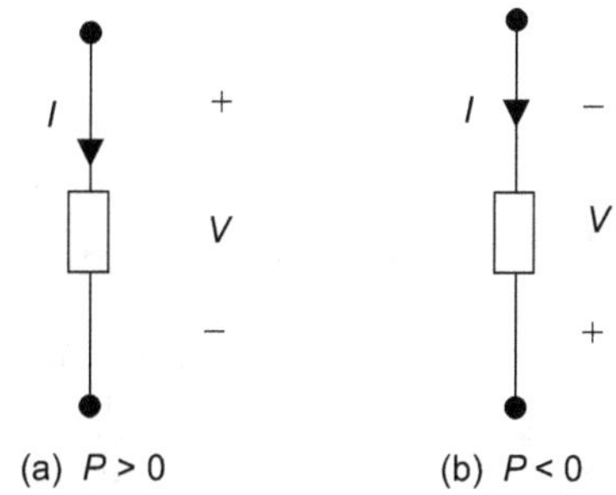

Figure 2.3 The reference direction of power

The reference direction of power

- If a circuit has mutually related reference polarity of current and voltage: $P > 0$ (absorption energy).
- If a circuit has non-mutually related reference polarity of current and voltage: $P < 0$ (releasing energy).

Example 2.3: Determine the reference direction of power in Figure 2.4(a and b).

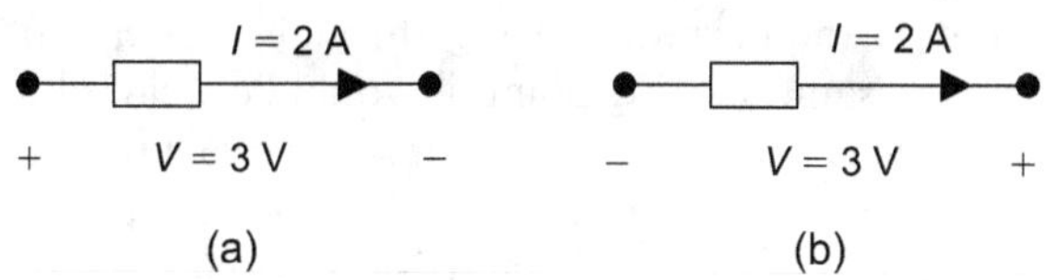

Figure 2.4 Illustrations for Example 2.3

Solution:

(a) $P = IV = (2\ \text{A})(3\ \text{V}) = 6\ \text{W}$ ($P > 0$, the resistor absorbs energy).
(b) $P = I(-V) = (2\ \text{A})(-3\ \text{V}) = -6\ \text{W}$ ($P < 0$, the resistor releases energy).

Example 2.4: $I = 2$ A, $V_1 = 6$ V, $V_2 = 14$ V and $E = 20$ V in a circuit as shown in Figure 2.5. Determine the powers dissipated on the resistors R_1, R_2, and R_1 and R_2 in series in this figure.

Solution:

Power for R_1 (a to b): $P_1 = V_1 I = (6\ \text{V})(2\ \text{A}) = 12\ \text{W}$ (absorption).
Power for R_2 (b to c): $P_2 = V_2 I = (14\ \text{V})(2\ \text{A}) = 28\ \text{W}$ (absorption).
Power for R_1 and R_2 (a to d): $P_3 = (-E)I = (-20\ \text{V})(2\ \text{A}) = -40\ \text{W}$ (releasing).
$P_1 + P_2 + P_3 = 12\ \text{W} + 28\ \text{W} - 40\ \text{W} = 0$ (energy conservation).

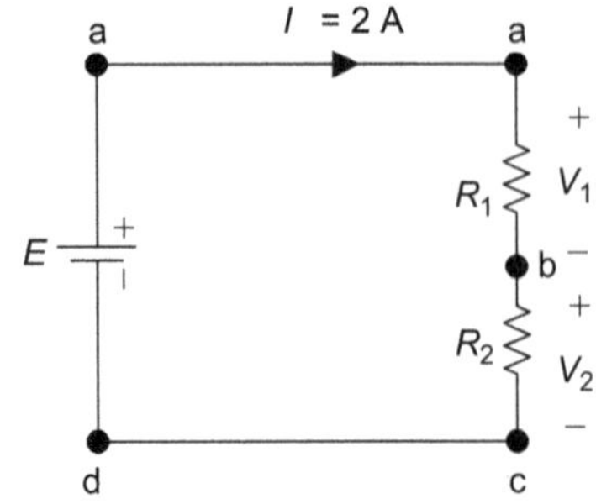

Figure 2.5 Circuit for Example 2.4

2.2 Kirchhoff's voltage law (KVL)

In 1847, a German physicist, physics professor Kirchhoff (Gustav Kirchhoff, 1824–1887) at Berlin University developed the two laws that established the relationship between voltage and current in an electric circuit. Kirchhoff's laws are the most important fundamental circuit laws for analysing and calculating electric circuits after Ohm's law.

2.2.1 Closed-loop circuit

A closed-loop circuit is a conducting path in a circuit that has the same starting and ending points. If the current flowing through a circuit from any point returns current to the same starting point, it would be a closed-loop circuit. As current flows through a closed-loop circuit, it is same as having a round trip, so the starting and ending points are the same, and they have the same potential positions. For example, Figure 2.6 is a closed-loop circuit, since current I starts at point a, passes through points b, c, d and returns to the starting point a.

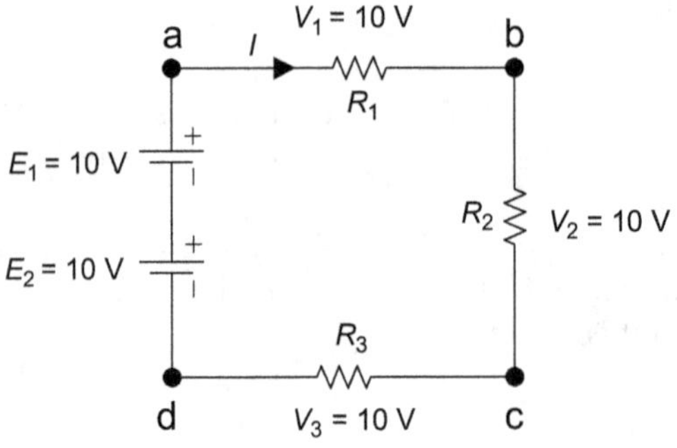

Figure 2.6 A closed-loop circuit

2.2.2 Kirchoff's voltage law #1

KVL #1 states that the algebraic sum of the voltage or potential difference along a closed-loop circuit is always equal to zero at any moment, or the sum of voltages in a closed-loop is always equal to zero, i.e. $\Sigma V = 0$. The voltage in KVL includes voltage rising from the voltage sources (E) and voltage dropping on circuit elements or loads.

The algebraic sum used in KVL #1 means that there are voltage polarities existing in a closed-loop circuit. It requires assigning a loop direction and it could be in either clockwise or counter-clockwise directions (usually clockwise).

- Assign a positive sign (+) for voltage (V or E) in the equation $\Sigma V = 0$, if the voltage reference polarity and the loop direction are the same, i.e. if the voltage reference polarity is from positive to negative and the loop direction is clockwise.
- Assign a negative sign (−) for voltage (V or E) in the equation $\Sigma V = 0$, if the voltage reference polarity and the loop direction are opposite, i.e. if the voltage reference polarity is from negative to positive, and the loop direction is clockwise.

Example 2.5a: Verify KVL #1 for the circuit of Figure 2.7.

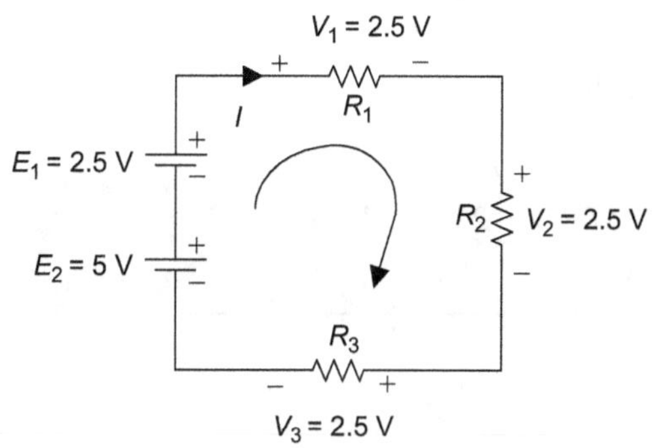

Figure 2.7 Circuit for Example 2.5

Solution:

Applying $\Sigma V = 0$ in Figure 2.7:

$$V_1 + V_2 + V_3 - E_2 - E_1 = 0$$
$$(2.5 + 2.5 + 2.5 - 5 - 2.5)\,\mathrm{V} = 0$$

KVL #1

$\Sigma V = 0$

- Assign a +ve sign for V or E if its reference polarity (+ to −) and loop direction (clockwise) are the same.
- Assign a −ve sign for V or E if its reference polarity (− to +) and loop direction (clockwise) are opposite.

2.2.3 KVL #2

KVL can also be expressed in another way: the sum of the voltage drops (V) around a closed loop must be equal to the sum of the voltage rises or voltage sources in a closed-loop circuit, i.e. $\Sigma V = \Sigma E$.

- Assign a positive sign (+) for V, if its reference and loop directions are the same; assign a negative sign (−) for V, if its reference and the loop directions are opposite.
- Assign a negative sign (−) for the voltage source (E) in the equation, if its reference polarity and the loop direction are the same, i.e. if its polarity is from +ve to –ve and the loop direction is clockwise. Assign a positive sign (+) for voltage source (E) in the equation if its reference polarity and loop direction are opposite, i.e. if its polarity is from negative to positive and the loop direction is clockwise.

Example 2.5b: Verify KVL #2 for the circuit of Figure 2.7.

Solution:

Applying $\Sigma V = \Sigma E$ in Figure 2.7:

$$V_1 + V_2 + V_3 = E_2 + E_1$$
$$(2.5 + 2.5 + 2.5)\text{V} = (2.5 + 5)\text{V}$$
$$7.5\,\text{V} = 7.5\,\text{V}$$

KVL #2

$\Sigma V = \Sigma E$

- Assign a +ve sign for V if its reference polarity and loop direction are the same; assign a −ve sign for V if its reference direction and loop direction are opposite.
- Assign a −ve sign for E if its reference polarity and loop direction are the same; assign a +ve sign for E if its polarity and loop direction are opposite.

2.2.4 Experimental circuit of KVL

KVL can be approved by an experimental circuit in Figure 2.8. If using a multimeter (voltmeter function) to measure voltages on all resistors and power supply in the circuit of Figure 2.8, the total voltage drops on all the resistors should be equal to the voltage for the DC power supply.

KVL #1, $\Sigma V = 0$: $(10 + 10 + 10 - 30)\ \text{V} = 0$
KVL #2, $\Sigma V = \Sigma E$: $(10 + 10 + 10)\ \text{V} = 30\ \text{V}$

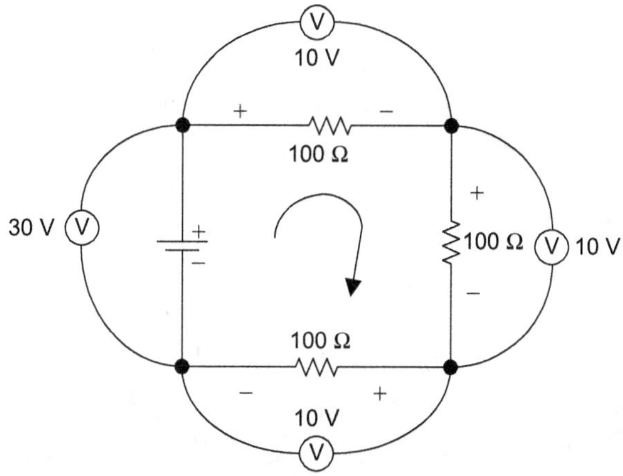

Figure 2.8 Experimental circuit of KVL

Example 2.6: Determine resistance R_3 in the circuit of Figure 2.9.

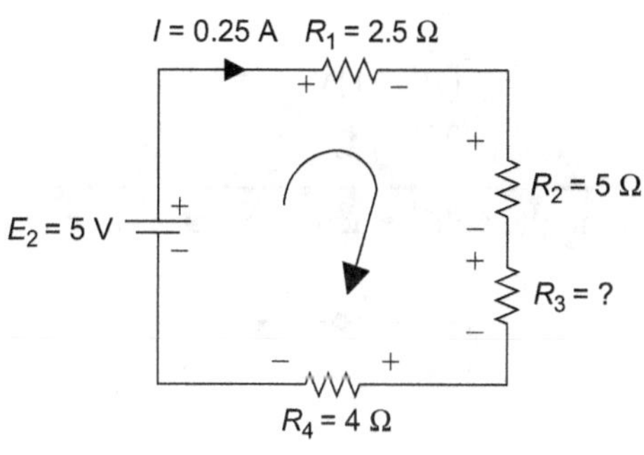

Figure 2.9 Circuit for Example 2.6

Solution:

$$R_3 \frac{V_3}{I}, \quad V_3 = ?$$

Applying KVL #1, $\Sigma V = 0$: $V_1 + V_2 + V_3 + V_4 - E = 0$
Therefore:

$$V_1 = I\,R_1 = (0.25\text{ A})(2.5\ \Omega) = 0.625\text{ V}$$
$$V_2 = I\,R_2 = (0.25\text{ A})(5\ \Omega) = 1.25\text{ V}$$
$$V_4 = I\,R_4 = (0.25\text{ A})(4\ \Omega) = 1\text{ V}$$

Solve for V_3 from $V_1 + V_2 + V_3 + V_4 - E = 0$:

$$V_3 = E - V_1 - V_2 - V_4 = (5 - 0.625 - 1.25 - 1)\text{V} = 2.125\text{ V}$$

Therefore,

$$R_3 = \frac{V_3}{I} = \frac{2.125\text{ V}}{0.25\text{ A}} = 8.5\ \Omega$$

2.2.5 KVL extension

KVL can be expanded from a closed-loop circuit to any scenario loop in a circuit, because voltage or potential difference in the circuit can exist between any two points in a circuit.

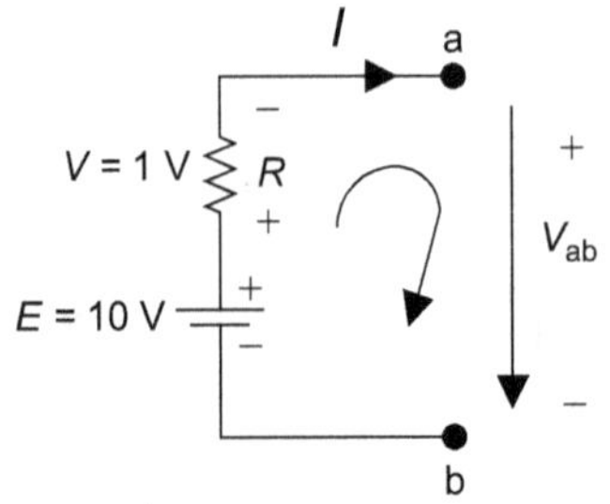

Figure 2.10 KVL extension

V_{ab} in the circuit of Figure 2.10 can be calculated using KVL #2 as follows:

$$\Sigma V = \Sigma E: \quad V + V_{ab} = E$$
$$V_{ab} = E - V$$
$$= (10 - 1)\text{V}$$
$$= 9\text{ V}$$

Example 2.7: Determine the voltage across points a to b (V_{ab}) in the circuit of Figure 2.11.

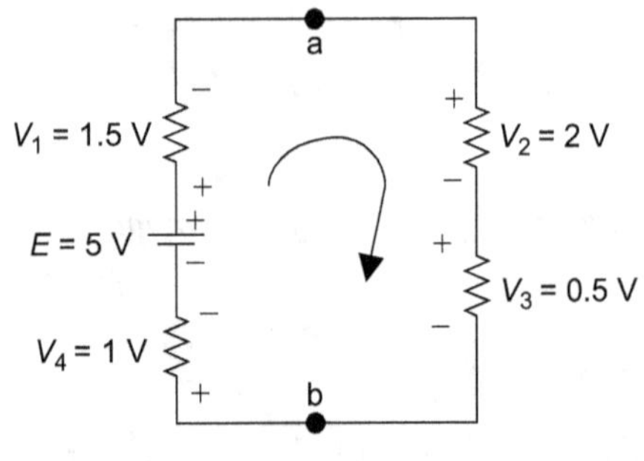

Figure 2.11 Circuit for Example 2.7

Solution:

V_{ab} can be solved by two methods as follows:

Method 1: $\Sigma V = 0$: $V_1 + V_{ab} + V_4 - E = 0$,
where $V_{ab} = E - V_1 - V_4 = (5 - 1.5 - 1)\text{V} = 2.5\text{ V}$.
Method 2: $\Sigma V = 0$: $V_2 + V_3 - V_{ab} = 0$,
where $V_{ab} = V_2 + V_3 = (2 + 0.5)\text{V} = 2.5\text{ V}$.

2.2.6 The physical property of KVL

The results from Example 2.7 show that voltage across two points a and b is the same, and it does not matter which path or branch is used to solve for voltage between these two points, the result should be the same. Therefore, the physical property of KVL is that *voltage does not depend on the path.*

2.3 Kirchhoff's current law (KCL)

2.3.1 KCL #1

KCL #1 states that the algebraic sum of the total currents entering and exiting a node or junction of the circuit is equal to zero, i.e. $\Sigma I = 0$.

- Assign a positive sign (+) to the current in the equation if current is entering the node.
- Assign a negative sign (−) to the current in the equation if current is exiting the node.

A *node* or junction is the intersectional point of two or more current paths where current has several possible paths to flow. A *branch* is a current path between two nodes with one or more circuit components in series. For instance, point A is a node in Figure 2.12, and it has six branches. I_1, I_2 and I_3 are the currents flowing into node A; I_4, I_5 and I_6 are the currents exiting the node A.

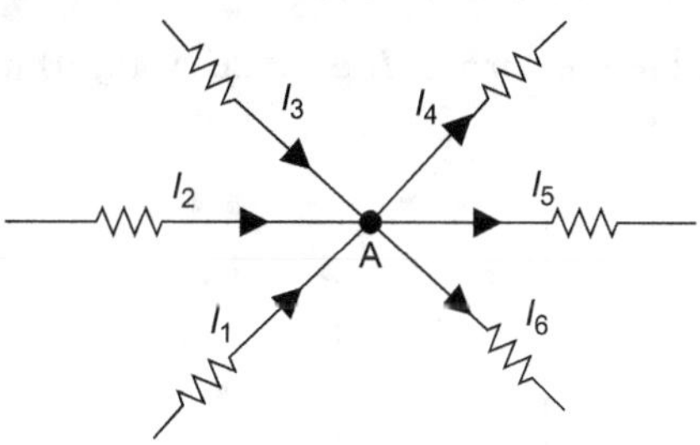

Figure 2.12 Nodes and branches

Applying KCL #1: $I_1 + I_2 + I_3 - I_4 - I_5 - I_6 = 0$

KCL #1

$\Sigma I = 0$

- Assign a +ve sign for current in KCL if I is entering the node.
- Assign a −ve sign for current in KCL if I is exiting the node.

2.3.2 KCL #2

KCL can also be expressed in another way: the total current flowing into a node is equal to the total current flowing out of the node, i.e. $\Sigma I_{in} = \Sigma I_{out}$

- Assign a positive sign (+) to current I_{in} in the equation if current is entering the node; assign a negative sign (−) for I_{in} if current is exiting the node.
- Assign a positive sign (+) to current I_{out} in the equation if current is exiting the node; assign a negative sign (−) for I_{out} if current is entering the node.

Example 2.8: Verify KVL #1 and #2 for the circuit of Figure 2.13.

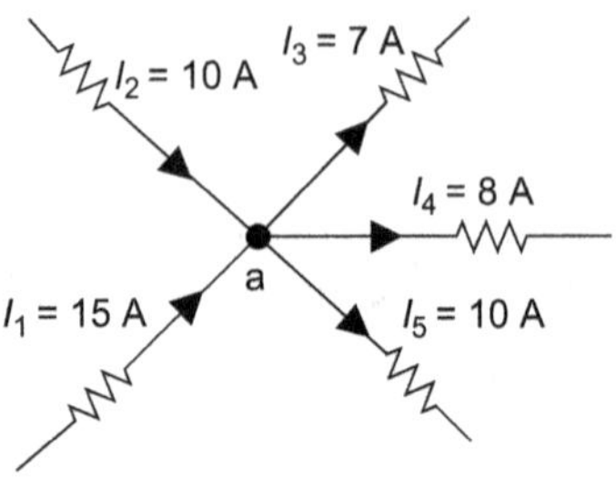

Figure 2.13 Circuit for Example 2.8

Solution:

KCL #2: $\Sigma I_{in} = \Sigma I_{out}$: $I_1 + I_2 = I_3 + I_4 + I_5$
Substituting I with its respective values, we get $(15 + 10)\text{A} = (7 + 8 + 10)\text{A}$
KCL #1: $\Sigma I = 0$: $I_1 + I_2 - I_3 - I_4 - I_5 = 0$
Substituting I with its respective values, we get $(15 + 10 - 7 - 8 - 10)\text{A} = 0$

Example 2.9: Determine the current I_1 at node A and B in Figure 2.14.

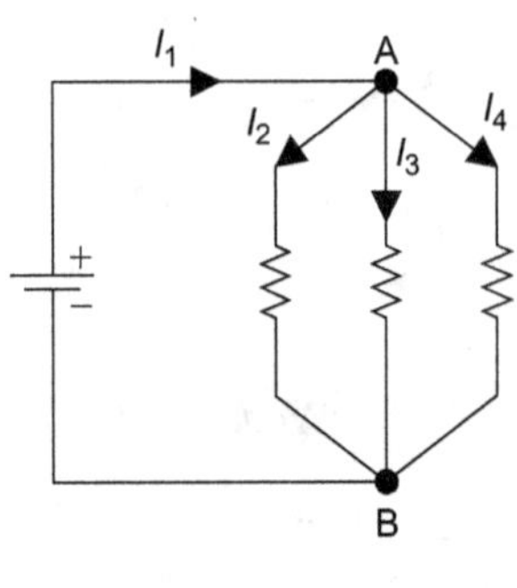

Figure 2.14 Circuit for Example 2.9

Solution:

Node A:

$$\Sigma I = 0: \quad I_1 - I_2 - I_3 - I_4 = 0$$
$$\Sigma I_{in} = \Sigma I_{out}: \quad I_1 = I_2 + I_3 + I_4$$

Node B:

$$\Sigma I = 0: \quad I_2 + I_3 + I_4 - I_1 = 0$$
$$\Sigma I_{in} = \Sigma I_{out}: \quad I_2 + I_3 + I_4 = I_1$$

KCL #2

$\Sigma I_{in} = \Sigma I_{out}$

- Assign a +ve sign for I_{in} if current is entering the node; assign a −ve sign for I_{in}, if current is exiting the node.
- Assign a +ve sign for I_{out} if current is exiting the node; assign a −ve sign for I_{out}, if current is entering the node.

Water flowing in a pipe can be analogized as current flowing in a conducting wire with KCL. Water flowing into a pipe should be equal to the water flowing out of the pipe. For example, in Figure 2.15, water flows in the three upstream creeks A, B and C merging together to a converging point and forms the main water flow out of the converging point to the downstream creek.

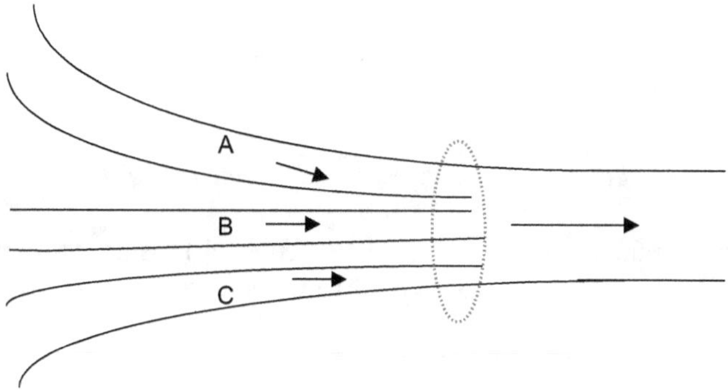

Figure 2.15 Creeks

Example 2.10: Determine current I_3 (you may calculate it by using one KCL, and prove it by using another one).

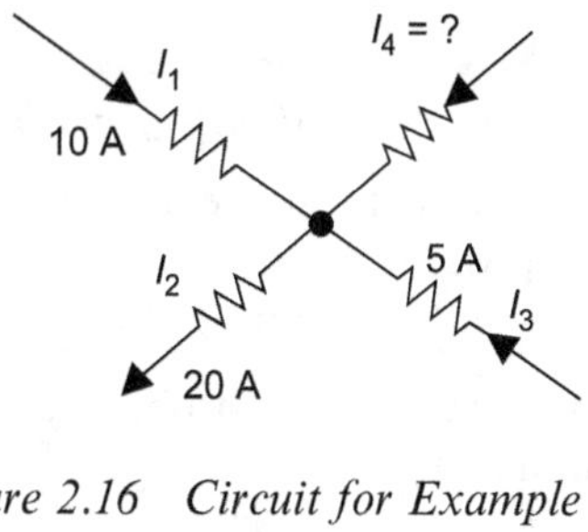

Figure 2.16 Circuit for Example 2.10

Solution:

$$\Sigma I = 0: \quad I_1 - I_2 + I_3 + I_4 = 0,$$
$$I_4 = I_2 - I_1 - I_3 = (20 - 10 - 5)\text{A} = 5\text{ A}$$

$$\Sigma I_{\text{in}} = \Sigma I_{\text{out}}: \quad I_1 + I_3 + I_4 = I_2$$
$$(10 + 5 + 5)\text{A} = 20\text{A}$$
$$20\text{A} = 20\text{A (hence proved)}$$

The KCL can be proved by an experimental circuit in Figure 2.17.

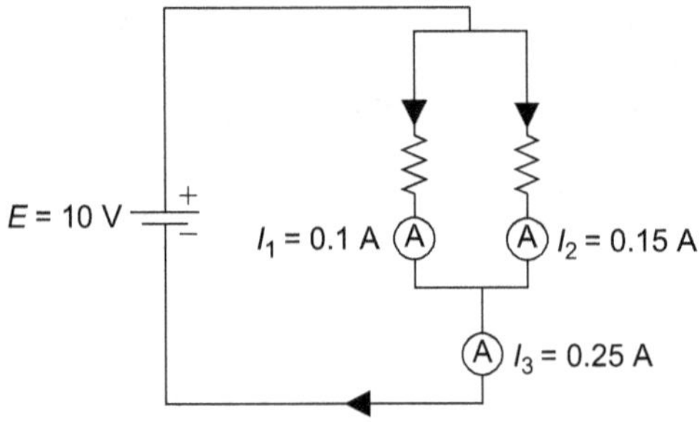

Figure 2.17 Experimental circuit for KCL

Measure branch currents I_1 and I_2 (entering) using two multimeters (ammeter function), and they are equal to the source branch current I_3 (exiting), $I_3 = I_1 + I_2 = 0.25$ A.

2.3.3 Physical property of KCL

The physical property of KCL is that charges cannot accumulate in a node; what arrives at a node is what leaves that node. This results from conservation of charges, i.e. charges can neither be created nor destroyed or the amount of charges that enter the node equals the amount of charges that exit the node. Another property of KCL is the continuity of current (or charges), which is similar to the continuity of flowing water, i.e. the water or current will never discontinue at any moment in a pipe or conductor.

2.3.4 Procedure to solve a complicated problem

It does not matter which field of natural science the problems belong to or how complicated they are, the procedure for analysing and solving them are all similar. The following steps outline the procedure:

1. Start from the unknown value in the problem and find the right equation that can solve this unknown.

2. Determine the new unknown of the equation in step 1 and find the equation to solve this unknown.
3. Repeat steps 1 and 2 until there are no more unknowns in the equation.
4. Substitute the solution from the last step into the previous equation, and solve the unknown. Repeat until the unknown in the original problem is solved.

Now let's try to use this method to solve I_1 in Example 2.11.

Example 2.11: $I_1 = ?$

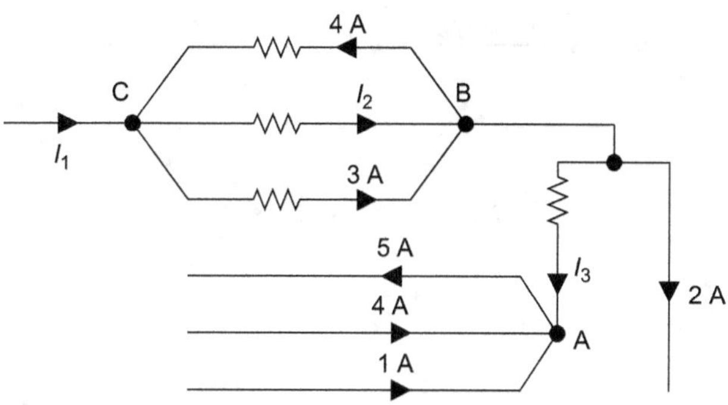

Figure 2.18 Circuit for Example 2.11

Solution:

The unknown in this problem is I_1. Find the right equation to solve I_1.

- At node C:

$$I_1 + 4\,\text{A} = I_2 + 3\,\text{A} \qquad I_2 = ? \tag{2.1}$$

  (Besides I_1, the unknown in this equation is I_2.)
- Find the right equation to solve I_2. At node B:

$$I_2 + 3\,\text{A} = 4\,\text{A} + I_3 + 2\,\text{A} \qquad I_3 = ? \tag{2.2}$$

  (Besides I_2, the unknown in this equation is I_3.)
- Find the right equation to solve I_3. At node A: $I_3 + 4$ A $= (5 + 1)$A, solve for I_3: $I_3 = 2$ A (there are no more unknown elements in this equation except for I_3).
- Substitute $I_3 = 2$ A into (2.2) and solve for I_2: $I_2 + 3$ A $= (4 + 2 + 2)$A, so $I_2 = 5$ A.
- Substitute $I_2 = 5$ A into (2.1) and solve for I_1: $I_1 + 4$ A $= (5 + 3)$A, therefore, $I_1 = 4$ A.

2.3.5 *Supernode*

The concept of the node can be extended to a circuit that contains several nodes and branches, and this circuit can be treated as a supernode. The circuit between nodes a and b in Figure 2.19 within the dashed circle can be treated as an extended node or supernode A; KCL can be applied to it:

$$\Sigma I_{in} = \Sigma I_{out} \quad \text{or} \quad I_1 = I_2$$

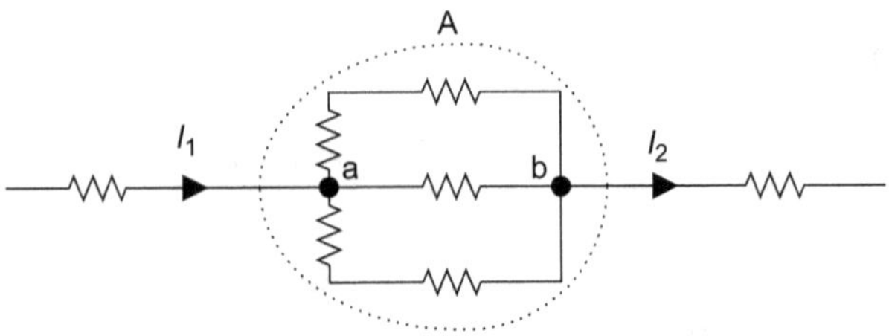

Figure 2.19 Supernode

Example 2.12: Determine the magnitudes and directions of I_3, I_4 and I_7 in the circuit of Figure 2.20.

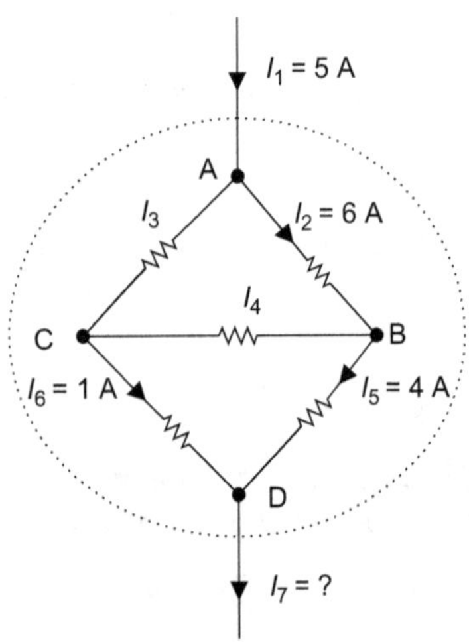

Figure 2.20 Circuit for Example 2.12

Solution: Treat the circuit between the nodes A and D (inside of the circle) as a supernode, and current entering the node A should be equal to current exiting the node D, therefore, $I_7 = I_1 = 5$ A.

- At node A: Since current entering node A is $I_1 = 5$ A, and current leaving node A is $I_2 = 6$ A, so $I_2 > I_1$. I_3 must be current entering node A to satisfy $\Sigma I_{in} = \Sigma I_{out}$, i.e. $I_1 + I_3 = I_2$ or 5 A + I_3 = 6 A, therefore, $I_3 = 1$ A.

- At node B: Since current entering node B is $I_2 = 6$ A and currents exiting node B is $I_5 = 4$ A, so $I_2 > I_5$. I_4 must be current exiting node B to satisfy $\Sigma I_{in} = \Sigma I_{out}$, i.e. $I_2 = I_4 + I_5$ or 6 A $= I_4 + 4$ A, therefore, $I_4 = 2$ A.
- Prove it at node C: $I_4 = I_3 + I_6$, 2 A = 1 A + 1 A, 2 A = 2 A (proved).

2.3.6 Several important circuit terminologies

- Node: The intersectional point of two or more current paths where current has several possible paths to flow.
- Branch: A current path between two nodes where one or more circuit components is in series.
- Loop: A complete current path where current flows back to the start.
- Mesh: A loop in the circuit that does not contain any other loops (non-redundant loop).

Note: A mesh is always a loop, but a loop is not necessary a mesh. A mesh can be analogized as a windowpane, and a loop may include several such windowpanes.

Example 2.13: List nodes, branches, meshes and loops in Figure 2.21.

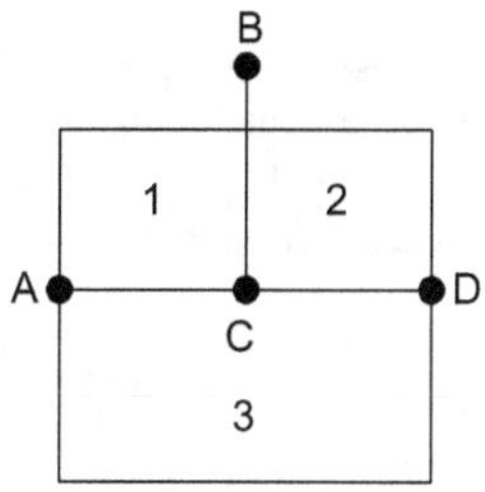

Figure 2.21 Illustration for Example 2.13

Solution:

- Node: four nodes – A, B, C and D
- Branch: six branches – AB, BD, AC, BC, CD and AD
- Mesh: 1, 2 and 3
- Loop: 1, 2, 3, A–B–D–C–A, A–B–D–A, etc.

2.4 Voltage source and current source

A power supply is a circuit device that provides electrical energy to drive the system, and it is a source that can provide EMF (electromotive force) and current to operate the circuit. The power supply can be classified into two categories: voltage source and current source.

2.4.1 Voltage source

2.4.1.1 Ideal voltage source

An ideal voltage source is a two-terminal circuit device that can provide a constant output voltage, V_{ab}, across its terminals, and is shown in Figure 2.22(a). Voltage of the ideal voltage source, V_S, will not change even if an external circuit such as a load, R_L, is connected to it as shown in Figure 2.22(b), so it is an independent voltage source. This means that the voltage of the ideal voltage source is independent of variations in its external circuit or load. The ideal voltage source has a zero internal resistance ($R_S = 0$), and it can provide maximum current to the load.

Current in the ideal voltage source is dependant on variations in its external circuit, so when the load resistance R_L changes, the current in the ideal voltage source also changes since $I = V/R_L$. The characteristic curve of an ideal voltage source is shown in Figure 2.22(c). The terminal voltage V_{ab} for an ideal voltage source is a constant, and same as the source voltage ($V_{ab} = V_S$), no matter what its load resistance R_L is.

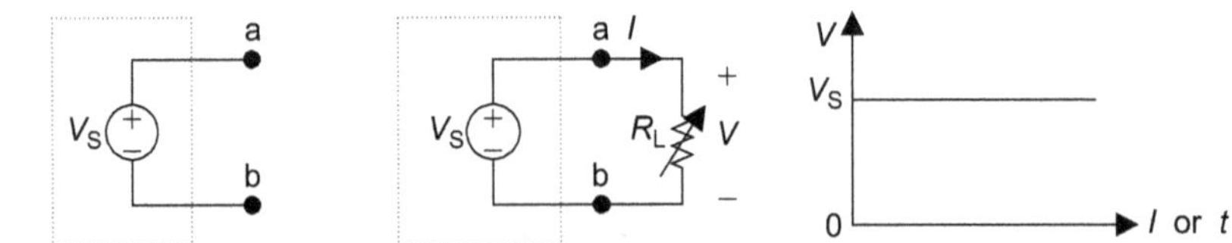

(a) Ideal voltage source (b) Ideal voltage source with a load (c) Characteristic curve

Figure 2.22 Ideal voltage source

Ideal voltage source

- It can provide a constant terminal voltage that is independent of the variations in its external circuit, $V_{ab} = V_S$.
- Its internal resistance, $R_S = 0$. Its current depends on variations in its external circuit.

2.4.1.2 Real voltage source

Usually a real-life application of a voltage source, such as a battery, DC generator or DC power supply, etc., will not reach a *perfect* constant output voltage after it is connected to an external circuit or load, since nothing is perfect. The real voltage sources all have a non-zero internal resistance R_S ($R_S \neq 0$).

The real voltage source (or voltage source) can be represented as an ideal voltage source V_S in series with an internal resistor R_S as shown in Figure 2.23(a). Once a load resistor R_L is connected to the voltage source (Figure 2.23(b)), the terminal voltage of the source V_{ab} will change if the load resistance R_L changes. Since the internal resistance R_S is usually very small, V_{ab} will be a little bit lower than the source voltage V_S ($V_{ab} = V_S - IR_S$).

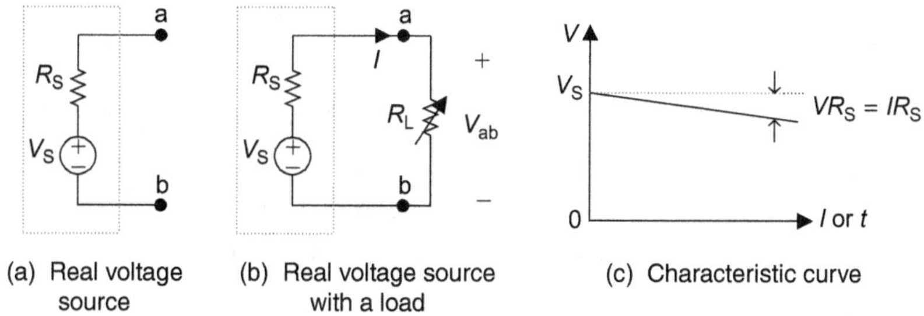

Figure 2.23 Real voltage source

A smaller internal resistance can also provide a higher current through the external circuit of the real voltage source because $I = V_S/(R_S + R_L)$. Once the load resistance R_L changes, current I in this circuit will change, and the terminal voltage V_{ab} also changes. This is why the terminal voltage of the real voltage source is not possible to keep at an ideal constant level ($V_{ab} \neq V_S$).

The internal resistance of a real voltage source usually is much smaller than the load resistance, i.e. $R_S \ll R_L$, so the voltage drop on the internal resistance (IR_S) is also very small, and therefore, the terminal voltage of the real voltage source (V_{ab}) is approximately stable:

$$V_{ab} = V_S - IR_S \approx V_S$$

When a battery is used as a real voltage source, the older battery will have a higher internal resistance R_S and a lower terminal voltage V_{ab}.

Real voltage source (voltage source)

It has a series internal resistance R_S, and $R_S \ll R_L$. The terminal voltage of the real voltage source is: $V_{ab} = V_S - IR_S$.

Example 2.14: Determine the terminal voltages of the circuit in Figure 2.24(a and b).

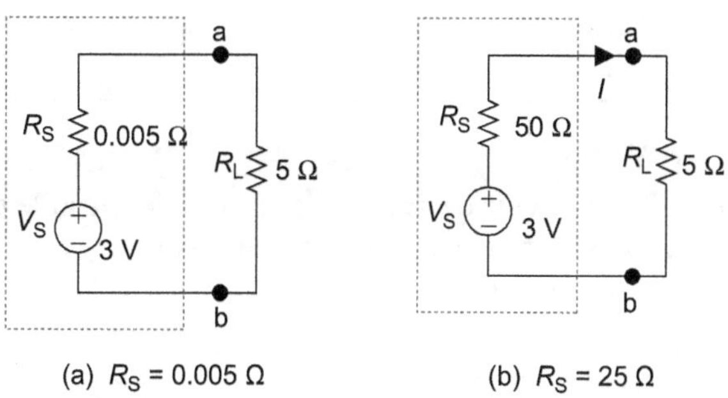

Figure 2.24 Circuits for Example 2.14

$$\text{When } R_S = 0.005\,\Omega, \quad I = \frac{V_S}{(R_S + R_L)} = \frac{3\text{ V}}{(0.005\,\Omega + 5\,\Omega)} \approx 0.5994\text{ A}$$

$$V_{ab} = IR_L = (0.5994\text{ A})(5\,\Omega) = 2.997\text{ V}$$

$$\text{When } R_S = 50\,\Omega, \quad I = \frac{V_S}{(R_S + R_L)} = \frac{3\text{ V}}{(50\,\Omega + 5\,\Omega)} \approx 0.055\text{ A}$$

$$V_{ab} = IR_L = (0.055\text{ A})(5\,\Omega) = 0.275\text{ V}$$

The above example indicates that the internal resistance has a great impact on the terminal voltage and current of the voltage source. Only when the internal resistance is very small, can the terminal voltage of the source be kept approximately stable, such as when $R_S = 0.005\ \Omega$, $V_{ab} = 2.997\text{ V} \approx V_S = 3\text{ V}$.

In this case, the terminal voltage V_{ab} is very close to the source voltage V_S. But when $R_S = 50\ \Omega$, $V_{ab} = 0.275\text{ V} \ll V_S = 3\text{ V}$, i.e. the terminal voltage V_{ab} is much less than the source voltage V_S.

A real voltage source has three possible working conditions:

- When an external load R_L is connected to a voltage source (Figure 2.25(a)): $V_{ab} = V_S - I R_S, \quad I = (V_S/R_S + R_L)$.
- Open circuit: when there is no external load R_L connected to a voltage source (Figure 2.25(b)): $V_{ab} = V_S, \quad I = 0$.
- Short circuit: when a jump wire is connected to the two terminals of a voltage source (Figure 2.25(c)): $V_{ab} = 0, \quad I = (V_S/R_S)$.

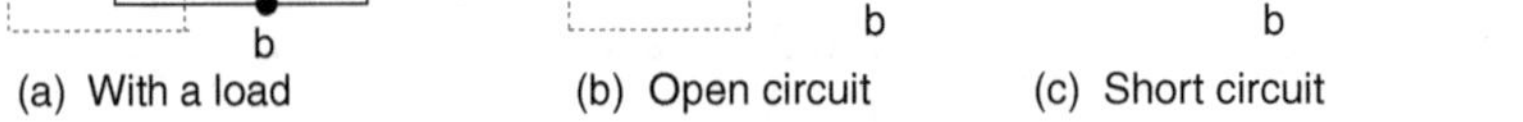

Figure 2.25 Three states of a voltage source

2.4.2 Current source

The current source is a circuit device that can provide a stable current to the external circuit. A transistor, an electronic element you may have heard, can be approximated as an example of a current source.

2.4.2.1 Ideal current source

An ideal current source is a two terminal circuit device that can provide a constant output current I_S through its external circuit. Current of the ideal current source will not change even an external circuit (load R_L) is connected to it, so it is an independent current source. This means the current of the ideal voltage source is independent of variations in its external circuit or load. The ideal current source has an infinite internal resistance ($R_S = \infty$), so it can provide a maximum current to the load. Its two-terminal voltage is determined by the external circuit or load.

The symbol of an ideal current source is shown in Figure 2.26(a), and its characteristic curve is shown in Figure 2.26(b). I_S represents the current for current source, and the direction of the arrow is the current direction of the source.

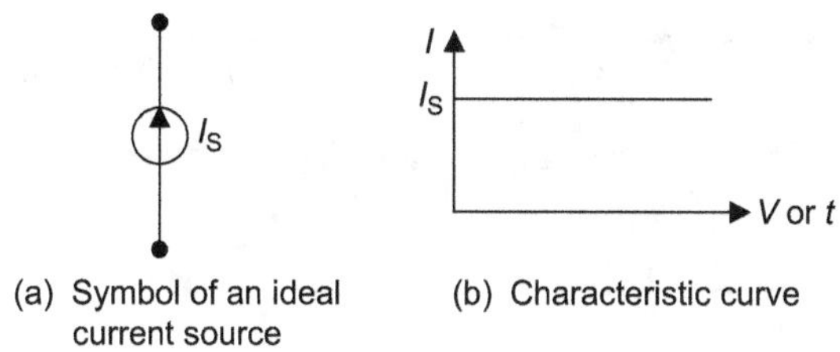

Figure 2.26 Ideal current source

Ideal current source

- It can provide a constant output current I_S that does not depend on the variations in its external circuit.
- Its internal resistance $R_S = \infty$.
- Its voltage depends on variations in its external circuit. $V_{ab} = I_S R_L$.

Example 2.15: The load resistance R_L is 1 000 and 50 Ω, respectively, in Figure 2.27. Determine the terminal voltage V_{ab} for the ideal current source in the circuit.

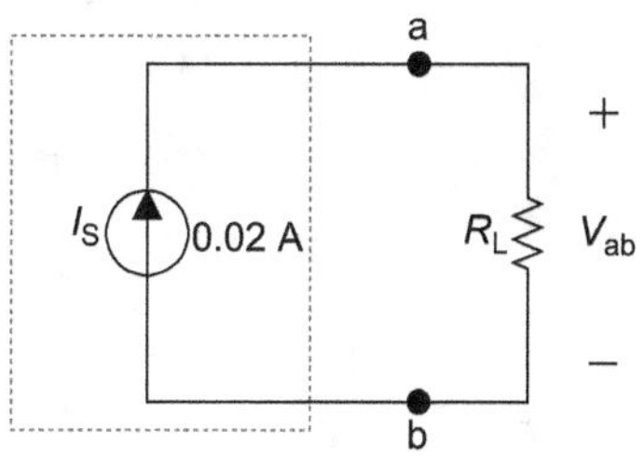

Figure 2.27 Circuit for Example 2.15

Solution:

When $R_L = 1000\,\Omega$, $V_{ab} = I_S\,R_L = (0.02\text{A})\,(1000\,\Omega) = 20\,\text{V}$

When $R_L = 50\,\Omega$, $V_{ab} = I_S\,R_L = (0.02\text{A})\,(50\,\Omega) = 1\,\text{V}$

The conditions of open circuit and short circuit of an ideal current source are as follows:

- Open circuit, $V_{ab} = \infty$, $I = 0$, as shown in Figure 2.28(a).
- Short circuit, $V_{ab} = 0$, $I = I_S$, as shown in Figure 2.28(b).

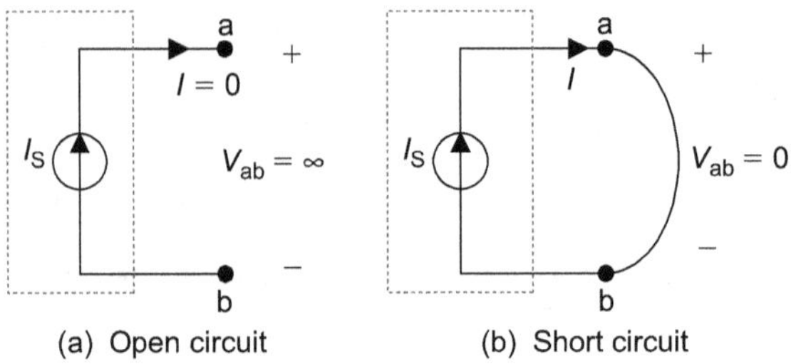

Figure 2.28 Open circuit and short circuit of an ideal current source

2.4.2.2 Real current source

Usually a real-life application of current source will not reach a *perfect* constant output current after it is connected to an external circuit or load, as the real current sources all have a non-infinite internal resistance R_S.

The real current source (or current source) can be represented as an ideal current source I_S in parallel with an internal resistor R_S. Once a load resistor R_L is connected to the current source as shown in Figure 2.29, the current of the source will change if the load resistance R_L changes. Since the internal resistance R_S of the current source usually is very large, the load current I will be a little bit lower than the source current I_S.

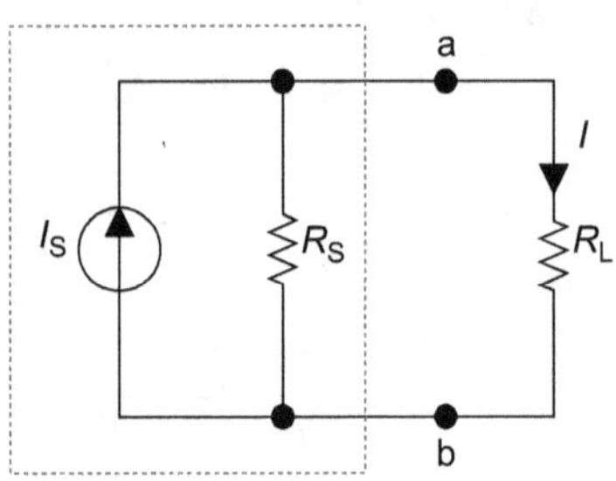

Figure 2.29 A real current source

Once the load resistance R_L changes, the current in the load will also changes. This is why the current of the real current source is not possible to keep at an ideal constant level.

A higher internal resistance R_S can provide a higher current through the external circuit of the real current source. The internal resistance of a real current source usually is much greater than the load resistance ($R_S \gg R_L$), and therefore the output current of the real current source is approximately stable.

Real current source (current source)

- It has an internal resistance R_S ($R_S \gg R_L$).
- R_S is in parallel with the current source.

2.5 International units for circuit quantities

2.5.1 International system of units (SI)

The international system of units (SI) was developed at the General Conference of the International Weight and Measures, which is the international authority that ensures dissemination and modifications of the SI units to reflect the latest advances in science and commerce. SI originates from the French '*Le Système International d'Unités*', which means the international system of units or the metric system to most people.

SI system is the world's most widely used modern metric system of measurement. Each physical quantity has a SI unit. There are seven basic units of the SI system and they are listed in Table 2.1.

Table 2.1 SI basic units

Quantity	**Quantity symbol**	**Unit**	**Unit symbol**
Length	l	Metre	m
Mass	M	Kilogram	kg
Time	t	Second	s
Current	I	Ampere	A
Temperature	T	Kelvin	K
Amount of substance	m	Mole	mol
Intensity of light	I	Candela	cd

SI units

International system of units (SI) is the world's most widely used modern metric system of measurement. There are seven base units of the SI system.

All other metric units can be derived from the seven SI basic units that are called 'derived quantities'. Some derived SI units for circuit quantities are given in Table 2.2.

Table 2.2 Some circuit quantities and their SI units

Quantity	**Quantity symbol**	**Unit**	**Unit symbol**
Voltage	V	Volt	V
Resistance	R	Ohm	Σ
Charge	Q	Coulomb	C
Power	P	Watt	W
Energy	W	Joule	J
Electromotive force	E or V_S	Volt	V
Conductance	G	Siemens	S
Resistivity	ρ	Ohm · metre	Ω · m

As you study the circuit theory more in-depth, you may use and add more circuit quantities and their derived SI units in this table.

2.5.2 Metric prefixes (SI prefixes)

Some time there are very large or small numbers when doing circuit analysis and calculation. A metric prefix (or SI prefix) is often used in the circuit calculation to reduce the number of zeroes. Large and small numbers are made by adding SI prefixes. A metric prefix is a modifier on the root unit that is in multiples of 10. In general science, the most common metric prefixes, such as milli, centi and kilo are used. In circuit analysis, more metric prefixes, such as nano and pico are used. Table 2.3 contains a complete list of metric prefixes.

Example 2.16:

(a) $47\,000\,\Omega = (?)\,\text{k}\Omega$
(b) $0.0505\,\text{A} = (?)\,\text{mA}$
(c) $0.0005\,\text{V} = (?)\,\mu\text{V}$
(d) $15\,000\,000\,000\,\text{C} = (?)\,\text{GC}$

Solution:

(a) $47\,000\,\Omega = 47 \times 10^3\,\Omega = 47\,\text{k}\Omega$
(b) $0.0505\,\text{A} = 50.5 \times 10^{-3}\,\text{A} = 50.5\,\text{mA}$
(c) $0.0005\,\text{V} = 500 \times 10^{-6}\,\text{V} = 500\,\mu\text{V}$
(d) $15\,000\,000\,000\,\text{C} = 15 \times 10^9 = 15\,\text{GC}$

Note:

- If a number is a *whole number*, move the decimal point to the *left*, and multiply the *positive* exponent of 10 (moving the decimal point three places each time). In Example 2.16(a), $47\,000\Omega = 47 \times 10^3\Omega$ (moving the decimal point three places to the left).

Table 2.3 Metric prefix table

Prefix	Symbol (abbreviation)	Exponential (power of 10)	Multiple value (in full)
Yotta	Y	10^{24}	1 000 000 000 000 000 000 000 000
Zetta	Z	10^{21}	1 000 000 000 000 000 000 000
Exa	E	10^{18}	1 000 000 000 000 000 000
Peta	P	10^{15}	1 000 000 000 000 000
Tera	**T**	$\mathbf{10^{12}}$	**1 000 000 000 000**
Giga	**G**	$\mathbf{10^{9}}$	**1 000 000 000**
Mega	**M**	$\mathbf{10^{6}}$	**1 000 000**
myria	my	10^{4}	10 000
kilo	**k**	$\mathbf{10^{3}}$	**1 000**
hecto	h	10^{2}	100
deka	da	10	10
deci	d	10^{-1}	0.1
centi	**c**	$\mathbf{10^{-2}}$	**0.01**
milli	**m**	$\mathbf{10^{-3}}$	**0.001**
micro	**μ (mu)**	$\mathbf{10^{-6}}$	**0.000 001**
nano	**n**	$\mathbf{10^{-9}}$	**0.000 000 001**
pico	**p**	$\mathbf{10^{-12}}$	**0.000 000 000 001**
femto	f	10^{-15}	0.000 000 000 000 001
atto	a	10^{-18}	0.000 000 000 000 000 001
zepto	z	10^{-21}	0.000 000 000 000 000 000 001
yocto	y	10^{-24}	0.000 000 000 000 000 000 000 001

Note: The most commonly used prefixes are shown in bold.

- If a number is a *decimal number*, move the decimal point to the *right*, and multiply the *negative* exponent of 10.
  In Example 2.16(b), 0.0505 A = 50.5×10^{-3} A (moving the decimal point three places to the right).
- If numbers have different prefixes, convert them to the same prefix first, then do the calculation.

Example 2.17: Determine the result of 30mA + 2000μA.

Solution:

$$\begin{aligned}30\,\text{mA} + 2000\,\mu\text{A} &= 30 \times 10^{-3}\,\text{A} + 2000 \times 10^{-6}\,\text{A}\\ &= 30 \times 10^{-3}\,\text{A} + 2 \times 10^{-3}\,\text{A}\\ &= (30 + 2)\text{mA}\\ &= 32\,\text{mA}\end{aligned}$$

Summary

Basic concepts

- Power: the speed of energy conversion, or work done per unit of time, $P = W/t$.

- Energy: the ability to do work.
- The reference direction of power:
  - If a circuit has mutually related reference polarity of current and voltage: $P > 0$ (absorption energy).
  - If a circuit has non-mutually related reference polarity of current and voltage: $P < 0$ (releasing energy).
- Branch: a current path between two nodes where one or more circuit components in series.
- Node: the intersectional point of two or more current paths where current has several possible paths to flow.
- Supernode: a part of the circuit that contains several nodes and branches.
- Loop: a complete current path where current flows back to the start.
- Mesh: a loop in the circuit that does not contain any other loops.
- Ideal voltage source: can provide a constant terminal voltage that does not depend on the variables in its external circuit. Its current depends on variables in its external circuit, $V_{ab} = V_S$, $R_S = 0$.
- Real voltage source: with a series internal resistance R_S ($R_S \ll R_L$), the terminal voltage of the real voltage source is: $V_{ab} = V_S - IR_S$.
- Ideal current source: can provide a constant output current Is that does not depend on the variations in its external circuit, $R_S = \infty$. Its voltage depends on variations in its external circuit.
- Real current source: with an internal resistance R_S in parallel with the ideal current source, $R_S \gg R_L$.

Formulas

- Work: $W = FS$
- Power: $P = W/t$
- Electrical power: $P = IV = I^2R = V^2/R$
- KVL #1: $\Sigma V = 0$
  - Assign a +ve sign for V or E if its reference polarity and loop direction are the same.
  - Assign a −ve sign for V or E if its reference polarity and loop direction are opposite.
- KVL #2: $\Sigma V = \Sigma E$
  - Assign a +ve sign for V if its reference polarity and loop direction are the same; assign a −ve sign for V if its reference direction and loop direction are opposite.
  - Assign a −ve sign for E if its reference polarity and loop direction are the same; assign a +ve sign for E if its polarity and loop direction are opposite.
- KCL #1: $\Sigma I_{in} = 0$
  - Assign a +ve sign for I if current is entering the node.
  - Assign a −ve sign for I if current is exiting the node.

- KCL #2: $\Sigma I_{in} = I_{out}$
  - Assign a +ve sign for I_{in} if current is entering the node; assign a −ve sign for I_{in} if current is exiting the node.
  - Assign a +ve sign for I_{out} if current is exiting the node; assign a −ve sign for I_{out} if current is entering the node.
- Some circuit quantities and their SI units

Quantity	Quantity symbol	Unit	Unit symbol
Voltage	V	Volt	V
Resistance	R	Ohm	Ω
Charge	Q	Coulomb	C
Power	P	Watt	W
Energy	W	Joule	J
Electro motive force	E or V_S	Volt	V
Conductance	G	Siemens	S
Resistivity	ρ	Ohm · metre	Ω · m

- The commonly used metric prefixes

Prefix	Symbol (abbreviation)	Exponential (power of 10)
pico	p	10^{-12}
nano	n	10^{-9}
micro	μ	10^{-6}
milli	m	10^{-3}
kilo	K	10^{3}
mega	M	10^{6}
giga	G	10^{9}
tera	T	10^{12}

Experiment 2: KVL and KCL

Objectives

- Construct and analyse series and parallel circuits.
- Apply Ohm's law and plot I–V characteristics.
- Experimentally verify KVL.
- Experimentally verify KCL.
- Analyse experimental data, circuit behaviour and performance, and compare them to theoretical equivalents.

Background information

- Ohm's law: $V = IR$

- I–V characteristics:

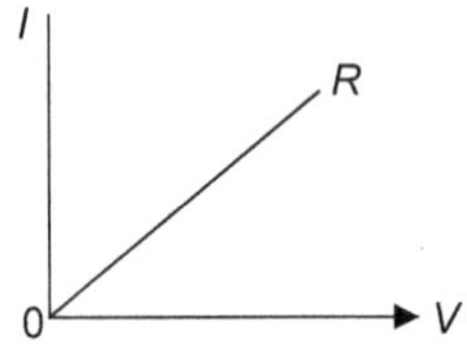

- KVL: $\Sigma V = 0$, or $\Sigma V = \Sigma E$
- KCL: $\Sigma I = 0$, or $\Sigma I_{in} = \Sigma I_{out}$

Equipment and components

- Digital multimeter (DMM)
- Breadboard
- DC power supply
- Switch
- Resistors: 240 Ω, 2.4 kΩ, 91 Ω, 2.7 kΩ, 3.9 kΩ and 910 Ω
- Some alligator clips
- Some wires and leads with banana-plug ends

Notes: (apply these notes to all experiments in this book)

- The ammeter (function) of the multimeter should be connected to the circuit after the power supply has been turned off.
- The voltmeter (function) of the multimeter should be connected in parallel with the component to measure voltage, and ammeter (function) should be connected in series with the component to measure current.
- Turn off the power supply before doing any circuit rearrangement, otherwise it will damage or harm experimental devices and components.
- Connect the negative terminal of the power supply to the ground using the black wire, and connect the positive terminal of the power supply to the component using the red wire. Connect other circuit components using different colour wires other than red and black.
- Use the actual resistance values to do the calculation.

Multimeter guide

- Recall: A multimeter is an electrical and electronic measuring instrument that combines functions of ammeter, voltmeter, ohmmeter, etc.
- Method for measuring *voltage* with a digital multimeter (voltmeter function):
  - Turn on the power supply after components have been connected in the circuit.

- Insert the multimeter's leads into the sockets labelled COM and V/Ω and turn on the multimeter (Figure L2.1).

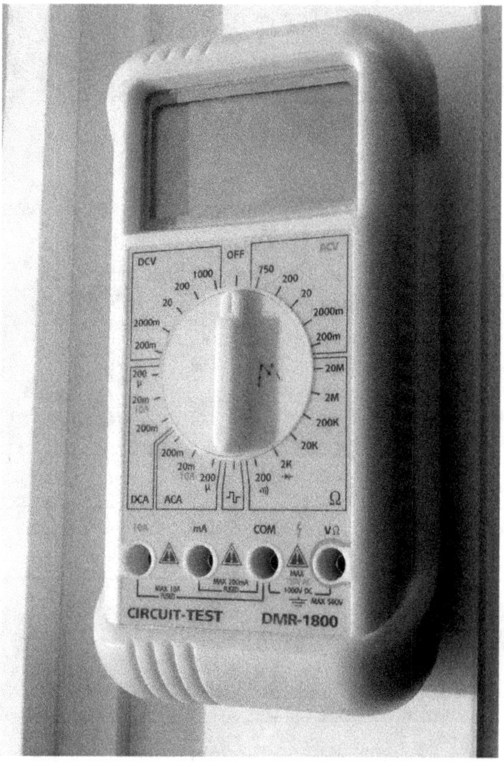

Figure L2.1 Multimeter

- Turn the central selector switch pointing to the voltage ranges with the DCV sign (DCV is for measuring DC voltage, and ACV is for measuring AC voltage), and where the estimated voltage value should be less than the maximum range.
- Make the measurement by connecting the component in parallel with the two leads of the multimeter. Connect or touch the red lead from the multimeter to terminal of the component, which is expected to have the more positive voltage, and connect or touch the black lead to the other terminal of the component.
- Read the displayed voltage value on the scale.
- Turn off the multimeter after the measurement.

- Method for measuring *current* with a digital multimeter (ammeter function):
  - Insert the multimeter's leads into the sockets labelled COM and A.
  - Connect the multimeter in series with the resistor branch that is going to make a measurement.

- Turn the central selector switch pointing to the current ranges (with mA or 10 A sign) where the estimated current value is closed to the maximum range.
- Turn on the power supply.
- Read the displayed current value on the scale.
- Turn off the multimeter after the measurement.

Procedure

Part I: Kirchhoff's voltage law (KVL)

1. Use the resistor colour code to choose three resistors with resistor values listed in Table L2.1. Measure each resistor using the multimeter (ohmmeter function). Record the values in Table L2.1.

Table L2.1

Resistance	R_1	R_2	R_3
Colour code resistor value	240 Ω	2.4 kΩ	91 Ω
Measured value			

2. Construct the series circuit shown in Figure L2.2 on the breadboard.

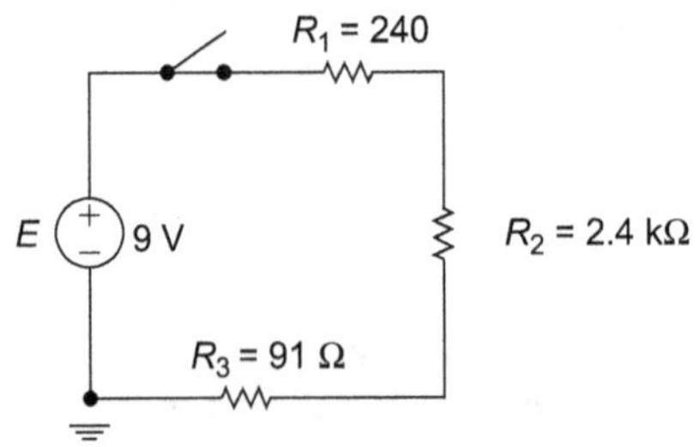

Figure L2.2 A series circuit

3. Calculate circuit current and voltages cross each resistor in Figure L2.2 (assuming the switch is turned on). Record the values in Table L2.2.

Table L2.2

	I	V_{R_1}	V_{R_2}	V_{R_3}	V_T
Formula for calculations					
Calculated value					
Measured value					

4. Set the power supply to 9 V, then turn on the switch, connect the multimeter (voltmeter function) in parallel with each resistor and power supply, and measure voltages across each resistor and power supply. Record the values in Table L2.2.
5. Use the direct method or indirect method to measure the circuit current. Record the value in Table L2.2.
   - Direct method: Connect the multimeter (ammeter function) in series with the circuit components, then turn on the switch and measure circuit current directly.
   - Indirect method: Apply Ohm's law to calculate the current with measured voltage and resistance.
6. Use measured values to plot I–V characteristics for 240 Ω resistor.
7. Substitute the measured voltage values from Table L2.2 into KVL equations to verify $\Sigma V = 0$ and $\Sigma V = \Sigma E$.

Part II: Kirchhoff's current law (KCL)

1. Construct a parallel circuit as shown in Figure L2.3 to the breadboard.

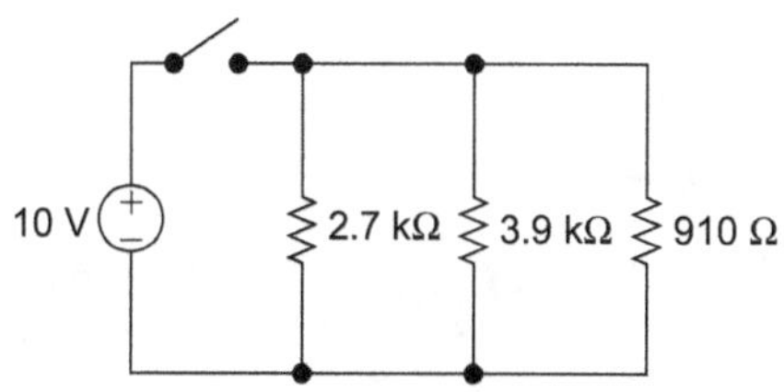

Figure L2.3 A parallel circuit

2. Calculate each branch current and total current in the circuit of Figure L2.3 (assuming the switch is turned on). Record the values in Table L2.3.

Table L2.3

Current	I_1	I_2	I_3	I_T
Formula for calculation				
Calculated values				
Measured values				

3. Set the power supply to 10 V, turn on the switch, measure each branch current and total current in the circuit by using direct or indirect methods (get the multimeter to function as an ammeter). Record the values in Table L2.3.
4. Substitute the measured current values from Table L2.3 into KCL equations to verify $\Sigma I = 0$ and $I_{in} = \Sigma I_{out}$.

Conclusion

Write your conclusions below:

Chapter 3
Series–parallel resistive circuits

Objectives

After completing this chapter, you will be able to:

- identify series circuits, parallel circuits and series–parallel circuits
- know how to determine the equivalent resistance for series, parallel and series–parallel resistive circuits
- calculate the resistance, voltage, current and power for series, parallel and series–parallel resistive circuits
- understand and apply the voltage-divider (VDR) and current-divider (CDR) rules
- identify the wye (Y) and delta (Δ) circuits
- know the method of wye (Y) and delta (Δ) conversions
- apply the method of Δ–Y conversions to simplify bridge circuits
- understand the method for measuring the unknown resistance of a balanced bridge circuit

3.1 Series resistive circuits and voltage-divider rule

Series, parallel and series–parallel resistive circuits are very often used electrical or electronic circuits. It is very important to construct electric circuits in different ways as to make practical use of them.

3.1.1 Series resistive circuits

A series circuit is the simplest circuit. It has all its elements connected in one loop of wire. It can be analogized by water flowing in a series of tanks connected by a pipe. The water flows through the pipe from tank to tank. The same amount of water will flow in each tank. The same is true of an electrical circuit. There is only one pathway by which charges can travel in a series circuit. The same amount of charges will flow in each component of the circuit, such as a light bulb, i.e. the current flow is the same throughout the circuit, so it has just one current in the series circuit.

Series circuit

- The components are connected one after the other.
- There is only one current path.
- The current flow through each component is always the same.

For example, Figure 3.1 illustrates an electrical circuit with three light bulbs (resistors) connected in series. If an ammeter is connected behind each light bulb, once the power turns on, the same current reading will be read on each ammeter. If we swap the position of the ammeter and the light bulb, the ammeter will still read the same current value. A practical example of a series circuit is a string of old Christmas lights.

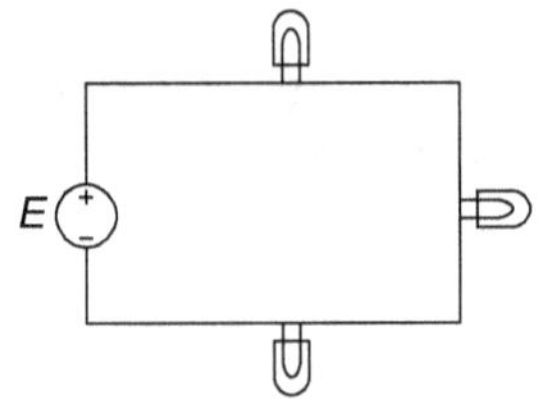

Figure 3.1 Series circuit

Many practical series circuits may not be as easily identifiable as Figures 3.1 and 3.2(a). Figure 3.2(b and c) are also series circuits but drawn in different ways. As long as the circuit elements are connected one after the other, and there

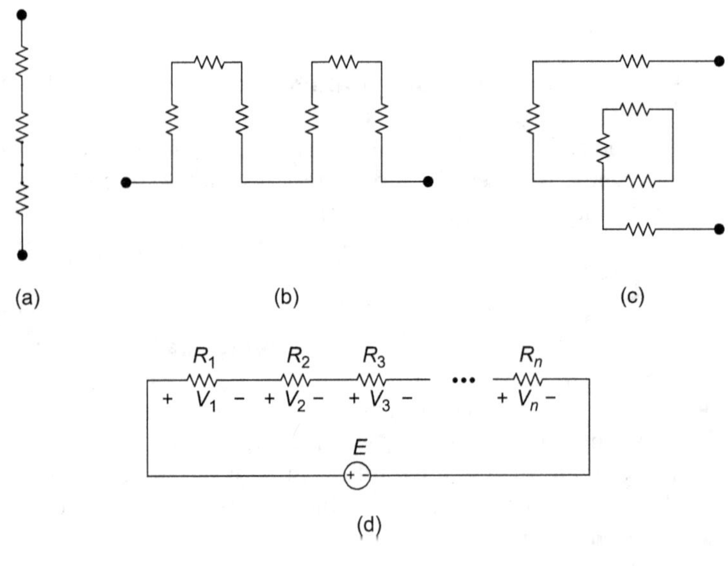

Figure 3.2 (a–c) Series resistive circuits and (d) series circuit

is only one current path for the circuit, it is said that they are connected in series. It does not matter if there is a different arrangement of the elements.

3.1.1.1 Total series voltage

The voltage across the source or power supply (total voltage) is equal to the sum of the voltage that drops across each resistor in a series circuit, i.e. the source voltage shared by each resistor. The terminal of the resistor connecting to the positive side of the voltage source is positive, and the terminal of the resistor connecting to the negative side of the voltage source is negative.

The total voltage V_T in the circuit of Figure 3.2(d) can be determined by Kirchhoff's voltage law (KVL) and Ohm's law. For n resistors connected in series, the total voltage will be as follows:

Total series voltage (V_T or E)

$$\begin{aligned} V_T &= E = V_1 + V_2 + \cdots + V_n \\ V_T &= IR_1 + IR_2 + \cdots + IR_n = IR_T \end{aligned} \tag{3.1}$$

For a series resistive circuit, (3.1) of the total voltage gives

$$V_T = I(R_1 + R_2 + \cdots + R_n)$$

and

$$R_T = R_1 + R_2 + \cdots + R_n$$

R_T is the mathematical equation for computing the total resistance (or equivalent resistance R_{eq}) of a series resistive circuit.

3.1.1.2 Total series resistance (or equivalent resistance)

The total resistance (R_T) of a series resistive circuit is the sum of all resistances in the circuit. It is also called the equivalent resistance (R_{eq}) because this resistance is *equivalent* to the sum of all resistances when you look through the two terminals of the series resistive circuit. The equivalent resistance of a series resistive circuit is the amount of resistance that a single resistor would need to equal the overall effect of the all resistors that are present in the circuit.

The total resistance of a series resistive circuit is always greater than any single resistance in that circuit.

Total series resistance (R_T) or equivalent resistance (R_{eq})

$$R_T = R_1 + R_2 + \cdots + R_n$$

3.1.1.3 Series current

From the definition of a series circuit we know that there is only one current path in a series circuit, that the current flowing through each element is always the same and that the current is always the same at any point in a series circuit. The current I flowing in a series resistive circuit, such as the one in Figure 3.2(d), can be determined from Ohm's law as follows:

Series current (I)

$$I = \frac{V_T}{R_T} = \frac{E}{R_T} = \frac{V_1}{R_1} = \frac{V_2}{R_2} = \cdots = \frac{V_n}{R_n}$$

3.1.1.4 Series power

Each of the resistors in a series circuit consumes power, which is dissipated in the form of heat. The total power (P_T) consumed by a series circuit is the sum of power dissipated by the individual resistor. Since this power must come from the source, the total power is actually the power supplied by the source. Multiply the current I on both sides of the total voltage equation $V_T = E = V_1 + V_2 + \cdots + V_n$, to get the total power $P_T = IE = IV_1 + IV_2 + \cdots + IV_n$.

Therefore, the total power in a series resistive circuit can be expressed as follows:

Total series power (P_T)

$$P_T = P_1 + P_2 + \cdots + P_n \quad \text{or} \quad P_T = IE = I^2 R_T = (E^2/R_T)$$

The power dissipated by the individual resistor in a series resistive circuit is as follows:

$$P_1 = I^2 R_1 = IV_1 = \frac{V_1^2}{R_1}$$

$$P_2 = I^2 R_2 = IV_2 = \frac{V_2^2}{R_2}$$

$$\cdots$$

$$P_n = I^2 R_n = IV_n = \frac{V_n^2}{R_n}$$

Example 3.1: A series resistive circuit is shown in Figure 3.3. Determine the following:

(a) Total resistance R_T
(b) Current I in the circuit
(c) Voltage across the resistor R_1
(d) Total voltage V_T
(e) Total power P_T

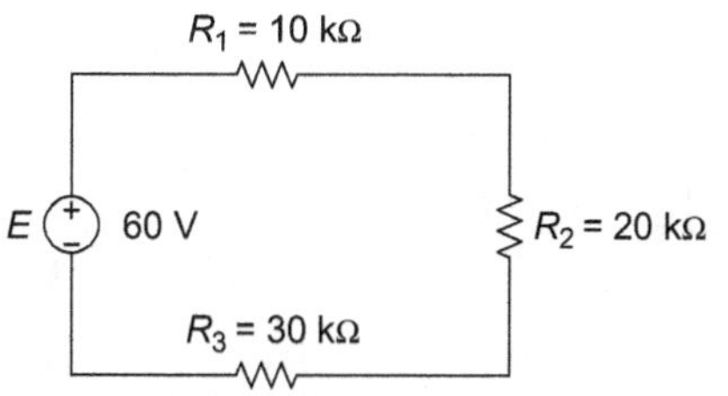

Figure 3.3 Circuit for Example 3.1

Solution:

(a) $R_T = R_1 + R_2 + R_3 = (10 + 20 + 30)\text{k}\Omega = 60\ \text{k}\Omega$
(b) $I = (E/R_T) = (60\ \text{V}/60\ \text{k}\Omega) = 1\ \text{mA}$
(c) $V_1 = IR_1 = (1\ \text{mA})(10\ \text{k}\Omega) = 10\ \text{V}$
(d) $V_T = IR_T = (1\ \text{mA})(60\ \text{k}\Omega) = 60\ \text{V}$, $V_T = E = 60\ \text{V}$ (checked)
(e) $P_T = IE = (1\ \text{mA})(60\ \text{V}) = 60\ \text{mW}$ or $P_T = I^2R_T = (1\ \text{mA})^2(60\ \text{k}\Omega) =$ 60 mW (checked)

3.1.2 Voltage-divider rule (VDR)

The VDR can be exhibited by using a pot (short for potentiometer). A pot is a variable resistor whose resistance across its terminals can be varied by turning a knob.

A pot is connected to a voltage source, as shown in Figure 3.4(a). Using a voltmeter to measure the voltage across the pot, the voltage relative to the negative side of the 100 V voltage source is ½ $E = 50$ V when the arrow (knob) is at the middle of the potentiometer. The voltage will increase when the arrow moves up, and the voltage will decrease when the arrow moves down. This is the principle of the voltage divider, i.e. the voltage divider is a design technique used to create different output voltages that is proportional to the input voltage. Actually, a potentiometer itself is an adjustable voltage divider.

The circuit in Figure 3.4(a) is equivalent to (b) since $R = R_1 + R_2 = 100\ \text{k}\Omega$. Therefore, the voltage divider means that the source voltage E or total voltage V_T is divided according to the value of the resistors in the series circuit. The output voltage from the divider changes when any of the resistor values change.

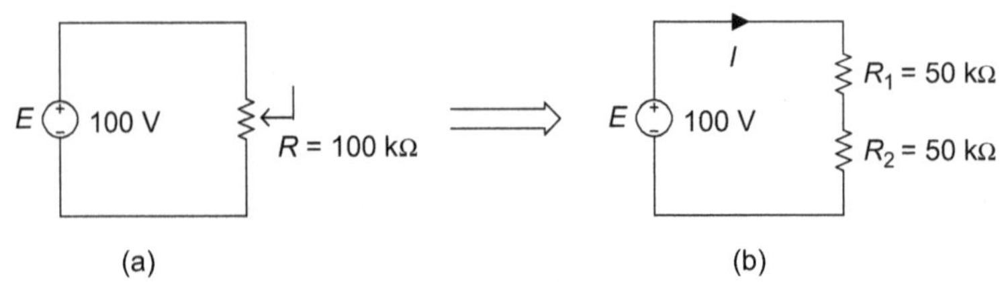

Figure 3.4 Voltage divider

In Figure 3.4(b),

$$I = \frac{E}{R_1 + R_2}$$

$$V_1 = IR_1 = \frac{E}{R_1 + R_2} R_1 = E\frac{R_1}{R_1 + R_2} = 100\ \text{V}\frac{50\ \text{k}\Omega}{(50 + 50)\text{k}\Omega} = 50\ \text{V}$$

$$V_2 = IR_2 = \frac{E}{R_1 + R_2} R_2 = E\frac{R_2}{R_1 + R_2} = 100\ \text{V}\frac{50\ \text{k}\Omega}{(50 + 50)\text{k}\Omega} = 50\ \text{V}$$

The above two equations are the VDRs for a series circuit of *two* resistors. When there are n resistors in series, using the same method we can obtain the general form of the VDR as follows:

$$I = \frac{E}{R_1 + R_2 + \cdots + R_n} = \frac{E}{R_T}$$

$$V_X = IR_X = \frac{E}{R_T} R_X = E\frac{R_X}{R_T} \quad \text{or} \quad V_X = V_T\frac{R_X}{R_T}$$

where R_X and V_X are the unknown resistance and voltage, and R_T and V_T are the total resistance and voltage in the series circuit.

VDR

- General form: $V_X = V_T\frac{R_X}{R_T}$ or $V_X = E\frac{R_X}{R_T}$ $(X = 1, 2, \ldots, n)$
- When there are only two resistors in series: $V_1 = V_T\frac{R_1}{R_1+R_2}$, $V_2 = V_T\frac{R_2}{R_1+R_2}$

Note: The *numerator* of the VDR is always the *unknown* resistance (this is worth memorizing).

Example 3.2: Use the VDR to determine the voltage drops across resistors R_2 and R_3 in the circuit of Figure 3.5.

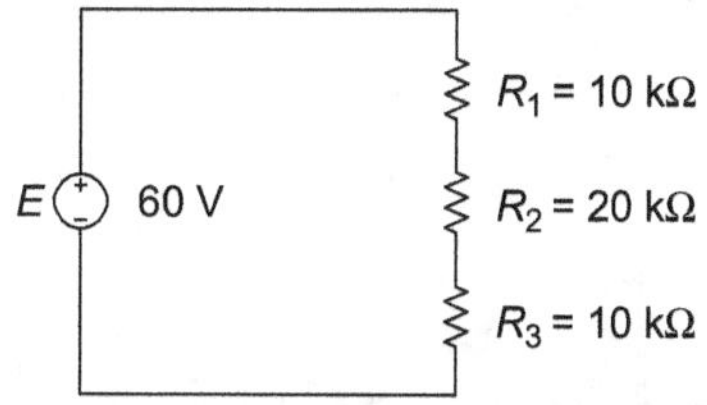

Figure 3.5 Circuit for Example 3.2

Solution:

Use the general form of the VDR $V_X = E(R_X/R_T)$

$$V_2 = E\frac{R_2}{R_T} = E\frac{R_2}{R_1 + R_2 + R_3} = 60\,\text{V}\frac{20\,\text{k}\Omega}{(10 + 20 + 10)\,\text{k}\Omega} = 30\,\text{V}$$

$$V_3 = E\frac{R_3}{R_T} = E\frac{R_3}{R_1 + R_2 + R_3} = 60\,\text{V}\frac{10\,\text{k}\Omega}{(10 + 20 + 10)\,\text{k}\Omega} = 15\,\text{V}$$

The practical application of the voltage divider can be the volume control of audio equipment. The knob of the pot in the circuit will eventually let you adjust the volume of the audio equipment.

3.1.3 Circuit ground

Usually, there is a ground for each electric circuit. Electrical circuit grounding is important because it is always at zero potential (0 V), and provides a reference voltage level in which all other voltages in a circuit are measured. There are two types of circuit grounds: one is the earth ground and another is the common ground (or chassis ground). Since an equal number of negative and positive charges are distributed throughout the earth at any given time, the earth is an electrically neutral body. So the earth is always at zero potential (0 V) and measurements can be made by using earth as a reference. An earth ground usually consists of a ground rod or a conductive pipe driven into the soil.

A chassis ground is a connection to the main chassis of a piece of electronic or electrical equipment, such as a metal plate. Chassis ground is also called common ground. All chassis grounds should lead to earth ground, so that it also provides a point that has zero voltage. The neutral point in the alternating circuit (AC) is an example of the common ground.

The difference between these two grounds can be summarized as follows:

- Earth ground: Connecting one terminal of the voltage source to the earth. The symbol for it is:

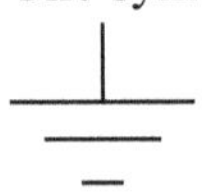

- Common ground or chassis ground: The common point for all elements in the circuit. All the common points are electrically connected together through metal plates or wires. The symbol for the common point is:

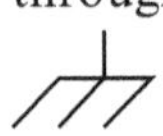

In a circuit, the voltage with the single-subscript notation (such as V_b) is the voltage drop from the point b with respect to ground. And the voltage with the double-subscripts notation (such as V_{bc}) is the voltage drop across the two points b and c (each point is represented by a subscript).

Example 3.3: Determine V_{bc}, V_{be} and V_b in the circuit of Figure 3.6.

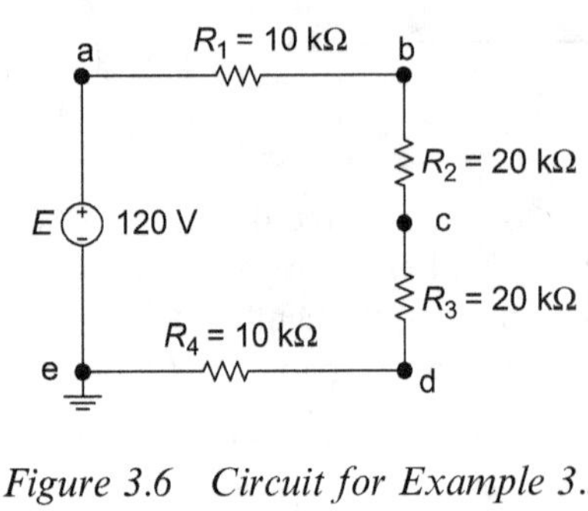

Figure 3.6 Circuit for Example 3.3

Solution:

$$V_{bc} = V_{R_2} = E\frac{R_2}{R_T} = E\frac{R_2}{R_1 + R_2 + R_3 + R_4}$$
$$= 120\,\text{V}\frac{20\,\text{k}\Omega}{(10 + 20 + 20 + 10)\text{k}\Omega} = 120\,\text{V}\frac{20\,\text{k}\Omega}{60\,\text{k}\Omega} = 40\,\text{V}$$

$$V_{be} = E\frac{R_2 + R_3 + R_4}{R_T} = 120\text{V}\frac{(20 + 20 + 10)\text{k}\Omega}{60\,\text{k}\Omega} = 100\,\text{V}$$

(Use the general form of the VDR $V_X = E\,(R_X/R_T)$. There the unknown voltage $V_X = V_{be}$, and the unknown resistance $R_X = R_2 + R_3 + R_4$).

$V_b = V_{be} = 100\,\text{V}$

Ground and voltage subscript notation

- Earth ground: connects to the earth ($V = 0$).
- Common ground (chassis ground): the common point for all components in the circuit ($V = 0$).
- Single-subscript notation: the voltage from the subscript with respect to ground.
- Double-subscript notation: the voltage across the two subscripts.

3.2 Parallel resistive circuits and the current-divider rule

3.2.1 Parallel resistive circuits

If any one of the light bulbs or resistors burns out or is removed in a series resistive circuit, the circuit is broken, no charges or current will move through the circuit and the entire circuit would stop operating. This is because there is only one current path in a series circuit. Old style Christmas lights were often wired in series, so if one light bulb burned out, the whole string of lights went off. This is the main disadvantage of a series circuit.

A parallel resistive circuit is composed of two or more series circuits connected to the same power source. Thus, it has more than one path for charges or current to follow. An obvious advantage of the parallel circuit is that burn out or removal of one bulb does not affect the other bulbs. They continue to operate because there is still a separate, independent path from the source to each of the other bulbs, so the other bulbs will stay lit.

Parallel resistive circuits also can be analogized by flowing water. When water flowing in a river across small islands, the one water path will be divided by the islands and split into many more water paths. When the water has passed the islands, it will become a single water path again. This scenario is illustrated in Figure 3.7(a). The circuit in Figure 3.7(b) is parallel resistive circuit equivalent to Figure 3.7(a).

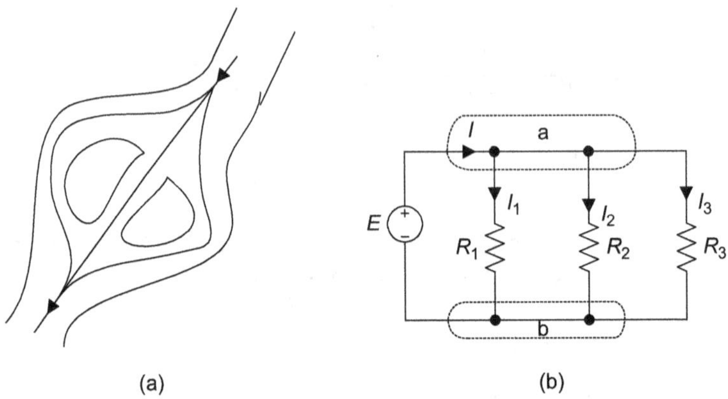

Figure 3.7 Parallel circuit

In the parallel resistive circuit, the source current divides among the available resistive branches in different paths. The total current I as illustrated in Figure 3.7(b) leaves the positive terminal of the voltage source and flows to node a (or supernode – chapter 2), which is a connecting point for the four branches. At node a, the total current I divides into three currents I_1, I_2 and I_3. These three currents flow through their resistors and rejoin at node b. The total current then flows from b back to the negative terminal of the source.

Note that the node does not have to physically be one single point. As long as several branches are connected together, then that part of the circuit is considered to be a node.

From the parallel circuit in Figure 3.7(b), we can see that circuit elements (resistors) are connected in parallel if the ends of one element are connected directly to the corresponding ends of the other.

Many practical series circuits may not be as easily identifiable as Figure 3.7(b), and circuits in Figure 3.8 are also parallel circuits but drawn in different ways. As long as the circuit elements are connected end to end and there are at least two current paths in the circuit, it is said that they are connected in parallel. It does not matter if there are different arrangements of the elements.

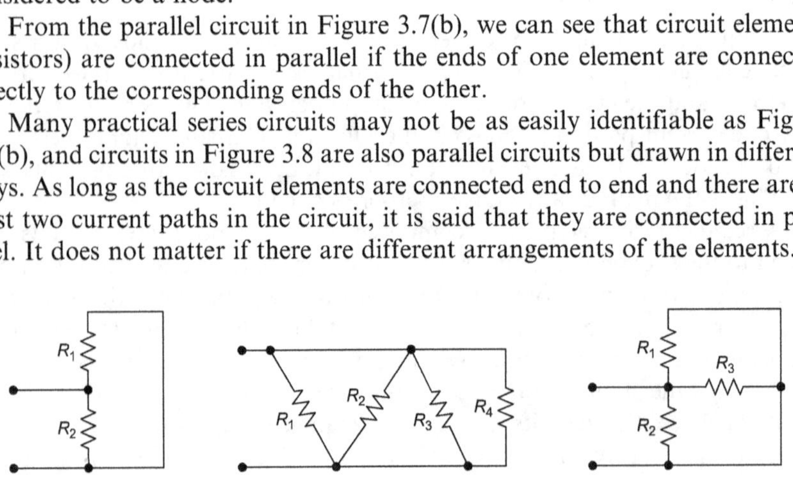

Figure 3.8 Parallel resistive circuits

A parallel circuit has two main advantages when compared with series circuits. The first is that a failure of one element does not lead to the failure of the other elements. The other is that more elements may be added in parallel without the need for increasing voltage.

Parallel circuit

- The components are connected end to end.
- There are at least two current paths in the circuit.
- The voltage across each component is the same.

3.2.1.1 Parallel voltage

Since all resistors in a parallel resistive circuit are connected between the two notes, the voltage between these two notes must be the same. In this case, the voltage drop across each resistor must be the same. All these must be equal to the supply voltage *E* for the parallel resistive circuit shown in Figure 3.9. If all the resistors are light bulbs and have the same resistances as in Figure 3.9, they will glow at the same brightness as they each receive the same voltage.

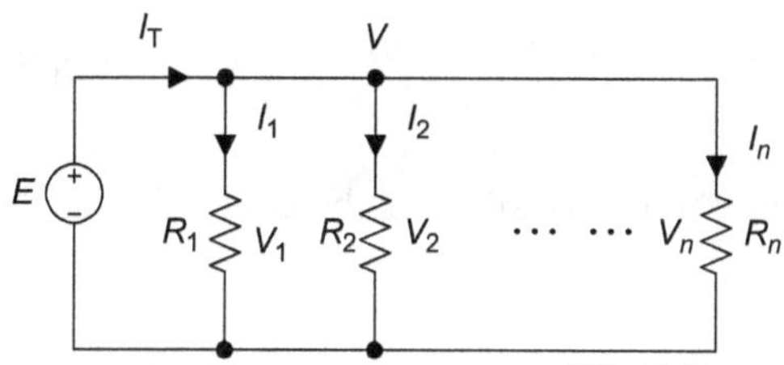

Figure 3.9 V and I in a parallel circuit

The voltage drop across each resistor must equal the voltage of the source in a parallel resistive circuit. This can be expressed in the following mathematical equation:

Parallel voltage

$$V = E = V_1 = V_2 = \cdots = V_n$$

3.2.1.2 Parallel current

If the parallel circuit was a river, the total volume of water in the river would be the sum of water in each branch (Figure 3.7(a)). This is the same with the current in the parallel resistive circuit. The total current is equal to the sum of currents in each resistive branch, and the total current entering and exiting parallel resistive circuit is the same.

If resistances are different in each branch of a parallel circuit, the branch currents will be different. The branch currents can be determined by Ohm's law, and the characteristics of the total current in a parallel circuit can be expressed in the following equations:

Parallel current

- Branch currents: $I_1 = \dfrac{V}{R_1}, \quad I_2 = \dfrac{V}{R_2}, \ldots, I_n = \dfrac{V}{R_n}$
- Total current: $I_T = \dfrac{V}{R_{eq}} = I_1 + I_2 + \cdots + I_n$

3.2.1.3 Equivalent parallel resistance

How much current is flowing through each branch in the parallel resistive circuit in Figure 3.9? It depends on the amount of resistance in each branch. The total resistance in a parallel circuit can be found by applying Ohm's law to the equation of the total current:

$$\frac{V}{R_{eq}} = I_1 + I_2 + \cdots + I_n = \frac{V}{R_1} + \frac{V}{R_2} + \cdots + \frac{V}{R_n} = V\left(\frac{1}{R_1} + \frac{1}{R_2} + \cdots + \frac{1}{R_n}\right)$$

Dividing the voltages on both sides of the equal sign in the above equation gives

$$\frac{1}{R_{eq}} = \frac{1}{R_1} + \frac{1}{R_2} + \cdots + \frac{1}{R_n}$$

Solving for R_{eq} from the above equation will give the equivalent resistance for the parallel circuit:

$$R_{eq} = \frac{1}{(1/R_1) + (1/R_2) + \cdots + (1/R_n)}$$

So the total resistance of a set of resistors in a parallel resistive circuit is found by adding up the reciprocals of the resistance values and then taking the reciprocal of the total.

It will be more convenient to use the conductance than the resistance in the parallel circuits. Since the conductance $G = 1/R$, therefore,

$$G_{eq} = \frac{1}{R_{eq}} = \frac{1}{R_1} + \frac{1}{R_2} + \cdots + \frac{1}{R_n} = G_1 + G_2 + \cdots + G_n$$

When there are only two resistors in parallel:

$$R_{eq} = \frac{1}{(1/R_1) + (1/R_2)} = \frac{1}{(R_1 + R_2)/(R_1R_2)} = \frac{R_1R_2}{R_1 + R_2}$$

Usually parallel can be expressed by a symbol of '//' such as: $R_1 \,//\, R_2 \,//\, \cdots \,//\, R_n$.

Equivalent parallel resistance and conductance

- $R_{eq} = \dfrac{1}{(1/R_1) + (1/R_2) + \cdots + (1/R_n)} = R_1//R_2// \cdots //R_n$
- $G_{eq} = G_1 + G_2 + \cdots + G_n$

When $n = 2$: $R_{eq} = \dfrac{R_1R_2}{R_1 + R_2} = R_1//R_2$

Note: The total resistance of the *series* resistive circuit is always *greater* than the individual resistance, and the total resistance of the *parallel* resistive circuit is always *less* than the individual resistance. So usually for parallel circuits the *equivalent* resistance is used instead of the *total* resistance.

3.2.1.4 Total parallel power

The total power is the sum of the power dissipated by the individual resistors in a parallel resistive circuit. By multiplying a voltage V to both sides of the equation for total current $I_T = I_1 + I_2 + \cdots + I_n$, the total power equation for the parallel circuits is obtained as follows:

$$I_T V = I_1 V + I_2 V + \cdots + I_n V = P_T$$

Total parallel power

$$P_T = P_1 + P_2 + \cdots + P_n \quad \text{or} \quad P_T = I_T V = I_T^2 R_{eq} = \frac{V^2}{R_{eq}}$$

The power consumed by each resistor in a parallel circuit is expressed as:

$$P_1 = I_1 V = I_1^2 R_1 = \frac{V^2}{R_1}, \quad P_2 = I_2 V = I_2^2 R_2 = \frac{V^2}{R_2}, \ldots, \quad P_n = I_n V = I_n^2 R_n = \frac{V^2}{R_n}$$

Example 3.4: A parallel circuit is shown in Figure 3.10. Determine (a) R_2, (b) I_T and (c) P_3, given $R_{eq} = 1.25$ kΩ, $R_1 = 20$ kΩ, $R_3 = 2$ kΩ and $I_3 = 18$ mA.

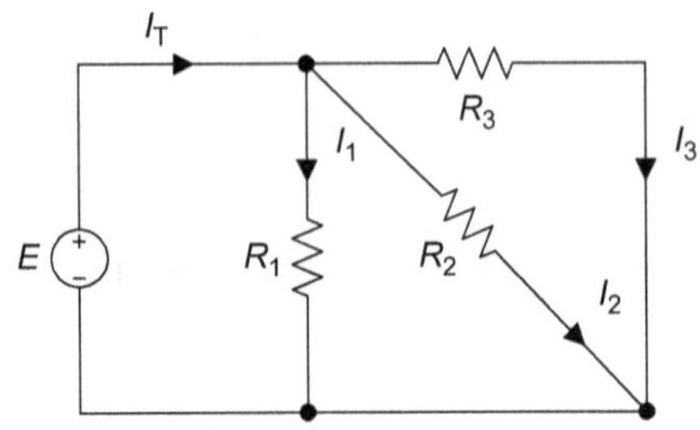

Figure 3.10 Figure for Example 3.4

Solution:

(a) Since $R_2 = 1/G_2$, determine G_2 first.

$$G_{eq} = G_1 + G_2 + G_3$$

$$G_2 = G_{eq} - G_1 - G_3 = \frac{1}{R_{eq}} - \frac{1}{R_1} - \frac{1}{R_3} = \frac{1}{1.25\,\text{k}\Omega} - \frac{1}{20\,\text{k}\Omega} - \frac{1}{2\,\text{k}\Omega}$$

$$= 0.25\,\text{mS}$$

$$\therefore \quad R_2 = \frac{1}{G_2} = \frac{1}{0.25\,\text{mS}} = 4\,\text{k}\Omega$$

$$R_{eq} = \frac{1}{(1/R_1) + (1/R_2) + (1/R_3)} = \frac{1}{(1/20\,\text{k}\Omega) + (1/4\,\text{k}\Omega) + (1/2\,\text{k}\Omega)}$$

$$= 1.25\,\text{k}\Omega \quad \text{(proved)}$$

(b) $I_T = \dfrac{E}{R_{eq}} = \dfrac{V_3}{R_{eq}} = \dfrac{I_3 R_3}{R_{eq}} = \dfrac{(18\,\text{mA})(2\,\text{k}\Omega)}{1.25\,\text{k}\Omega} = 28.8\,\text{mA}$

(c) $P_3 = I^2{}_3 R_3 = (18\ \text{mA})^2(2\ \text{k}\Omega) = 648\ \text{mW}$

3.2.2 Current-divider rule (CDR)

The VDR can be used for series circuits, and the CDR can be used for parallel circuits. As previously mentioned, parallel circuits can be analogized by flowing water. It is like a river, the water flowing through the river will be divided by small islands and the flow is split creating more water paths. This was shown in Figure 3.7(a).

The equations of the current divider can be derived by the following:

A parallel resistive circuit with n resistors was shown in Figure 3.9. In this circuit:

$$I_1 = \frac{E}{R_1}, \quad I_2 = \frac{E}{R_2}, \ldots, I_n = \frac{E}{R_n}$$

Inserting $E = I_T R_{eq}$ into the above equations gives:

$$I_1 = I_T \frac{R_{eq}}{R_1}, \quad I_2 = I_T \frac{R_{eq}}{R_2}, \ldots, I_n = I_T \frac{R_{eq}}{R_n}$$

These are the general form current-divider equations.

When there are only two resistors in parallel:

$$I_1 = I_T \frac{R_{eq}}{R_1} = I_T \frac{(R_1 R_2/(R_1 + R_2))}{R_1} = I_T \frac{R_2}{R_1 + R_2}$$

$$I_2 = I_T \frac{R_{eq}}{R_2} = I_T \frac{(R_1 R_2/(R_1 + R_2))}{R_2} = I_T \frac{R_1}{R_1 + R_2}$$

CDR

- General form: $I_X = I_T \frac{R_{eq}}{R_X}$ or $I_X = I_T \frac{G_X}{G_{eq}}$
- When there are two resistors in parallel: $I_1 = I_T \frac{R_2}{R_1 + R_2}$, $I_2 = I_T \frac{R_1}{R_1 + R_2}$

There I_X and R_X are unknown current and resistance, and I_T is the total current in the parallel resistive circuit.

Note: The CDR is similar in form to the VDR. The difference is that the *denominator* of the general form current divider is the *unknown* resistance. When there are two resistors in parallel, the *numerator* is the other resistance (other than the unknown resistance).

Recall the VDR:

$$V_X = V_T \frac{R_X}{R_T}, \quad V_1 = V_T \frac{R_1}{R_1 + R_2}, \quad V_2 = V_T \frac{R_2}{R_1 + R_2}$$

Example 3.5: Determine the current I_1, I_2 and I_3 in the circuit of Figure 3.11.

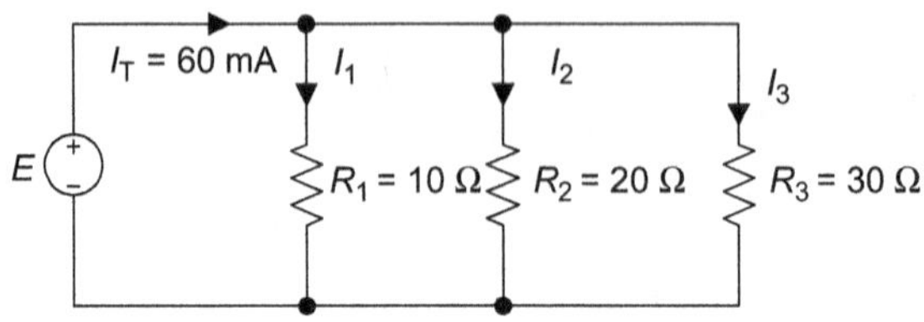

Figure 3.11 Circuit for Example 3.5

Solution:

$$R_{eq} = R_1//R_2//R_3 = \frac{1}{(1/R_1) + (1/R_2) + (1/R_3)}$$

$$= \frac{1}{(1/(10\,\Omega)) + (1/(20\,\Omega)) + (1/(30\,\Omega))} \approx 5.455\,\Omega$$

$$I_1 = I_T \frac{R_{eq}}{R_1}$$
$$= 60\,\text{mA} \frac{5.455\,\Omega}{10\,\Omega}$$
$$= 32.73\,\text{mA}$$

$$I_2 = I_T \frac{R_{eq}}{R_2}$$
$$= 60\,\text{mA} \frac{5.455\,\Omega}{20\,\Omega}$$
$$\approx 16.37\,\text{mA}$$

$$I_3 = I_T \frac{R_{eq}}{R_3}$$
$$= 60\,\text{mA} \frac{5.455\,\Omega}{30\,\Omega}$$
$$= 10.91\,\text{mA}$$

The conclusion that can be drawn from the above example is that the greater the branch resistance, the less the current flows through that branch, or the less the share of the total current.

Example 3.6: Determine the resistance R_2 for the circuit in Figure 3.12.

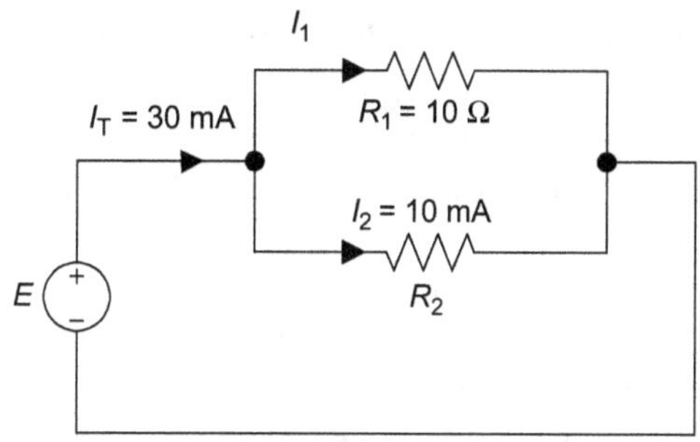

Figure 3.12 Circuit for Example 3.6

Solution:

Solve R_2 from the current-divider formula

$$I_2 = I_T[R_1/(R_1 + R_2)]$$

$$I_2(R_1 + R_2) = I_T R_1$$

$$I_2 R_2 = I_T R_1 - I_2 R_1, \quad R_2 = \frac{R_1(I_T - I_2)}{I_2}$$

$$R_2 = \frac{10\,\Omega(30 - 10)\text{mA}}{10\,\text{mA}}$$
$$= 20\,\Omega$$

3.3 Series–parallel resistive circuits

The most practical electric circuits are not simple series or parallel configurations, but combinations of series and parallel circuits, or the series–parallel configurations. Many circuits have various combinations of series and parallel components, i.e. circuit elements are series-connected in some parts and parallel in others.

The series–parallel configurations have a variety of circuit forms, and some of them may be very complex. However, the same principles and rules or laws that have been introduced in the previous chapters are applied. The key to solving series–parallel circuits is to identify which parts of the circuit are series and which parts are parallel and then simplify them to an equivalent circuit and find an equivalent resistance.

Series–parallel circuit
The series–parallel circuit is a combination of series and parallel circuits.

3.3.1 Equivalent resistance

Method for determining the equivalent resistance of series–parallel circuits:

- Determine the equivalent resistance of the parallel part of the series–parallel circuits.
- Determine the equivalent resistance of the series part of the series–parallel circuits.
- Plot the equivalent circuit if necessary.
- Repeat the above steps until the resistances in the circuit can be simplified to a single equivalent resistance R_{eq}.

Note: Determine R_{eq} step by step from the far end of the circuit to the terminals of the R_{eq}.

Example 3.7: Analysis of the series–parallel circuit in Figure 3.13.

In Figure 3.13(a), the resistor R_5 is in series with R_6 and in parallel with R_4 and R_3. This can be expressed by the equivalent circuit in Figure 3.13(b). R_2 is in series with $(R_5 + R_6) // R_4 // R_3$ and in parallel with R_1. That is the equivalent resistance R_{eq} for the series–parallel circuit, i.e. $R_{eq} = \{[(R_5 + R_6) // R_4 // R_3] + R_2\} // R_1$.

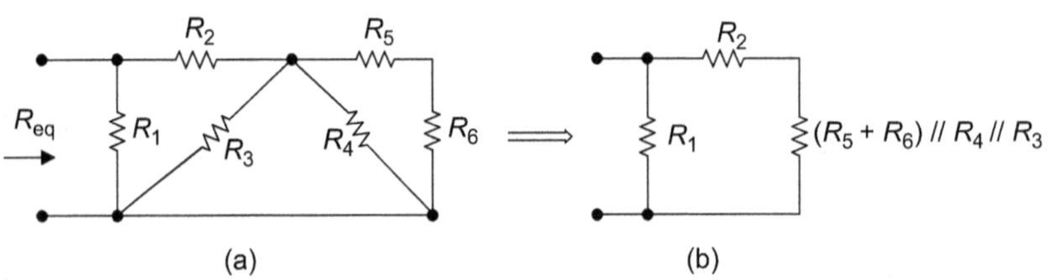

Figure 3.13 Circuits for Example 3.7

Example 3.8: Determine the equivalent resistance R_{eq} for the circuit shown in Figure 3.14(a).

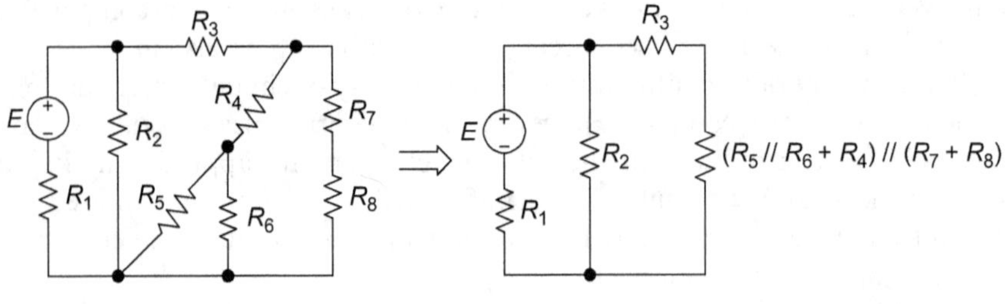

Figure 3.14 Circuits for Example 3.8

Solution: $R_{eq} = [(R_5 // R_6 + R_4) // (R_7 + R_8) + R_3] // R_2 + R_1$

Example 3.9: Determine the R_{eq} for the circuit shown in Figure 3.15(a).

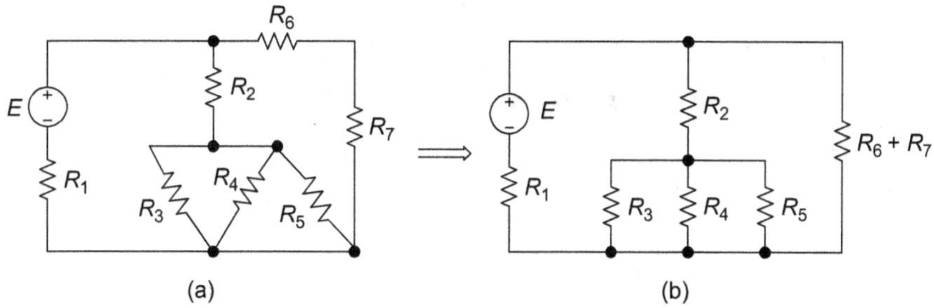

Figure 3.15 Circuits for Example 3.9

Solution: $R_{eq} = [(R_3 \,//\, R_4 \,//\, R_5) + R_2\,] \,//\, (R_6 + R_7) + R_1$

3.3.2 Method for analysing series–parallel circuits

After determining the equivalent resistance of the series–parallel circuit, the total current as well as currents and voltages for each resistor can be determined by using the following steps:

- Apply Ohm's law with the equivalent resistance solved from the previous section to determine the total current in the equivalent circuit $I_T = E/R_{eq}$.
- Apply the VDR, CDR, Ohm's law, KCL and KVL to determine the unknown currents and voltages.

Example 3.10: Determine the currents and voltages for each resistor in the circuit of Figure 3.16.

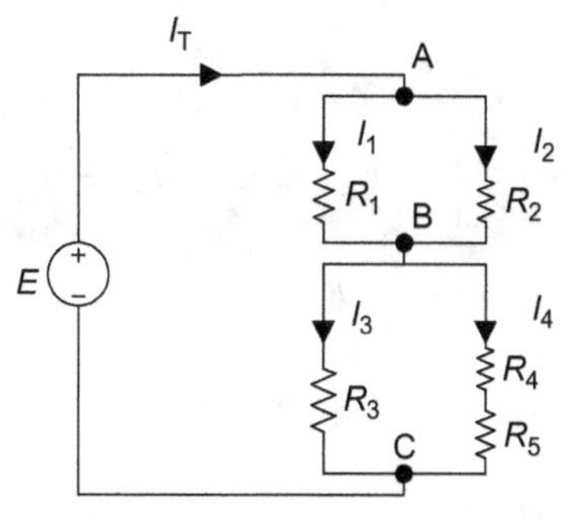

Figure 3.16 Circuit for Example 3.10

Solution:

- $R_{eq} = (R_1 \,//\, R_2) + [(R_4 + R_5) \,//\, R_3]$
  $I_T = \dfrac{E}{R_{eq}}$
- $V_{R_1} = V_{R_2} = V_{AB} = I_T(R_1//R_2), \quad I_1 = \dfrac{V_{AB}}{R_1}, \quad I_2 = \dfrac{V_{AB}}{R_2}$
  or $I_1 = I_T \dfrac{R_2}{R_1 + R_2}, \quad I_2 = I_T \dfrac{R_1}{R_1 + R_2}$ (the CDR)
- $V_{R_3} = V_{R_4} + V_{R_5} = V_{BC} = I_T[(R_4 + R_5)//R_3]$
  $I_3 = \dfrac{V_{BC}}{R_3}, \quad I_{4,5} = \dfrac{V_{BC}}{R_4 + R_5}$

Check: $I_T = I_1 + I_2$ or $I_T = I_3 + I_{4,5}$ (KCL)
$V_{AB} + V_{BC} = E$ (KVL)

Example 3.11: Determine the current I_T in the circuit of Figure 3.17.

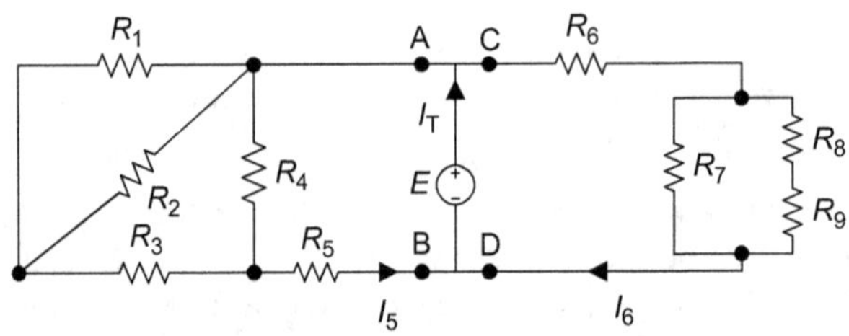

Figure 3.17 Circuit for Example 3.11

Solution:

$I_T = I_5 + I_6$

The method of analysis: $I_T = ?$

$$I_T = I_5 + I_6 \qquad (I_5 = ?, I_6 = ?)$$

$$I_5 = \frac{V_{AB}}{R_{AB}} \qquad \left(I_5 = \frac{V_{AB}}{R_{AB}}, \quad V_{AB} = E, \quad R_{AB} = ?\right)$$

$$R_{AB} = [(R_1//R_2 + R_3)//R_4] + R_5$$

$$I_6 = \frac{V_{CD}}{R_{CD}} \qquad \left(I_6 = \frac{V_{CD}}{R_{CD}}, \quad V_{CD} = E, \quad R_{CD} = ?\right)$$

$$R_{CD} = [(R_8 + R_9)//R_7] + R_6$$

3.4 Wye (Y) and delta (Δ) configurations and their equivalent conversions

3.4.1 Wye and delta configurations

Sometimes the circuit configurations will be neither in series nor in parallel, and the analysis method for series–parallel circuits described in previous chapters may not apply. For example, the configuration of three resistors R_a, R_b and R_c in the circuit of Figure 3.18(a) are neither in series nor in parallel. So how do we determine the equivalent resistance R_{eq} for this circuit? If we convert this to the configuration of resistors R_1, R_2 and R_3 in the circuit of Figure 3.18(b), the problem can be easily solved, i.e. $R_{eq} = [(R_1 + R_d) \,//\, (R_2 + R_e)] + R_3$.

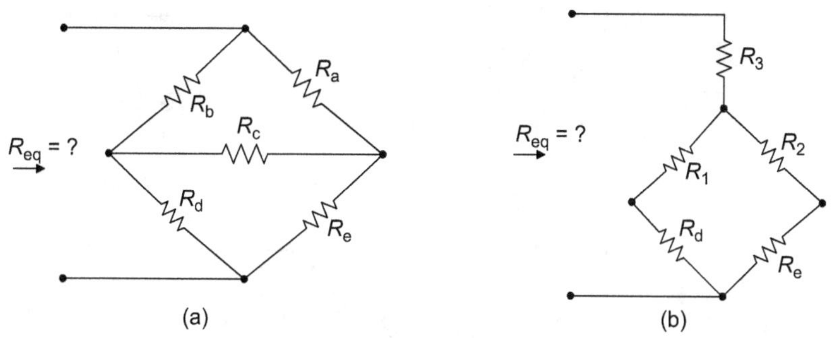

Figure 3.18 Delta (Δ) and wye (Y) conversions

The resistors of R_a, R_b and R_c in the circuit of Figure 3.18(a) are said to be in the delta (Δ) configuration; R_1, R_2 and R_3 in the circuit of Figure 3.18(b) is called the wye (Y) configuration. The delta and wye designations are from the fact that they look like a triangle and the letter Y, respectively, in electrical drawings. They are also referred to as tee (T) and pi (π) circuits as shown in Figure 3.19.

Wye (Y) and delta (Δ) configurations

Y or T configuration: 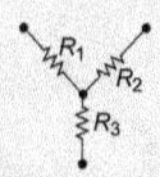

Δ or π configuration: 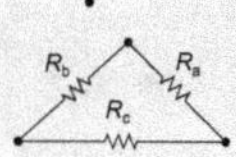

Wye (Y) and delta (Δ) configurations are often used in three phase AC circuits. They can also be used in the bridge circuit that will be discussed later. It is very important to know the conversion method of the two circuits and be able to convert back and forth between the wye (Y) and delta (Δ) configurations.

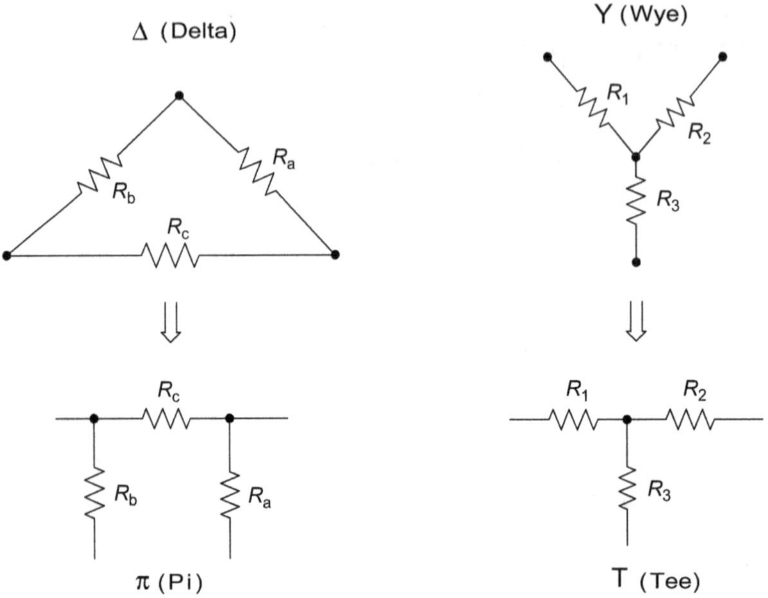

Figure 3.19 π and T configurations

3.4.2 Delta to wye conversion (Δ→Y)

There are three terminals in the delta (Δ) or wye (Y) configurations that can be connected to other circuits. The delta or wye conversion is used to establish equivalence for the circuits with three terminals, meaning that the resistors of the circuits between any two terminals must have the same values for both circuits as shown in Figure 3.20.

i.e. $R_{ac\,(Y)} = R_{ac\,(\Delta)} \quad R_{ab\,(Y)} = R_{ab\,(\Delta)} \quad R_{bc\,(Y)} = R_{bc\,(\Delta)}$

The following equations can be obtained from Figure 3.20:

$$R_{ac} = R_1 + R_3 = R_b//(R_a + R_c) = \frac{R_b(R_a + R_c)}{R_b + (R_a + R_c)} \tag{3.2}$$

$$R_{ab} = R_1 + R_2 = R_c//(R_a + R_b) = \frac{R_c(R_a + R_b)}{R_c + (R_a + R_b)} \tag{3.3}$$

$$R_{bc} = R_2 + R_3 = R_a//(R_b + R_c) = \frac{R_a(R_b + R_c)}{R_a + (R_b + R_c)} \tag{3.4}$$

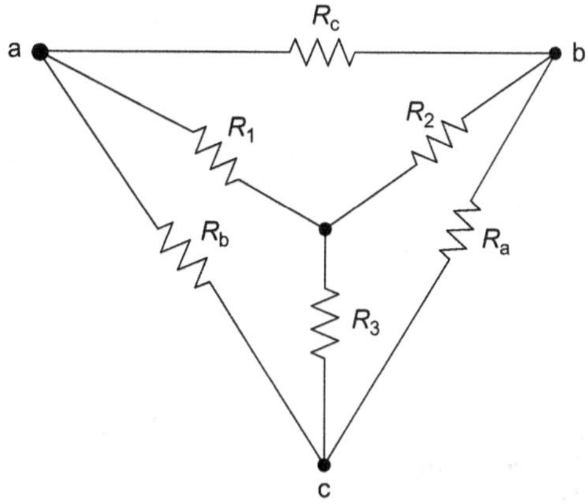

Figure 3.20 Delta and wye configurations

Subtracting equation (3.3) from the sum of equations (3.2) and (3.4) gives

$$(R_1 + R_3) + (R_2 + R_3) - (R_1 + R_2) = \frac{R_bR_a + R_bR_c}{R_b + (R_a + R_c)} + \frac{R_aR_b + R_aR_c}{R_a + (R_b + R_c)} - \frac{R_cR_a + R_cR_b}{R_c + (R_a + R_b)}$$

$$2R_3 = \frac{2R_aR_b}{R_a + R_b + R_c} \quad R_3 = \frac{R_aR_b}{R_a + R_b + R_c} \tag{3.5}$$

Similarly, subtracting equation (3.4) from the sum of equations (3.2) and (3.3) gives

$$R_1 = \frac{R_bR_c}{R_a + R_b + R_c} \tag{3.6}$$

And subtracting equation (3.2) from the sum of equations (3.3) and (3.4) gives

$$R_2 = \frac{R_aR_c}{R_a + R_b + R_c} \tag{3.7}$$

The circuit in delta configuration is converted to wye configuration as shown in Figure 3.21.

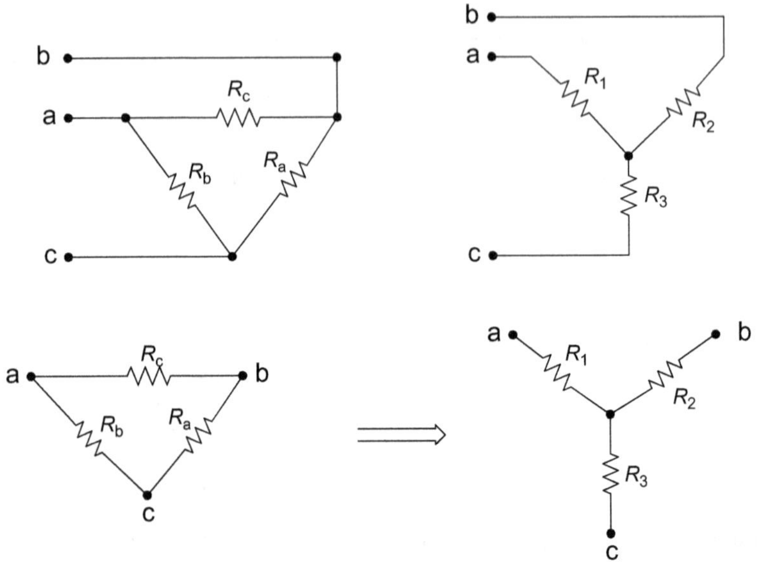

Figure 3.21 Delta converted to wye configuration

Equations for Δ→Y

$$R_1 = \frac{R_b R_c}{R_a + R_b + R_c}, \quad R_2 = \frac{R_a R_c}{R_a + R_b + R_c}, \quad R_3 = \frac{R_a R_b}{R_a + R_b + R_c}$$

3.4.3 Wye to delta conversion (Y→Δ)

The equations to convert wye to delta configuration can be derived based on (3.5–3.7).

Adding the products of equations (3.7) and (3.5), (3.5) and (3.6), and (3.6) and (3.7) gives

$$R_2R_3 + R_3R_1 + R_1R_2 = \frac{R_aR_cR_aR_b}{(R_a + R_b + R_c)^2} + \frac{R_aR_bR_bR_c}{(R_a + R_b + R_c)^2} + \frac{R_bR_cR_aR_c}{(R_a + R_b + R_c)^2}$$

$$R_2R_3 + R_3R_1 + R_1R_2 = \frac{R_aR_bR_c(R_a + R_b + R_c)}{(R_a + R_b + R_c)^2} = \frac{R_aR_bR_c}{R_a + R_b + R_c} \qquad (3.8)$$

Using (3.8) to divide (3.6), (3.7) and (3.5) individually will give the equations to convert wye to delta configurations as follows:

Equations for Y→Δ

$$R_a = \frac{R_1R_2 + R_2R_3 + R_3R_1}{R_1}, \quad R_b = \frac{R_1R_2 + R_2R_3 + R_3R_1}{R_2},$$

$$R_c = \frac{R_1R_2 + R_2R_3 + R_3R_1}{R_3}$$

For example, (3.8) divided by (3.6) gives

$$\frac{R_2R_3 + R_3R_1 + R_1R_2}{R_1} = \frac{R_aR_bR_c/(R_a + R_b + R_c)}{R_bR_c/(R_a + R_b + R_c)} = R_a$$

The circuit in wye configuration is converted to delta as shown in Figure 3.22.

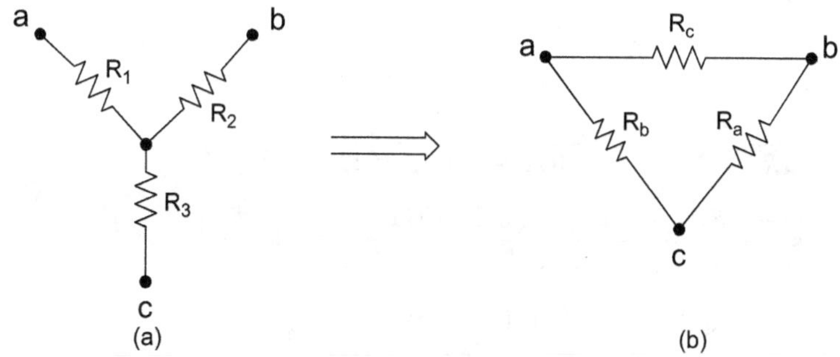

Figure 3.22 Wye converted to delta configuration

3.4.3.1 R_Y and R_Δ

If all resistors in the wye (Y) configuration have the same values, i.e. $R_1 = R_2 = R_3 = R_Y$, then all the resistances in the delta (Δ) configuration will also be the same, i.e. $R_a = R_b = R_c = R_\Delta$. This can be obtained from Y→Δ or Δ→Y conversion equations. Use Y→Δ as an example.

$$R_\Delta = R_a = \frac{R_1R_2 + R_2R_3 + R_3R_1}{R_1} = \frac{3R_YR_Y}{R_Y} = 3R_Y$$

$$R_\Delta = R_b = \frac{R_1R_2 + R_2R_3 + R_3R_1}{R_2} = \frac{3R_YR_Y}{R_Y} = 3R_Y$$

$$R_\Delta = R_c = \frac{R_1R_2 + R_2R_3 + R_3R_1}{R_3} = \frac{3R_YR_Y}{R_Y} = 3R_Y$$

So all delta resistances (R_Δ) have the same values.

$$R_a = R_b = R_c = R_\Delta = 3R_Y$$

In the above condition, the delta resistance R_Δ and wye resistance R_Y has the following relationship:

$$\text{If } R_a = R_b = R_c = R_\Delta, \quad R_1 = R_2 = R_3 = R_Y$$

$$R_Y = \frac{1}{3}R_\Delta \quad \text{or} \quad R_\Delta = 3R_Y$$

Example 3.12: Convert Δ to Y in the circuit of Figure 3.22, then Y to Δ to prove the accuracy of the equations. There the delta resistances $R_a = 30\ \Omega$, $R_b = 20\ \Omega$ and $R_c = 10\ \Omega$.

Solution:

$\Delta \to Y$:

$$R_3 = \frac{R_a R_b}{R_a + R_b + R_c} = \frac{(30\,\Omega)(20\,\Omega)}{30\,\Omega + 20\,\Omega + 10\,\Omega} = 10\,\Omega$$

$$R_2 = \frac{R_a R_c}{R_a + R_b + R_c} = \frac{(30\,\Omega)(10\,\Omega)}{30\,\Omega + 20\,\Omega + 10\,\Omega} = 5\,\Omega$$

$$R_1 = \frac{R_b R_c}{R_a + R_b + R_c} = \frac{(20\,\Omega)(10\,\Omega)}{30\,\Omega + 20\,\Omega + 10\,\Omega} \approx 3.33\,\Omega$$

$Y \to \Delta$:

$$R_a = \frac{R_1R_2 + R_2R_3 + R_3R_1}{R_1} = \frac{[(3.33)(5) + (5)(10) + (10)(3.33)]\,\Omega^2}{3.33\,\Omega}$$

$$= \frac{99.95\Omega^2}{3.33\Omega} \approx 30\,\Omega$$

$$R_b = \frac{R_1R_2 + R_2R_3 + R_3R_1}{R_2} = \frac{99.95\,\Omega^2}{5\,\Omega} \approx 20\,\Omega$$

$$R_c = \frac{R_1R_2 + R_2R_3 + R_3R_1}{R_3} = \frac{99.95\,\Omega^2}{10\,\Omega} \approx 10\,\Omega$$

The calculated delta resistances $R_a = 30\ \Omega$, $R_b = 20\ \Omega$ and $R_c = 10\ \Omega$ are the same with the resistances that were given (proved).

3.4.4 Using Δ → Y conversion to simplify bridge circuits

Sir Charles Wheatstone (1802–1875), a British physicist and inventor, is most famous for the Wheatstone bridge circuit. He was the first person who implemented the bridge circuit when he 'found' the description of the device. The bridge was invented by Samuel Hunter Christie (1784–1865), a British scientist. The Wheatstone bridge circuit can be used to measure unknown resistors. A basic Wheatstone bridge circuit is illustrated in Figure 3.23(a).

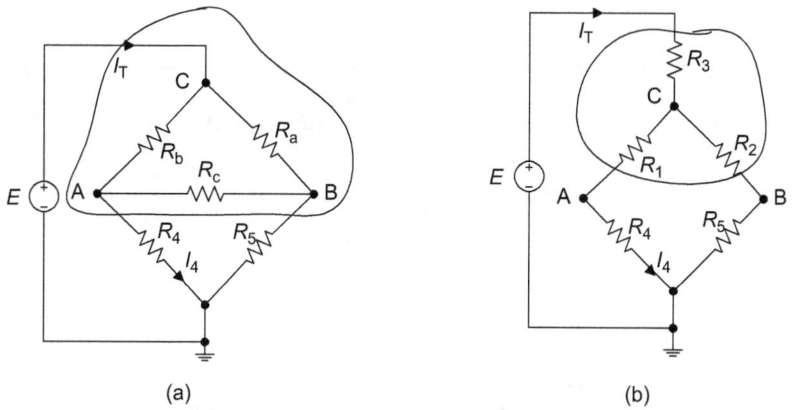

Figure 3.23 Wheatstone bridge circuit. (a) Delta (Δ) and (b) wye (Y)

Example 3.13: Determine the equations to calculate the total current I_T and branch current I_4 for the bridge circuit in Figure 3.23(a).

To determine the equation for the total current I_T of the bridge circuit, the equivalent resistance R_{eq} has to be determined first. The normal series–parallel analysis methods cannot be used to determine this resistance. However, Figure 3.23(a) can be converted to Figure 3.23(b) using the Δ → Y equivalent conversion, and R_1, R_2 and R_3 in Figure 3.23(b) can be determined by the equations of Δ → Y conversion.

$$R_1 = \frac{R_b R_c}{R_a + R_b + R_c}, \quad R_2 = \frac{R_a R_c}{R_a + R_b + R_c}, \quad R_3 = \frac{R_a R_b}{R_a + R_b + R_c}$$

The equivalent resistance R_{eq} of the bridge can be determined as follows:

$$R_{eq} = R_3 + [(R_1 + R_4)//(R_2 + R_5)]$$

The total current can be solved as $I_T = E/R_{eq}$.

The branch current $I_4 = I_T((R_2 + R_5)/[(R_1 + R_4) + (R_2 + R_5)])$ (the CDR).
If the wire between A and B in the circuit of Figure 3.23(a) is open, the equivalent resistance R_{eq} will be

$$R_{eq} = (R_b + R_4)//(R_a + R_5)$$

3.4.5 Balanced bridge

When the voltage across points A and B in a bridge circuit shown in Figure 3.24 is zero, i.e. $V_{AB} = 0$, the Wheatstone bridge is said to be balanced. A balanced Wheatstone bridge circuit can accurately measure an unknown resistor.

First determine the voltage V_{AB} in the points A and B.

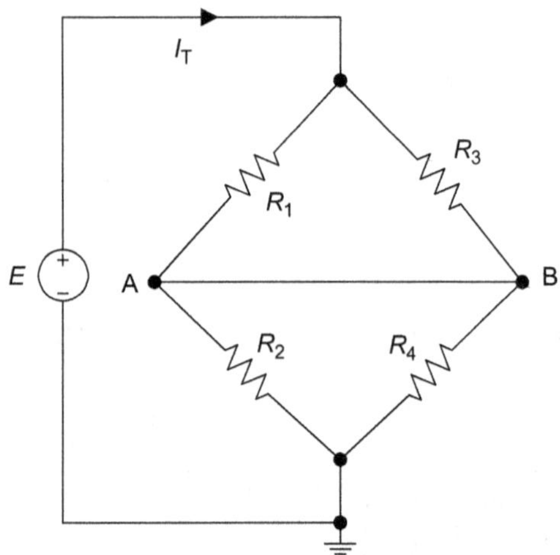

Figure 3.24 A balanced bridge

The voltage V_{AB} is the voltage from point A to ground (V_A) and then from ground to point B, i.e.

$$V_{AB} = V_A + (-V_B)$$

$$V_{AB} = E\frac{R_2}{R_1 + R_2} - E\frac{R_4}{R_3 + R_4} \quad \text{(the VDR)}$$

$$V_{AB} = E\frac{R_2(R_3 + R_4) - R_4(R_1 + R_2)}{(R_1 + R_2)(R_3 + R_4)} = E\frac{R_2R_3 - R_4R_1}{(R_1 + R_2)(R_3 + R_4)}$$

When $V_{AB} = 0$, or when the bridge is balanced, the numerator of the above equation will be zero, i.e.

$$R_2R_3 - R_4R_1 = 0, \quad \text{this gives:} \quad R_2R_3 = R_4R_1$$

Balanced bridge
When $V_{AB} = 0$, $R_2R_3 = R_4R_1$

3.4.6 Measure unknown resistors using the balanced bridge

The method of using the balanced bridge to measure an unknown resistor is as follows:

If the unknown resistor is in the position of R_4 in the circuit of Figure 3.24, using a variable (adjustable) resistor to replace R_2 and connecting a Galvanometer in between terminals A and B can measure the small current I_G in terminals A and B as shown in Figure 3.25. (A galvanometer is a type of ammeter that can measure small current accurately.)

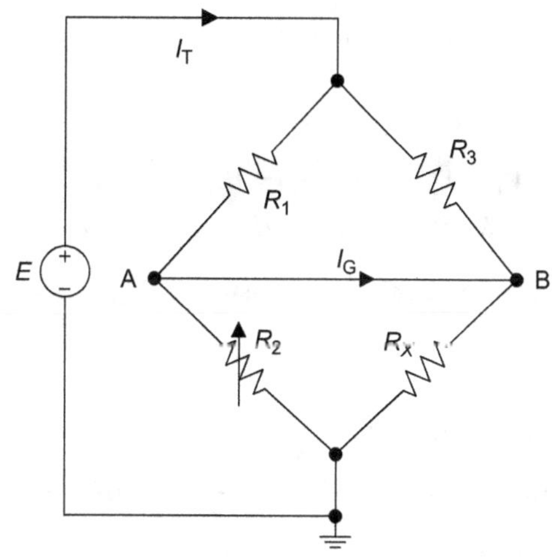

Figure 3.25 Measure an unknown R using a balanced bridge

Adjust R_2 until the current I_G measured by the Galvanometer or current in the A and B branch is zero ($I_G = 0$). This means $V_{AB} = 0$, or the bridge is balanced. $R_X = R_4$ at this time can be determined by the equation of the balanced bridge as follows:

From: $R_2R_3 = R_4R_1$

Solving for R_4: $R_X = R_4 = \dfrac{R_2R_3}{R_1}$

So the unknown resistor value can be determined from the ratios of the resistances in a balanced Wheatstone bridge.

Example 3.14: $R_1 = 100\ \Omega$, $R_2 = 330\ \Omega$ and $R_3 = 470\ \Omega$ in a balanced bridge circuit as shown in Figure 3.25. Determine the unknown resistance R_X.

Solution:

From $R_2R_3 = R_4R_1$ solving for R_4

$$R_X = R_4\frac{R_2R_3}{R_1} = \frac{(330\,\Omega)(470\,\Omega)}{100\,\Omega} = 1.551\ \text{k}\Omega$$

Summary

Series circuits

- Series circuits: All components are connected one after the other, there is only one circuit path, and the current flow through each component is always the same.
- Total series voltage:

$$V_T = E = V_1 + V_2 + \cdots + V_n = IR_T$$
$$V_T = IR_1 + IR_2 + \cdots + IR_n = IR_T$$

- Total series resistance (equivalent resistance R_{eq}): $R_T = R_1 + R_2 + \cdots + R_n$
- Series current:

$$I = \frac{V_T}{R_T} = \frac{E}{R_T} = \frac{V_1}{R_1} = \frac{V_2}{R_2} = \cdots = \frac{V_n}{R_n}$$

- Total series power: $P_T = P_1 + P_2 + \cdots + P_n = IE = I^2R_T = \frac{E^2}{R_T}$
- The VDR

  General form:

$$V_X = V_T\frac{R_X}{R_T} \quad \text{or} \quad V_X = E\frac{R_X}{R_T} \quad (X = 1, 2, \ldots, n)$$

  When there are only two resistors in series:

$$V_1 = V_T\frac{R_1}{R_1 + R_2}, \quad V_2 = V_T\frac{R_2}{R_1 + R_2}$$

- The earth ground: Connects to the earth ($V = 0$).
- Common ground or chassis ground: the common point for all components in the circuit ($V = 0$).

- Single-subscript notation: the voltage from the subscript with respect to ground.
- Double-subscript notation: the voltage across the two subscripts.

Parallel circuits

- Parallel circuits: The components are connected end to end, there are at least two current paths in the circuit, and the voltage across each component is the same.
- Parallel voltage: $V = E = V_1 = V_2 = \cdots = V_n$
- Parallel currents:

$$I_1 = \frac{V}{R_1}, \quad I_2 = \frac{V}{R_2}, \ldots, I_n = \frac{V}{R_n}$$

$$I_T = \frac{V}{R_{eq}} = I_1 + I_2 + \cdots + I_n$$

- Equivalent parallel resistance:

$$R_{eq} = \frac{1}{(1/R_1) + (1/R_2) + \cdots + (1/R_n)} = R_1//R_2//\cdots//R_n$$

When $n = 2$:

$$R_{eq} = \frac{R_1 R_2}{R_1 + R_2} = R_1//R_2$$

- Equivalent parallel conductance: $G_{eq} = G_1 + G_2 + \cdots + G_n$
- Total parallel power:

$$P_T = P_1 + P_2 + \cdots + P_n = I_T^2 R_{eq} = \frac{V^2}{R_{eq}} = I_T V$$

- The CDR
General form:

$$I_X = I_T \frac{R_{eq}}{R_X} \text{ or } I_X = I_T \frac{G_X}{G_{eq}}$$

(I_X and R_X are unknown current and resistance.)

When there are two resistors in parallel:

$$I_1 = I_T \frac{R_2}{R_1 + R_2}, \quad I_2 = I_T \frac{R_1}{R_1 + R_2}$$

Series–parallel circuits

- Series–parallel circuits are a combination of series and parallel circuits.
- Method for determining the equivalent resistance of series–parallel circuits:
  - Determine the equivalent resistance of the parallel part of the series–parallel circuits.
  - Determine the equivalent resistance of the series part of the series–parallel circuits.
  - Plot the equivalent circuit if necessary.
  - Repeat the above steps until the resistance in the circuit can be simplified to a single equivalent resistance R_{eq}.
- Method for analysing series–parallel circuits:
  - Apply Ohm's law to determine the total current:

$$I_T = \frac{E}{R_{eq}}$$

  - Apply VDR, CDR, Ohm's law, KCL and KVL to determine the unknown currents and voltages.

Wye and delta configurations and their conversions

- Y or T circuit: 

, Δ or π circuit: 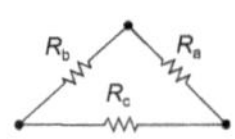

- $\Delta \rightarrow$ Y:

$$R_1 = \frac{R_b R_c}{R_a + R_b + R_c}, \quad R_2 = \frac{R_a R_c}{R_a + R_b + R_c}, \quad R_3 = \frac{R_a R_b}{R_a + R_b + R_c}$$

- Y $\rightarrow \Delta$:

$$R_a = \frac{R_1R_2 + R_2R_3 + R_3R_1}{R_1}, \quad R_b = \frac{R_1R_2 + R_2R_3 + R_3R_1}{R_2},$$

$$R_c = \frac{R_1R_2 + R_2R_3 + R_3R_1}{R_3}$$

- If $R_a = R_b = R_c = R_\Delta$ and $R_1 = R_2 = R_3 = R_Y$:

$$R_Y \frac{1}{3} R_\Delta \quad \text{or} \quad R_\Delta = 3R_Y$$

- The balanced bridge: When $V_{AB} = 0$, $R_2R_3 = R_4R_1$

Experiment 3: Series–parallel resistive circuits

Objectives

- Review series and parallel resistive circuits.
- Construct and analyse series–parallel resistive circuits.
- Measure voltages and currents for series–parallel resistive circuits.
- Review the applications of KCL and KVL.
- Verify the theoretical analysis, and compare the experimental results with theory calculations.
- Apply the CDR to circuit analysis.
- Design and test a voltage divider.
- Measure unknown resistors using a Wheatstone bridge circuit.

Background information

- Equivalent (or total) series resistance: $R_{eq} = R_T = R_1 + R_2 + \cdots + R_n$
- Equivalent parallel resistance:

$$R_{eq} = \frac{1}{(1/R_1) + (1/R_2) + \cdots + (1/R_n)} = R_1//R_2// \cdots //R_n$$

- When there are only two resistors in parallel:

$$R_{eq} = \frac{R_1 R_2}{R_1 + R_2} = R_1//R_2$$

- For a balanced Wheatstone bridge: when $V_{AB} = 0$, $R_2 R_3 = R_4 R_1$
- CDR:

$$I_X = I_T \frac{R_{eq}}{R_X}$$

When there are only two resistors in parallel:

$$I_1 = I_T \frac{R_2}{R_1 + R_2}, \quad I_2 = I_T \frac{R_1}{R_1 + R_2}$$

- VDR:

$$V_X = V_T \frac{R_X}{R_T} \quad \text{or} \quad V_X = E \frac{R_X}{R_T}$$

When there are only two resistors in parallel:

$$V_1 = V_T \frac{R_1}{R_1 + R_2}, \quad V_2 = V_T \frac{R_2}{R_1 + R_2}$$

Equipment and components

- Digital multimeter
- Breadboard
- DC power supply
- Switch
- Resistors:
  - four resistors with any values,
  - one 10 kΩ variable resistor,
  - 360 Ω, 510 Ω, 5.1 kΩ, 750 Ω, 1.2 kΩ, 2.4 kΩ, 5.1 kΩ, 910 Ω, 2.4 kΩ, 6.2 and 9.1 kΩ each and
  - two 1.1 kΩ.

Procedure

Part I: Equivalent series and parallel resistance

1. Take four unknown resistors and record their colour code resistor values in Table L3.1.
2. Get the multimeter to function as an ohmmeter and measure the values of these four resistors. Record the values in Table L3.1.

Table L3.1

Resistance	R_1	R_2	R_3	R_4
Colour code resistor value				
Measured value				

3. Connect four resistors in series with the DC power supply as shown in Figure L3.1. Adjust the source voltage to the suitable value according to the value of the resistors, and then connect the DC power supply to the circuit in Figure L3.1 (consult your instructor before you turn on the switch).

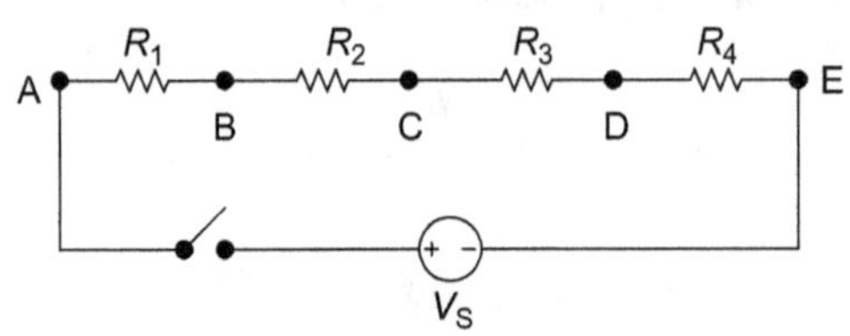

Figure L3.1 A series circuit

4. Calculate the equivalent series resistance R_{eq}, voltage across each resistor V_{AB}, V_{BC}, V_{CD}, V_{DE}, current I, and record in Table L3.2. Use direct method or indirect method to measure current I.

Recall:

- Direct method: Connect the multimeter (ammeter function) in series with the circuit components, turn on the switch and measure circuit current directly.
- Indirect method: Applying Ohm's law to calculate current by using the measured voltage and resistance.

5. Turn on the switch for the circuit in Figure L3.1, get the multimeter function as an ohmmeter, voltmeter and ammeter, respectively, and measure R_{eq}, V_{AB},, V_{BC}, V_{CD}, V_{DE}, current I. Record the values in Table L3.2.

Table L3.2

	R_{eq}	V_{AB}	V_{BC}	V_{CD}	V_{DE}	I
Formula for calculations						
Calculated value						
Measured value						

6. Connect four resistors in parallel with the DC power supply as shown in Figure L3.2.

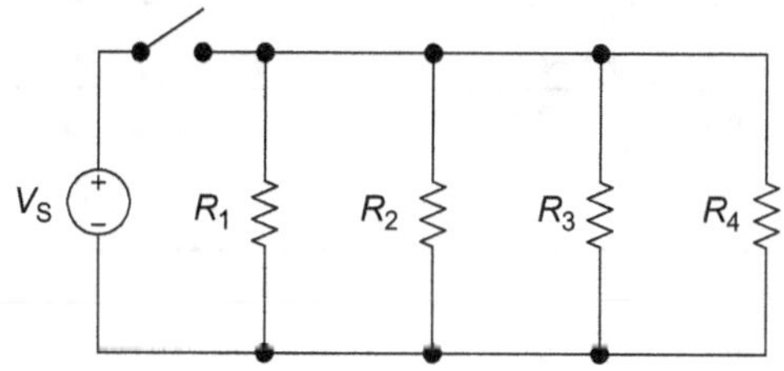

Figure L3.2 A parallel circuit

7. Calculate the equivalent parallel resistance R_{eq}, currents $I_{R_1}, I_{R_2}, I_{R_3}, I_{R_4}$ and I_T. Record the values in Table L3.3.
8. Turn on the switch for the circuit in Figure L3.2, measure R_{eq}, $I_{R_1}, I_{R_2}, I_{R_3}, I_{R_4}$ and I_T using the multimeter (ohmmeter and ammeter functions). Record the values in Table L3.3.

Table L3.3

	R_{eq}	I_T	I_{R_1}	I_{R_2}	I_{R_3}	I_{R_4}
Formula for calculations						
Calculated value						
Measured value						

Part II: Series–parallel resistive circuit

1. Connect a series–parallel circuit as shown in Figure L3.3 on the breadboard.

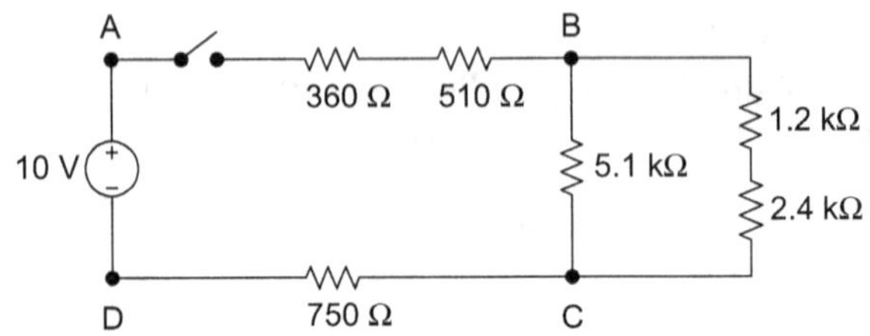

Figure L3.3 A series–parallel circuit

2. Calculate the equivalent resistance R_{eq}, currents I_T, $I_{5.1\ k\Omega}$ (current flowing through the branch of 5.1 kΩ resistor), and voltages V_{AB}, V_{BC} and V_{CD} for the circuit in Figure L3.3. Record the values in Table L3.4.
3. Turn on the switch for the circuit in Figure L3.3, measure R_{eq}, I_T, $I_{5.1\ k\Omega}$, V_{AB}, V_{BC} and V_{CD}. Record the values in Table L3.4.

Table L3.4

	R_{eq}	I_T	V_{AB}	V_{BC}	V_{CD}	$I_{5.1\ k\Omega}$
Formula for calculations						
Calculated value						
Measured value						

Part III: Voltage divider

1. Design and construct a voltage divider as shown in Figure L3.4. When $E = 12$ V, $V_A = 6$ V (voltage across the resistor R_2), and $I = 6$ mA, calculate resistance R_1 and R_2. Record the values in Table L3.5.
2. Measure resistance R_1 and R_2 using the multimeter (ohmmeter function) for the circuit in Figure L3.4. Record the values in Table L3.5.

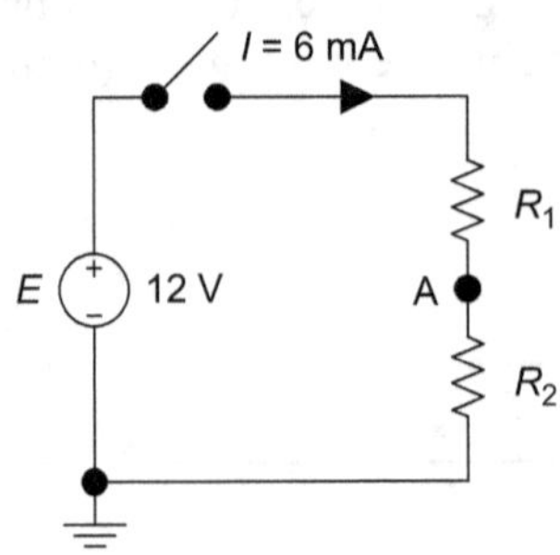

Figure L3.4 Voltage divider circuit

Table L3.5

	R_1	R_2	V_{R_1}	V_{R_2}
Formula for calculations				
Calculated value				
Measured value				

3. Calculate voltages V_{R_1} and V_{R_2} for the circuit in Figure L3.4. Record the values in Table L3.5.
4. Measure voltages V_{R_1} and V_{R_2} using the multimeter (voltmeter function) for the circuit in Figure L3.4. Record the values in Table L3.5.

Part IV: Wheatstone bridge

1. Measure the value of each resistor of R_X using the multimeter (ohmmeter function) in Table L3.6. Record the values in Table L3.6.

Table L3.6

Colour code value for R_X	**910 Ω**	**2.4 kΩ**	**6.2 kΩ**	**9.1 kΩ**
Multimeter measured R_X value				
Formula to calculate R_3				
Multimeter measured R_3 value				
Bridge measured R_X value				

2. Construct a bridge circuit as shown in Figure L3.5 on the breadboard, and connect the 910 Ω R_X resistor ($R_4 = R_X$) to the circuit.

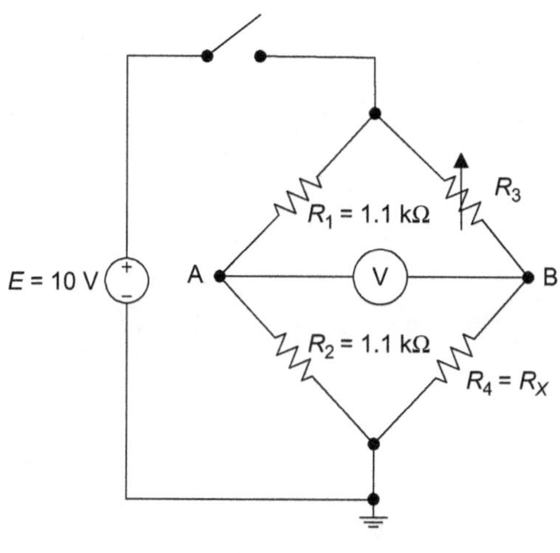

Figure L3.5 Bridge circuit

3. Calculate the value of the variable resistor R_3 for the balanced bridge circuit (when $R_X = 910\ \Omega$, and $V_{AB} = 0$) in Figure L3.5. Record the value in Table L3.6.
4. Turn on the switch, and carefully adjust the variable resistor R_3 when using the multimeter (voltmeter function) to measure the voltage across terminals A and B until the multimeter voltage is approximately zero.
5. Turn off the switch, use the multimeter (ohmmeter function) to measure the value of the variable resistor R_3. Record the value in Table L3.6.
6. Calculate the value of R_X when $V_{AB} = 0$ using the formula of the balanced bridge (use measured R_3 value). Record the values in Table L3.6.
7. Turn off the switch for the circuit in Figure L3.5, then replace the other three R_X resistors listed in Table L3.6 one by one to the circuit, and repeat steps 3 to 6.
8. Compare the measured and calculated R_X values. Are there any significant differences? If so, explain the reasons.

Conclusion

Write your conclusions below:

Chapter 4

Methods of DC circuit analysis

Objectives

After completing this chapter, you will be able to:

- convert voltage source to an equivalent current source and vice versa
- know the methods of voltage sources in series and parallel
- know the methods of current sources in series and parallel
- understand the branch current analysis method and apply it to circuit analysis
- understand the mesh analysis method and apply it to circuit analysis
- understand the node voltage analysis method and apply it to circuit analysis

4.1 Voltage source, current source and their equivalent conversions

4.1.1 Source equivalent conversion

It is sometimes easier to convert a current source to an equivalent voltage source or vice versa to analyse and calculate the circuits. The source *equivalent* conversion means that if loads are connected to both the terminals of the two sources after conversion, the load voltage V_L and current I_L of the two sources should be the same (Figure 4.1). So the source equivalent conversion actually means that the source terminals are equivalent, though the internal characteristics of each source circuit are not equivalent.

If the internal resistance R_S in Figure 4.1(a and b) is equal, the source voltage is $E = I_S R_S$ in Figure 4.1(a) and the source current is $I_S = E/R_S$ in Figure 4.1(b), then the current source and voltage source can be equivalently converted.

When performing the source equivalent conversion, we need to pay attention to the polarities of the sources. The reference polarities of voltage and current of the sources should be the same before and after the conversion as shown in Figures 4.1 and 4.2 (notice the polarities of sources E and I_S in the two figures).

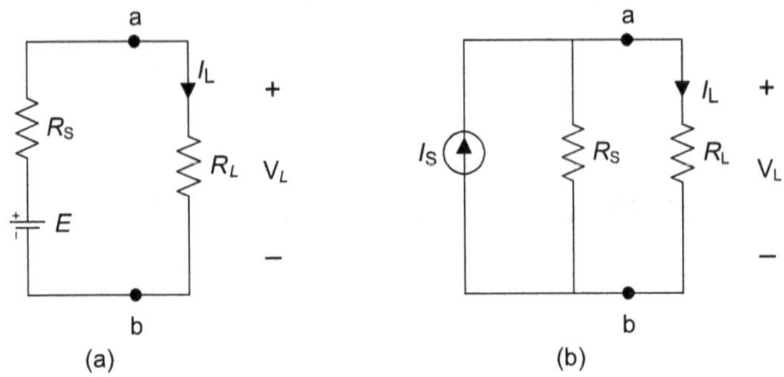

Figure 4.1 Sources equivalent conversion

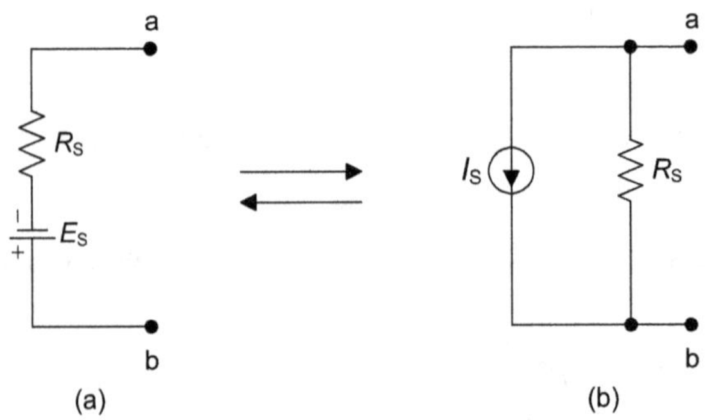

Figure 4.2 Polarity of conversion

Source equivalent conversion

- Voltage source → Current source $R_S = R_S$, $I_S = E/R_S$
- Current source → Voltage source $R_S = R_S$, $E = I_S/R_S$

The following procedure can verify that the load voltage V_L and load current I_L in two circuits of Figure 4.1(a and b) are equal after connecting a load resistor R_L to the two terminals of these circuits.

- The voltage source in Figure 4.1(a):

$$I_L = \frac{E}{R_S + R_L}$$
$$V_L = E\frac{R_L}{R_S + R_L} = I_S R_S \frac{R_L}{R_S + R_L}$$

(Applying the voltage-divider rule and $E = I_S R_S$)

- The current source in Figure 4.1(b):

$$I_L = I_S \frac{R_S}{R_S + R_L} = \frac{E}{R_S + R_L}$$
$$V_L = I_L R_L = \left(I_S \frac{R_S}{R_S + R_L}\right) R_L$$

(Applying the current-divider rule and $E = I_S R_S$)

So the load voltages and currents in the two circuits of Figure 4.1(a and b) are the same, and the source conversion equations have been proved.

Example 4.1: Convert the voltage source in Figure 4.3(a) to an equivalent current source and calculate the load current I_L for the circuit in Figure 4.3(a and b).

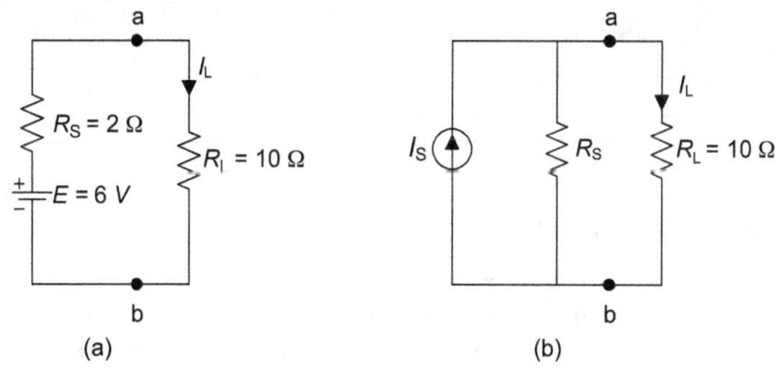

Figure 4.3 Circuit for Example 4.1

Solution: The equivalent current source after the source conversion is shown in Figure 4.3(b); R_S is still 2 Ω in Figure 4.3(b).

For Figure 4.3(b):

$$I_S = \frac{E}{R_S} = \frac{6\,\text{V}}{2\,\Omega} = 3\,\text{A}$$
$$I_L = I_S \frac{R_S}{R_S + R_L} = 3\,\text{A}\frac{2\,\Omega}{(2+10)\Omega} = 0.5\,\text{A}$$

For Figure 4.3(a):

$$I_L = \frac{E}{R_S + R_L} = \frac{6\,\text{V}}{(2+10)\Omega} = 0.5\,\text{A}$$

Example 4.2: Convert the current source in Figure 4.4(a) to an equivalent voltage source, and determine the voltage source E_S and internal resistance R_S in Figure 4.4(b).

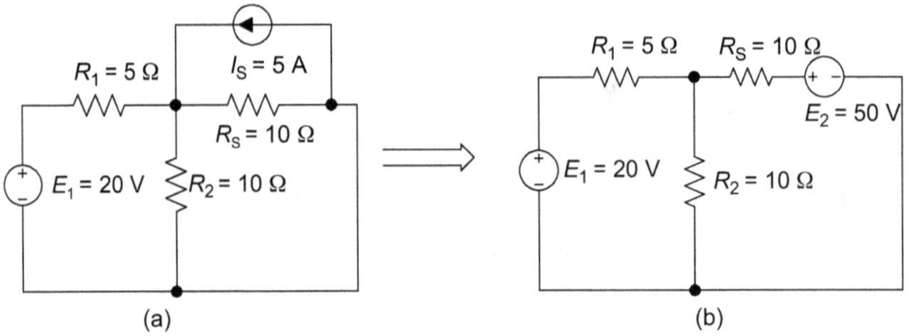

Figure 4.4 Circuit for Example 4.2

Solution:

$$R_S = 10\,\Omega$$
$$E_S = I_S R_S = (5\,\text{A})(10\,\Omega) = 50\,\text{V}$$

4.1.2 Sources in series and parallel

4.1.2.1 Voltage sources in series

A circuit of voltage sources in series and its equivalent circuit are shown in Figure 4.5. Voltage sources connected in series are similar with the resistors connected in series, that is the equivalent internal resistance R_S for series voltage sources is the sum of the individual internal resistances:

$$R_S = R_{S1} + R_{S2} + \cdots + R_{Sn}$$

and the equivalent voltage E or V_S for series voltage sources is the algebraic sum of the individual voltage sources:

$$E = E_1 + E_2 + \cdots + E_n$$

or

$$V_S = V_1 + V_2 + \cdots + V_n$$

Assign a positive sign (+) if the individual voltage has the same polarity as the equivalent voltage E (or V_S); assign a negative sign (−) if the individual voltage has a different polarity from the equivalent voltage E (or V_S) as shown in Figure 4.5.

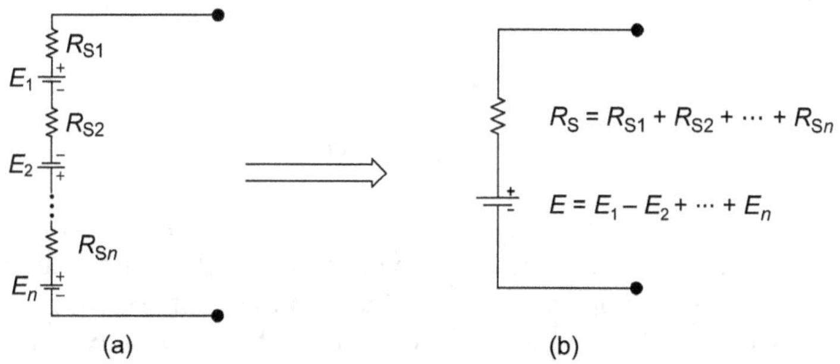

Figure 4.5 Voltage sources in series

A flashlight is an example of voltage sources in series, where batteries are connected in series to increase the total equivalent voltage.

Voltage sources in series

$R_S = R_{S1} + R_{S2} + \cdots + R_{Sn}$

$E = E_1 + E_2 + \cdots + E_n$ or $V_S = V_1 + V_2 + \cdots + V_n$

- Assign a +ve sign if E_n has same polarity as E (or V_S)
- Assign a −ve sign if E_n has different polarity from E (or V_S)

4.1.2.2 Voltage sources in parallel

A circuit of voltage sources in parallel and its equivalent circuit are shown in Figure 4.6. The equivalent voltage for the parallel voltage sources is the same as the voltage for each individual voltage source:

$$E = E_1 = E_2 = \cdots = E_n \quad \text{or} \quad V_S = V_{S1} = V_{S2} = \cdots = V_{Sn}$$

and the equivalent internal resistance R_S is the individual internal resistances in parallel:

$$R_S = R_{S1} \,//\, R_{S2} \,//\, \cdots \,//\, R_{Sn}$$

Note: Only voltage sources that have the same values and polarities can be connected in parallel by using the method mentioned above. If voltage sources

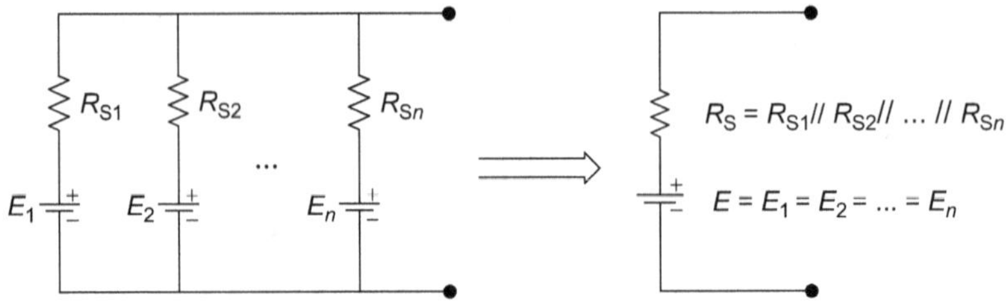

Figure 4.6 Voltage sources in parallel

having different values and polarities are connected in parallel, it can be solved by using Millman's theory, which will be discussed in chapter 5 (section 5.4).

An example of an application for voltage sources connected in parallel is for boosting (or jump starting) a 'dead' vehicle. You may have experienced using jumper cables by connecting the dead battery in parallel with a good car battery or with a booster (battery charger) to recharge the dead battery. It is the process of using the power from a charged battery to supplement the power of a discharged battery. It can provide twice the amount of current to the battery of the 'dead' vehicle and successfully start the engine.

Voltage sources in parallel

$R_S + R_{S1} \; // \; R_{S2} \; // \cdots // \; R_{Sn}$

$E = E_1 = E_2 = \cdots = E_n$ or $V_S = V_{S1} = V_{S2} = \cdots = V_{S_n}$

Only voltage sources that have the same values and polarities can be in parallel.

4.1.2.3 Current sources in parallel

A circuit of current sources in parallel and its equivalent circuit are shown in Figure 4.7. Current sources connected in parallel can be replaced by a single equivalent resistance R_S in parallel with a single equivalent current I_S.

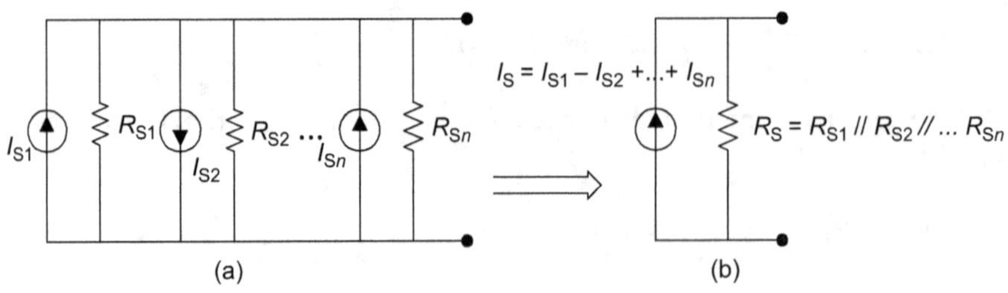

Figure 4.7 Current sources in parallel

The equivalent resistance R_S is the individual internal resistances in parallel:

$$R_S = R_{S1} \,//\, R_{S2} \,//\cdots//\, R_{Sn}$$

The equivalent current I_S is the algebraic sum of the individual current sources:

$$I_S = I_{S1} + I_{S2} + \cdots + I_{Sn}$$

Assign a positive sign (+) if the individual current is in the same direction as the equivalent current I_S; assign a negative sign (−) if the individual current is in a different direction from the equivalent current I_S.

Current sources in parallel
$R_S = R_{S1} \,//\, R_{S2} \,//\ldots//\, R_{Sn}$

$I_S = I_{S1} + I_{S2} + \cdots + I_{Sn}$

- Assign a +ve sign for I_{Sn} if it has the same polarity as I_S
- Assign a −ve sign for I_{Sn} if it has different polarity from I_S

4.1.2.4 Current sources in series

Only current sources that have the same polarities and same values can be connected in series. There is only one current path in a series circuit, so there must be only one current flowing through it. This is the same concept as Kirchhoff's current law (KCL), otherwise if the current entering point A does not equal the current exiting point A in Figure 4.8, KCL would be violated at point A.

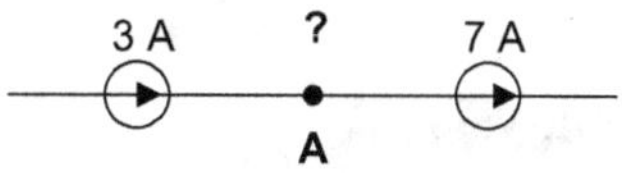

Figure 4.8 KCL is violated at point A

Current sources in series
Only current sources that have the same polarities and values can be connected in series.

Example 4.3: Determine the load voltage V_L in Figure 4.9.

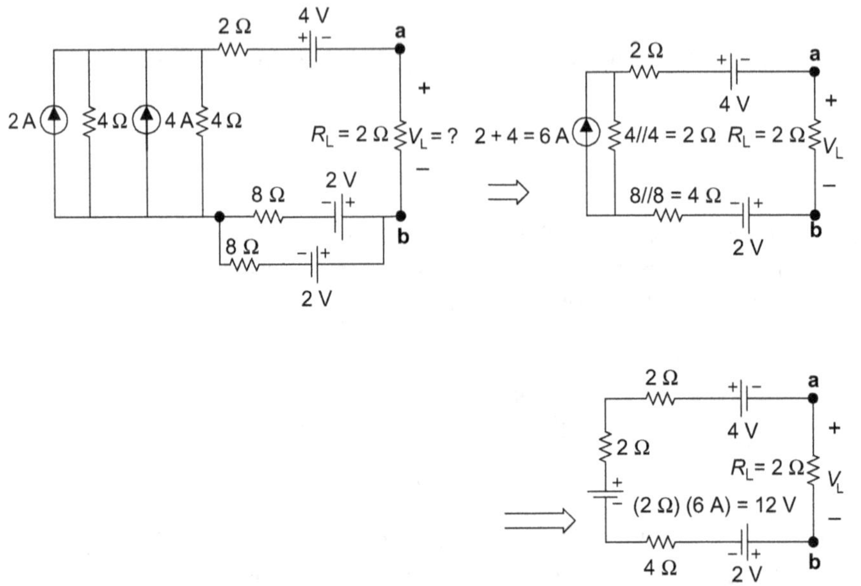

Figure 4.9 Circuit for Example 4.3

Solution: The process of source equivalent conversion is shown in the circuit of Figure 4.9. Determine V_L by using the voltage-divider rule as follows:

$$V_L = V_{ab} = IR_L = \frac{E}{R_T} R_L$$
$$= \frac{(-4 + 12 - 2)\text{V}}{(2 + 2 + 4 + 2)\Omega}(2\,\Omega) = 1.2\,\text{V}$$

4.2 Branch current analysis

The methods of analysis stated in chapter 3 are limited to an electric circuit that has a single power source. If an electric circuit or network has more than one source, it can be solved by the circuit analysis techniques that are discussed in chapters 4 and 5. The branch current analysis is one of several basic methods for analysing electric circuits.

The branch current analysis is a circuit analysis method that writes and solves a system of equations in which the unknowns are the branch currents. This method applies Kirchhoff's laws and Ohm's law to the circuit and solves the branch currents from simultaneous equations. Once the branch currents have been solved, other circuit quantities such as voltages and powers can also be determined.

The branch current analysis technique will use the terms node, branch and independent loop (or mesh); let us review the definitions of these terms.

- Node: The intersectional point of two or more current paths where current has several possible paths to flow.
- Branch: A current path between two nodes where one or more circuit components is in series.
- Loop: A complete current path that allows current to flow back to the start.
- Mesh: A loop in the circuit that does not contain any other loop (it can be analysed as a windowpane).

The circuits in Figure 4.10 have three meshes (or independent loops) and different number of nodes (the dark dots).

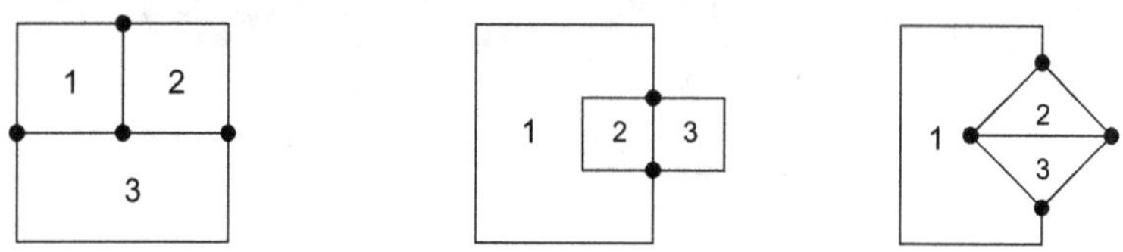

Figure 4.10 Nodes and meshes

Branch current analysis
A circuit analysis method that writes and solves a system of Kirchhoff's current law (KCL) and voltage law (KVL) equations in which the unknowns are the branch currents (it can be used for a circuit that has more than one source).

4.2.1 Procedure for applying the branch circuit analysis

1. Label the circuit.
   - Label all the nodes.
   - Assign an arbitrary reference direction for each branch current.
   - Assign loop direction for each mesh (choose clockwise direction).
2. Apply KCL to numbers of independent nodes ($n-1$), where n is the number of nodes.
3. Apply KVL to each mesh (or windowpane), and the number of KVL equations should be equal to the number of meshes, or Equation # = branch # – (nodes # – 1).

4. Solve the simultaneous equations resulting from steps 2 and 3, using determinant or substitution methods to determine each branch current.
5. Calculate the other circuit unknowns from the branch currents in the problem if necessary.

The procedure of applying the branch current analysis method is demonstrated in the following example.

Example 4.4: Use the branch current analysis method to determine each branch current, power on resistor R_2 and also the voltage across the resistor R_1 in the circuit of Figure 4.11.

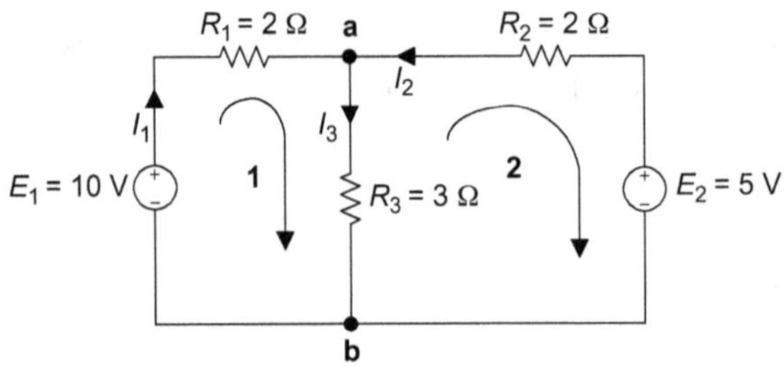

Figure 4.11 Circuit for Example 4.4

Solution: This circuit contains two voltage sources, and cannot be solved by using the methods we have learned in chapter 3; let us try to use the branch current analysis method.

1. Label the circuit as shown in Figure 4.11.
   - Label the nodes a and b.
   - Assign an arbitrary reference direction for each branch current as shown in Figure 4.11.
   - Assign clockwise loop direction for each mesh as shown in Figure 4.11.
2. Apply KCL to $(n-1) = (2-1) = 1$ number of independent nodes (there are two nodes a and b, and $n = 2$):

$$I_1 + I_2 = I_3 \tag{4.1}$$

3. Apply KVL to each mesh (windowpane). The number of KVL equations should be equal to the number of meshes. As there are two meshes in Figure 4.11, we should write two KVL equations.

Or Equation # = branch # – (nodes # – 1) = 3 – (2 – 1) = 2

Mesh 1: $I_1R_1 + I_3R_3 - E_1 = 0$ (4.2)

Mesh 2: $-I_2R_2 - I_3R_3 + E_2 = 0$ (4.3)

Recall KVL #1 ($\Sigma V = 0$): Assign a positive sign (+) for E or $V = IR$ if its reference polarity and loop direction are the same; otherwise assign a negative sign (−).

4. Solve the simultaneous equations resulting from steps 2 and 3, and determine branch currents I_1, I_2 and I_3 (three equations can solve three unknowns).
   - Rewrite the above three equations in standard form:

$$\begin{aligned} I_1 + I_2 - I_3 &= 0 \\ I_1R_1 + 0 + I_3R_3 &= E_1 \\ 0 - I_2R_2 - I_3R_3 &= -E_2 \end{aligned}$$

   - Substitute the values into equations:

$$\begin{aligned} I_1 + I_2 - I_3 &= 0 \\ 2I_1 + 0 + 3I_3 &= 10\,\text{V} \\ 0 - 2I_2 - 3I_3 &= -5\,\text{V} \end{aligned}$$

   - Solve simultaneous equations using determinant method:

$$\begin{aligned} \Delta &= \begin{vmatrix} 1 & 1 & -1 \\ 2 & 0 & 3 \\ 0 & -2 & -3 \end{vmatrix} \\ &= (1)(0)(-3) + (2)(-2)(-1) + (0)(1)(3) - (-1)(0)(0) \\ &\quad - (3)(-2)(1) - (-3)(2)(1) \\ &= 4 - (-6) - (-6) = 16 \end{aligned}$$

$$\begin{aligned} I_1 &= \frac{\begin{vmatrix} 0 & 1 & -1 \\ 10 & 0 & 3 \\ -5 & -2 & -3 \end{vmatrix}}{\Delta} = \frac{(10)(-2)(-1) + (-5)(3)(1) - (-3)(10)(1)}{16} \\ &\approx 2.19\,\text{A} \end{aligned}$$

$$\begin{aligned} I_2 &= \frac{\begin{vmatrix} 1 & 0 & -1 \\ 2 & 10 & 3 \\ 0 & -5 & -3 \end{vmatrix}}{\Delta} = \frac{(1)(10)(-3) + (2)(-5)(-1) - (-3)(-5)(1)}{16} \\ &\approx 0.31\,\text{A} \end{aligned}$$

$$I_3 = \frac{\begin{vmatrix} 1 & 1 & 0 \\ 2 & 0 & 10 \\ 0 & -2 & -5 \end{vmatrix}}{\Delta} = \frac{-(10)(-2)(1) - (-5)(2)(1)}{16} \approx 1.88\text{A}$$

$$I_1 \approx 2.19\,\text{A}, \quad I_2 \approx -0.31\,\text{A}, \quad I_3 \approx 1.88\,\text{A}$$

(Negative sign (–) for I_2 indicates that the actual direction of I_2 is opposite with its assigned reference direction.)

5. Calculate the other circuit unknowns from the branch currents:

$$P_2 = I_2^2 R_2 = (-0.31)^2(2) \approx 0.19\,\text{W}$$
$$V_1 = I_1 R_1 = (2.19)(2) = 4.38\,\text{V}$$

Example 4.5: Determine current I_3 in Figure 4.12 using the branch current analysis.

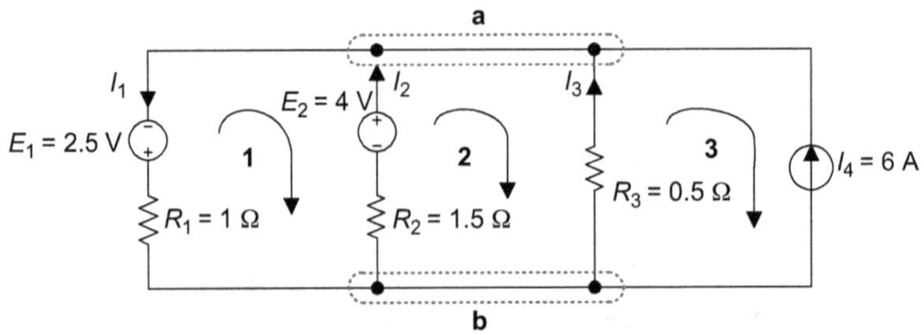

Figure 4.12 Circuit for Example 4.5

Solution:

1. Label the nodes, reference direction for branch currents and loop directions in the circuit as shown in Figure 4.12.
2. Apply KCL to $(n-1) = (2-1) = 1$ number of independent nodes (there are two nodes or supernodes a and b, so $n = 2$): $-I_1 + I_2 + I_3 + I_4 = 0$ ($I_4 = 6\,\text{A}$).
3. Apply KVL around each mesh (or windowpanes); there are three meshes in Figure 4.12, so you should write three KVL equations.

   There is no need to write KVL for mesh 1 since mesh 1 current is already known to be equal to the source current I_1 ($I_1 = 5$ A); therefore, the number of loop equations can be reduced from 3 to 2:

$$\text{Mesh 1:} \quad -I_1R_1 + E_1 + E_2 - I_2R_2 = 0$$
$$\text{Mesh 2:} \quad I_2R_2 - E_2 - I_3R_3 = 0$$

4. Solve the simultaneous equations resulting from steps 2 and 3, and determine the branch current I_2.

$$
\begin{aligned}
-I_1 - 1.5I_2 &= -2.5 - 4 \quad & -I_1 - 1.5I_2 + 0I_3 &= -6.5 \\
0I_1 + 1.5I_2 - 0.5I_3 &= 4 \quad & 0I_1 + 1.5I_2 - 0.5I_3 &= 4 \\
-I_1 + I_2 + I_3 &= -6 \quad & -I_1 + I_2 + I_3 &= -6
\end{aligned}
$$

Solve the above simultaneous equations using the determinant method:

$$
\Delta = \begin{vmatrix} -1 & -1.5 & 0 \\ 0 & 1.5 & -0.5 \\ -1 & 1 & 1 \end{vmatrix}
$$

$$
= (-1)(1.5)(1) + (-1)(-0.5)(-1.5) - (-0.5)(1)(-1) = -2.75
$$

$$
I_2 = \frac{\begin{vmatrix} -1 & -6.5 & 0 \\ 0 & 4 & -0.5 \\ -1 & -6 & 1 \end{vmatrix}}{\Delta}
$$

$$
= \frac{(-1)(4)(1) + (-1)(-0.5)(-6.5) - (-0.5)(-6)(-1)}{-2.75} \approx 1.55\ \text{A}
$$

$I_2 \approx 1.55\ \text{A}$

4.3 Mesh current analysis

The branch current analysis in section 4.2 is a circuit analysis method that writes and solves a system of KCL and KVL equations in which the unknowns are the *branch currents*. Mesh current analysis is a circuit analysis method that writes and solves a system of KVL equations in which the unknowns are the *mesh currents* (a current that circulates in the mesh). It can be used for a circuit that has more than one source.

The branch current analysis is a fundamental method for understanding mesh current analysis; mesh analysis is more practical and easier to use. Mesh current analysis uses KVL and does not need to use KCL. Applying KVL to get the mesh equations and solve unknowns implies that it will have less unknown variables, less simultaneous equations and therefore less calculation than branch current analysis. After solving mesh currents, the branch currents of the circuit will be easily determined.

Mesh current analysis

A circuit analysis method that writes and solves a system of KVL equations in which the unknowns are the *mesh currents* (it can be used for a circuit that has more than one source).

4.3.1 Procedure for applying mesh current analysis

1. Identify each mesh, and label all the nodes and reference directions for each mesh current (a current that circulates in the mesh) clockwise.
2. Apply KVL to each mesh of the circuit, and the number of KVL equations should be equal to the number of meshes (windowpanes).

   Or Equation # = branch # – (nodes # – 1)

   Assign a positive sign (+) for each self-resistor voltage, and a negative sign (–) for each mutual-resistor voltage in KVL equations.
   - Self-resistor: A resistor that is located in a mesh where only one mesh current flows through it.
   - Mutual resistor: A resistor that is located in a boundary of two meshes and has two mesh currents flowing through it.
3. Solve the simultaneous equations resulting from step 2 using determinant or substitution methods, and determine each mesh current.
4. Calculate the other circuit unknowns such as branch currents from the mesh currents in problem if necessary (choose the reference direction of branch currents first).

Note:
- Convert the current source to the voltage source first in the circuit, if there is any.
- If the circuit has a current source, the source current will be the same as the mesh current, so the number of KVL equations can be reduced.

The procedure for applying the mesh current analysis method is demonstrated in the following examples.

Example 4.6: Use the mesh current analysis method to determine each mesh current and branch currents I_{R_1}, I_{R_2} and I_{R_3} in the circuit of Figure 4.13.

Solution:

1. Label all the reference directions for each mesh current I_1 and I_2 (clockwise) as shown in Figure 4.13.

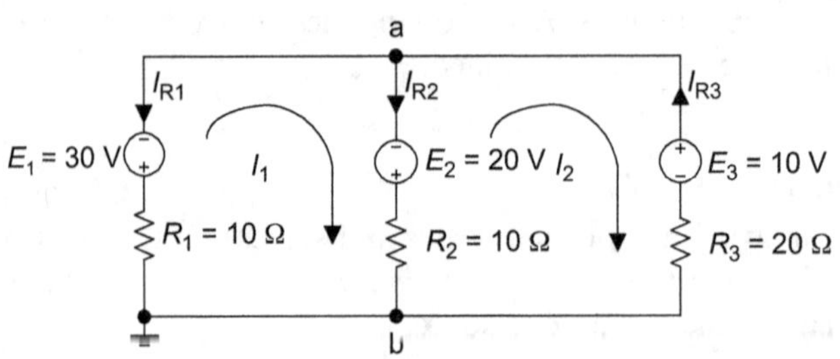

Figure 4.13 Circuit for Example 4.6

2. Apply KVL around each mesh (windowpane), and the number of KVL equations is equal to the number of meshes (there are two meshes in Figure 4.13). Alternatively, use the number of KVL: [branch # − (nodes # − 1)] = 3 − (2 − 1) = 2.

   Assign a positive sign (+) for each self-resistor voltage, and a negative sign (−) for each mutual-resistor voltage in KVL ($\Sigma V = \Sigma E$).

$$\text{Mesh 1: } (R_1 + R_2)\, I_1 - R_2\, I_2 = -E_1 + E_2$$

$$\text{Mesh 2: } -R_2\, I_1 + \;(R_2 + R_3)\, I_2 = -E_2 - E_3$$

Note: These equations were written by *inspection* of the circuit (inspection method):

	First column I_1		Second column I_2		Source E
Mesh 1:	(Self-resistor)I_1	−	(Mutual resistor)I_2	=	$-E_1 + E_2$
Mesh 2:	(Mutual resistor)I_1	+	(Self-resistor)I_2	=	$-E_2 - E_3$

3. Solve the simultaneous equations resulting from step 2, and determine the mesh currents I_1 and I_2:

$$\begin{aligned} &(10 + 10)I_1 - 10I_2 = -30 + 20 \\ &20I_1 - 10I_2 = -10 \end{aligned} \tag{4.4}$$

$$\begin{aligned} &-10I_1 + (10 + 20)I_2 = -20 - 10 \\ &-10I_1 + 30I_2 = -30 \end{aligned} \tag{4.5}$$

Solve for I_1 and I_2 using the substitution method as follows:

- Solve for I_1 from (4.4):

$$\begin{aligned} 20I_1 &= -10 + 10I_2 \\ I_1 &= -\frac{1}{2} + \frac{1}{2}I_2 \end{aligned} \tag{4.6}$$

- Substitute I_1 into (4.5) and solve for I_2:

$$-10\left(-\frac{1}{2} + \frac{1}{2}I_2\right) + 30I_2 = -30$$

$$I_2 = -1.4\,\text{A}$$

- Substitute I_2 into (4.6) and solve for I_1:

$$I_1 = \frac{1}{2} + \frac{1}{2}(-1.4)$$

$$I_1 = -0.2\,\text{A}$$

4. Assuming the reference direction of unknown branch current I_{R_2} as shown in Figure 4.13, calculate I_{R_2} from the mesh currents by applying KCL at node a:

$$\Sigma I = 0: -I_{R_1} - I_{R_2} + I_{R_3} = 0 \quad \text{or} \quad I_1 - I_{R_2} - I_2 = 0$$
$$(\text{since } I_1 = -I_{R_1} \text{ and } I_2 = -I_{R_3})$$
$$I_{R_2} = I_1 - I_2 = -0.2 - (-1.4) = 1.2\,\text{A}$$
$$I_{R_1} = -I_1 = 0.2\,\text{A}, \quad I_{R_3} = -I_2 = 1.4\,\text{A}$$

Example 4.7: Write the mesh equations using the mesh current analysis method for the circuit in Figure 4.14.

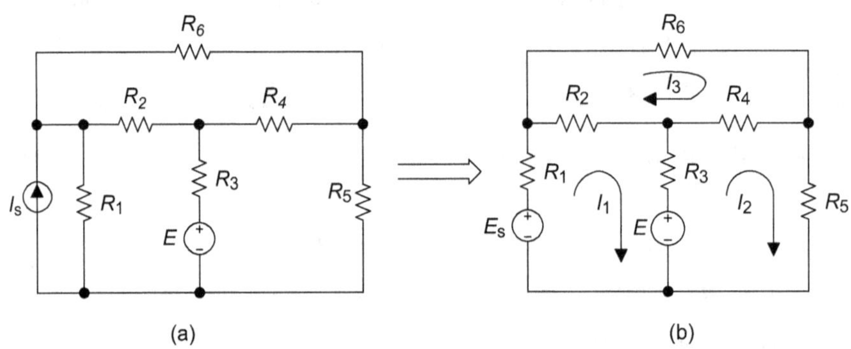

Figure 4.14 Circuit for Example 4.7 ($E_S = I_S R_1$)

Solution: Convert the current source to a voltage source as shown in Figure 4.14.

1. Label all the nodes and the reference directions for each mesh current (clockwise), as shown in Figure 4.14(b);
2. Apply KVL for each mesh (windowpane), and the number of KVL equations is equal to the number of meshes (there are three meshes in Figure 4.14(b))

$$\text{Mesh 1: } (R_1 + R_2 + R_3)I_1 - R_3I_2 - R_2I_3 = E_s - E$$
$$\text{Mesh 2: } -R_3I_1 + (R_3 + R_4 + R_5)I_2 - R_4I_3 = E$$
$$\text{Mesh 3: } -R_2I_1 - R_4I_2 + (R_2 + R_4 + R_6)I_3 = 0$$

4.4 Nodal voltage analysis

The node voltage analysis is another method for analysis of an electric circuit with two or more sources. The node voltage analysis is a circuit analysis method that writes and solves a set of simultaneous KCL equations in which the unknowns are the *node voltages*. Recall that node is the intersectional point of two or more current paths. Node voltage is voltage between a node and the reference node.

Node voltage analysis
A circuit analysis method that writes and solves a set of simultaneous KCL equations in which the unknowns are the *node voltages* (it can be used for a circuit that has more than one source).

4.4.1 Procedure for applying the node voltage analysis

1. Label the circuit.
   - Label all the nodes and choose one of them to be the reference node. Usually ground or the node with the most branch connections should be chosen as the reference node (at which voltage is defined as zero).
   - Assign an arbitrary reference direction for each branch current (this step can be skipped if using the inspection method).
2. Apply KCL to all $n - 1$ nodes except for the reference node (n is the number of nodes).
   - Method 1: Write KCL equations and apply Ohm's law to the equations; either resistance or conductance can be used. Assign a positive sign (+) for the self- resistor or self-conductance voltage and a negative sign (−) for the mutual-resistor or mutual-conductor voltage.
   - Method 2: Convert voltage sources to current sources and write KCL equations using the inspection method.
3. Solve the simultaneous equations and determine each nodal voltage.
4. Calculate the other circuit unknowns such as branch currents from the nodal voltages in the problem, if necessary.

The procedure to apply node voltage analysis method is demonstrated in the following example.

Example 4.8: Write the node voltage equations for the circuit shown in Figure 4.15(a) using node voltage analysis method.

Solution:

1. Label nodes a, b and c, and choose ground c to be the reference node; assign the reference current directions for each branch as shown in Figure 4.15(a).
2. Apply KCL to $n - 1 = 3 - 1 = 2$ nodes (nodes a and b).

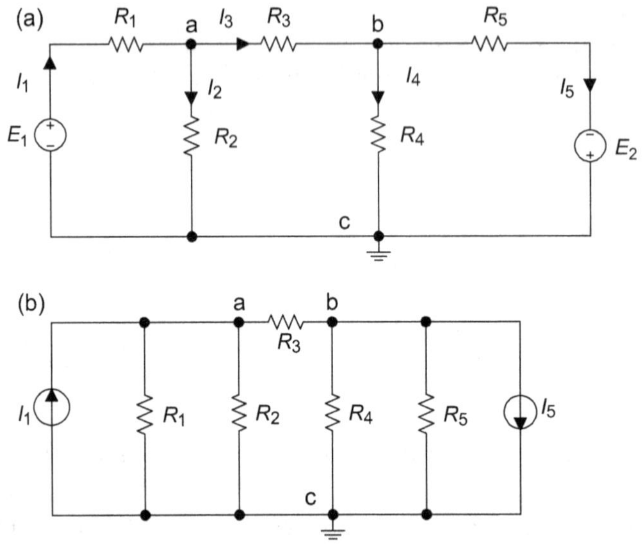

Figure 4.15 (a) Circuit for Example 4.8. (b) Circuit for method 2

- Method 1: Write KCL equations and apply Ohm's law to the equations.

$$\text{Node a:}\quad I_1 - I_2 - I_3 = 0,\quad \frac{E_1 - V_a}{R_1} - \frac{V_a}{R_2} - \frac{V_a - V_b}{R_3} = 0$$

$$\text{Node b:}\quad I_3 - I_4 - I_5 = 0,\quad \frac{V_a - V_b}{R_3} - \frac{V_b}{R_4} - \frac{V_b + E_2}{R_5} = 0$$

Or use conductance ($G = 1/\mathrm{R}$)

$$(E_1 - V_a)G_1 - V_aG_2 - (V_a - V_b)G_3 = 0$$
$$(V_a - V_b)G_3 - V_bG_4 - (V_b + E_2)G_5 = 0$$

- Method 2: Convert two voltage sources to current sources from Figure 4.15(a) to Figure 4.15(b), and write KCL equations by inspection.

– Use conductance:

First column (V_a)		Second column (V_b)		Source I_S
Node a: $(G_1 + G_2 + G_3)V_a$	$-$	G_3V_b	$=$	I_1
Node b: $-G_3V_a$	$+$	$(G_3 + G_4 + G_5)V_b$	$=$	$-I_5$

– Use resistance:

$$\left(\frac{1}{R_1}+\frac{1}{R_2}+\frac{1}{R_3}\right)V_\mathrm{a}-\frac{1}{R_3}V_\mathrm{b}=I_1$$

$$-\frac{1}{R_3}V_\mathrm{a}+\left(\frac{1}{R_3}+\frac{1}{R_4}+\frac{1}{R_5}\right)V_\mathrm{b}=-I_5$$

Note: The inspection method is similar with the one in mesh current analysis. The difference is that mesh current analysis uses *mesh currents* in each column, and node voltage analysis uses *node voltage* in each column. (Assign a positive sign (+) for the self-resistance/conductance voltage and entering node current, and a negative sign (–) for the mutual-conductor or mutual-resistor voltage and exiting node current.)

3. Two equations can solve two unknowns, which are the node voltages V_a and V_b.

Example 4.9: Use the node voltage analysis to calculate resistances R_1 and R_2, and current I_1 and I_2 for the circuit shown in Figure 4.16(a).

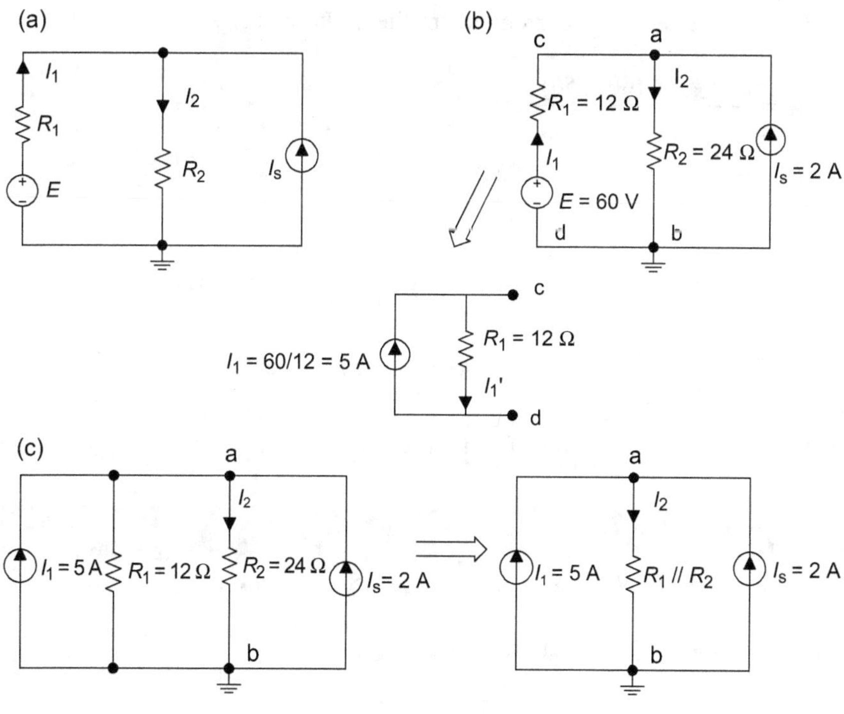

Figure 4.16 Circuits for Example 4.9

Solution:

- Label nodes a and b, and choose b to be the reference node, and assign the reference current direction for each branch as shown in Figure 4.16(b).
- Apply KCL to $n-1=2-1=1$ node (node a):
  - Use method 1: Write KCL equations and apply Ohm's law to the equations:

$$I_1 - I_2 + I_S = 0, \quad \frac{E-V_a}{R_1} - \frac{V_a}{R_2} + I_S = 0$$

  - Or use conductance: $(E-V_a)G_1 - V_aG_2 + I_s = 0$
- Solve the above equation and determine the node voltage V_a:

$$\frac{E}{R_1} - \frac{V_a}{R_1} - \frac{V_a}{R_2} + I_S = 0$$

$$\frac{E}{R_1} + I_S = V_a\left(\frac{1}{R_1} + \frac{1}{R_2}\right)$$

$$V_a = \frac{(E/R_1) + I_S}{(1/R_1) + (1/R_2)} = \frac{[(60/12) + 2]\mathrm{A}}{[(1/12) + (1/24)]\mathrm{S}} = \frac{7\,\mathrm{A}}{0.125\,\mathrm{S}} = 56\,\mathrm{V}$$

- Calculate the branch currents from the nodal voltages:

$$I_1 = \frac{E-V_a}{R_1} = \frac{(60-56)\mathrm{V}}{12\,\Omega} = 0.33\,\mathrm{A}$$

$$I_2 = -\frac{V_a}{R_2} = -\frac{56\,\mathrm{V}}{24\,\Omega} \approx -2.33\,\mathrm{A}$$

  - Use method 2: Convert voltage source to current source from the circuit of Figure 4.16(a) to the circuit of Figure 4.16(c):

$$I_1 = \frac{E}{R_1} = \frac{60}{12} = 5\,\mathrm{A} \quad R_1//R_2 = 12//24 = 8\,\Omega$$

Write KCL equation to node a using the inspection method:

$$\frac{V_a}{R_1//R_2} = I_1 + I_S$$

$$V_a = (I_1 + I_s)(R_1//R_2) = (5\,\mathrm{A} + 2\,\mathrm{A})(8\,\Omega) = 56\,\mathrm{V}$$

(V_a is the same as that from method 1)

Example 4.10: Write node voltage equations with resistances and conductances in the circuit of Figure 4.17 using the inspection method.

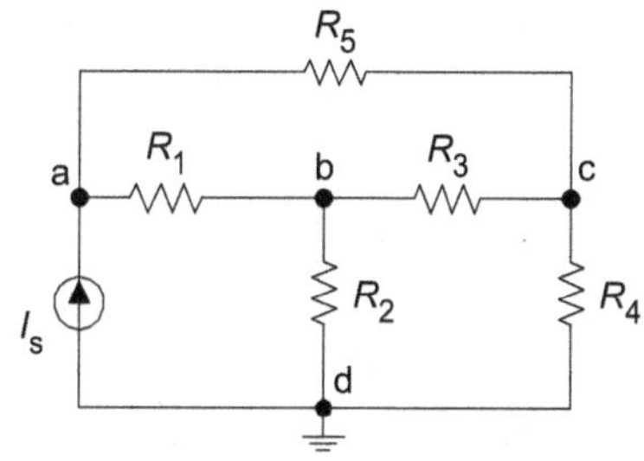

Figure 4.17 Circuit for Example 4.10

Solution:

1. Label all nodes a, b, c and d ($n = 4$) in the circuit as shown in Figure 4.17, and choose d to be the reference node. (The step to assign each branch current with reference direction can be skipped since it is used for the inspection method.)
2. Write KCL equations to $n - 1 = 4 - 1 = 3$ nodes using the inspection method.
   - Use resistance:

$$\text{Node a:} \quad \left(\frac{1}{R_1} + \frac{1}{R_5}\right) V_a - \frac{1}{R_1} V_b - \frac{1}{R_5} V_c = I_S$$

$$\text{Node b:} \quad -\frac{1}{R_1} V_a + \left(\frac{1}{R_1} + \frac{1}{R_2} + \frac{1}{R_3}\right) V_b - \frac{1}{R_3} V_C = 0$$

$$\text{Node c:} \quad -\frac{1}{R_5} V_a - \frac{1}{R_3} V_b + \left(\frac{1}{R_3} + \frac{1}{R_4} + \frac{1}{R_5}\right) V_C = 0$$

   - Use conductance:

$$\text{Node a:} \quad (G_1 + G_5)V_a - G_1 V_b - G_5 V_c = I_S$$

$$\text{Node b:} \quad -G_1 V_a + (G_1 + G_2 + G_3)V_b - G_3 V_c = 0$$

$$\text{Node c:} \quad -G_5 V_a - G_3 V_b + (G_3 + G_4 + G_5)V_c = 0$$

3. Three equations can solve three unknowns (node voltages V_a, V_b and V_c)

4.5 Node voltage analysis vs. mesh current analysis

The choice between mesh current analysis and node voltage analysis is often made on the basis of the circuit structure:

- The node voltage analysis is preferable for solving a circuit that is a parallel circuit, with current source(s), less nodes and more branches, and thus it is more convenient to solve the circuit unknowns.

- The mesh current analysis is preferable for solving a circuit that has fewer meshes, more nodes, with voltage sources and requires solving circuit branch currents.

Summary

Source equivalent conversions and sources in series and parallel

- Voltage source → Current source: $R_S = R_S, I_S = \dfrac{E}{R_S}$
- Current source → Voltage source: $R_S = R_S, E = I_S R_S$
- Voltage sources in series: $R_S = R_{S1} + R_{S2} + \cdots + R_{S}n$
  Assign a positive sign (+) if it has the same polarity with E (or V_S), otherwise assign a negative sign (−).
- Voltage sources in parallel: $\begin{matrix} R_S = R_{S1}//R_{S2}//\cdots//R_{Sn} \\ E = E_1 = E_2 = \cdots E_n \end{matrix}$

  Only voltage sources that have the same values and polarities can be in parallel.
- Current sources in series: Only current sources that have the same polarities and values can be in series.

Branch current analysis

A circuit analysis method that writes and solves a system of KCL and KVL equations in which the unknowns are the branch currents.

The procedure for applying the branch current analysis is given in Section 4.2.1.

Mesh current analysis

A circuit analysis method that writes and solves a system of KVL equations in which the unknowns are the *mesh currents* (it can be used for a circuit that has more than one source).

The procedure for applying the mesh current analysis is given in Section 4.3.1.

Nodal voltage analysis

A circuit analysis method that writes and solves a set of simultaneous of KCL equations in which the unknowns are the *node voltages*.

The procedure for applying the nodal voltage analysis is given in Section 4.4.1.

Note: The branch current analysis, mesh current analysis and node voltage analysis can be used for a circuit that has more than one source.

Experiment 4: Mesh current analysis and nodal voltage analysis

Objectives

- Construct circuits with two voltage sources.
- Experimentally verify the methods of solving a circuit with two power supplies.
- Experimentally verify the mesh current analysis method.
- Experimentally verify the node voltage analysis method.
- Analyse the experimental data, circuit behaviour and performance, and compare them to the theoretical equivalents.

Equipment and components

- Multimeter
- Breadboard
- Dual-output DC power supply
- Switches (2)
- Resistors: 1.8 kΩ (2), 3 kΩ, 8.2 kΩ, 9.1 kΩ and 3.9 kΩ

Background information

- Mesh current analysis: A circuit analysis method that writes and solves a system of KVL equations in which the unknowns are the *mesh currents*. It can be used for a circuit that has more than one source.
- Nodal voltage analysis: A circuit analysis method that writes and solves a set of simultaneous KCL equations in which the unknowns are the *node voltages*. It can be used for a circuit that has more than one source.

Lab procedure

Part I: Experimentally verify the mesh current analysis method

1. Construct a circuit as shown in Figure L4.1 on the breadboard.
2. Calculate voltage V_{R_2} and current I_{R_2} using the mesh current analysis method (assuming the switches are turned on). Record the values in Table L4.1.

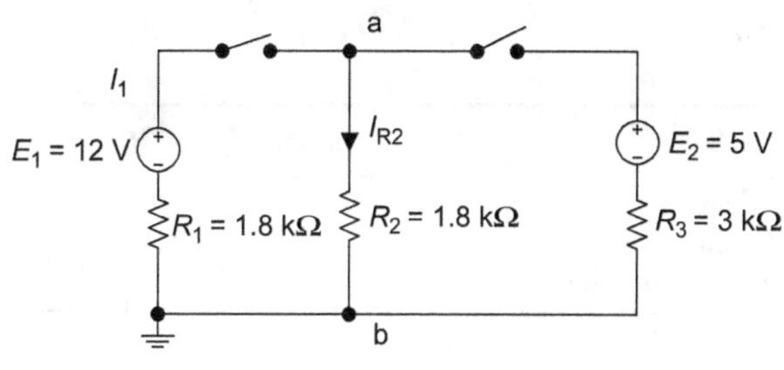

Figure L4.1 Circuit for mesh current analysis

Table L4.1 Circuit for mesh current analysis

Quantity	I_{R_2}	V_{R_2}
Formula for calculations		
Calculated value		
Measured value		

3. Set outputs of the dual-output power supply to 12 and 5 V, respectively, and turn on the two switches. Connect the multimeter (voltmeter function) in parallel to resistor R_2 and measure V_{R_2}. Record the values in Table L4.1.
4. Use direct method or indirect method to measure current I_{R_2}. Record the value in Table L4.1.
5. Compare the measured values and calculated values; are there any significant differences? If so, explain the reasons.

Part II: Experimentally verify the nodal voltage analysis method

1. Construct a circuit shown in Figure L4.2 on the breadboard.

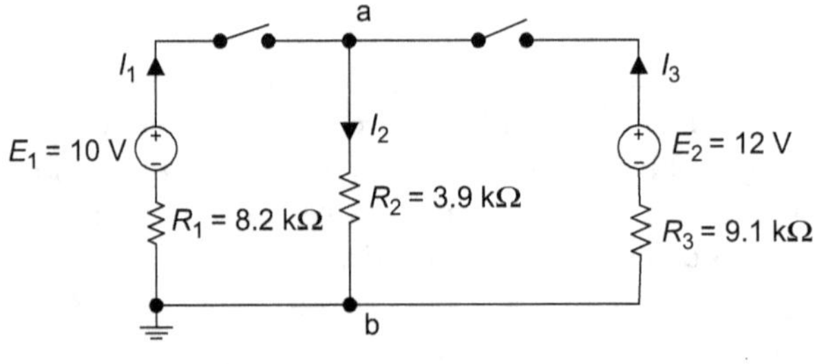

Figure L4.2 Circuit for nodal voltage analysis

2. Calculate the nodal voltage V_a using the nodal voltage analysis method (assuming the two switches are turned on). Record the value in Table L4.2.

Table L4.2 Circuit for nodal voltage analysis

	V_a	I_1	I_2	I_3
Formula for calculations				
Calculated value				
Measured value				

3. Calculate branch currents I_1, I_2 and I_3. Record the values in Table L4.2.

4. Set outputs of the dual-output power supply to 10 and 12 V, respectively, and then turn on the two switches. Connect the multimeter (voltmeter function) in parallel to resistor R_2 and measure V_a. Record the values in Table L4.2.
5. Measure branch currents I_1, I_2 and I_3 using either the direct method or indirect method. Record the values in Table L4.2.
6. Compare the measured values and calculated values; are there any significant differences? If so, explain the reasons.

Conclusion

Write your conclusions below:

Chapter 5

The network theorems

Objectives

After completing this chapter, you will be able to:

- understand the concept of linear circuits
- determine currents or voltages of networks using the superposition theorem
- understand Thevenin's and Norton's theorems and know how to convert their equivalent circuits
- determine currents or voltages of networks using Thevenin's and Norton's theorems
- understand the maximum power transfer theorem, and the method of transferring maximum power to the load
- determine currents or voltages of the parallel voltage source circuits using Millman's theorem
- determine currents or voltages of networks using the substitution theorem

The main methods for analysing series and parallel circuits in chapter 3 are Kirchhoff's laws. The branch current method, mesh or loop analysis method and node voltage analysis method also use KCL and KVL as the main backbone. When the practical circuits are more and more complex, especially in multi-loop electric circuits, the applications of the above methods solving for currents and voltages can be quite complicated. This is because you need to solve the higher-order mathematic equations when using these methods, i.e. you have to use complex algebra to handle multiple circuit unknowns.

The scientists working in the field of electrical engineering have developed more simplified theorems to analyse these kinds of complex circuits (the complicated circuit is also called the network). This chapter presents several theorems useful for analysing such complex circuits or networks. These theorems include the superposition theorem, Thevenin's theorem, Norton's theorem, Millman's theorem and the substitution theorem. In electrical network analysis, the fundamental rules are still Ohm's law and Kirchhoff's laws.

Network
A network is a complicated circuit.

Linearity property

The linearity property of a component describes a linear relationship between cause and effect. The pre-requirement of applying some of the above network theorems is that the analysed network must be a linear circuit. The components of a linear circuit are the linear components.

An example of linear component is a linear resistor. The voltage and current (input/output) of this linear resistor have a directly proportional (a straight line) relationship. A linear circuit has an output that is directly proportional to its input. The linear circuit can also be defined as follows: as long as the input/output signal timing does not depend on any characteristic of the input signal, it will be a linear circuit.

5.1 Superposition theorem

5.1.1 Introduction

When several power sources are applied to a single circuit or network at the same time, the superposition theorem can be used to separate the original network into several individual circuits for each power source working separately. Then, use series/parallel analysis to determine voltages and currents in the modified circuits. The actual unknown currents and voltages with all power sources can be determined by their algebraic sum; this is the meaning of the theorem's name – 'superimposed'. This method can avoid complicated mathematical calculations.

Superposition theorem
The unknown voltages or currents in a network are the sum of the voltages or currents of the individual contributions from each single power supply, by setting the other inactive sources to zero.

5.1.2 Steps to apply the superposition theorem

1. Turn off all power sources except one, i.e. replace the voltage source with the short circuit (placing a jump wire), and replace the current source with an open circuit. Redraw the original circuit with a single source.
2. Analyse and calculate this circuit by using the single source series–parallel analysis method.

3. Repeat steps 1 and 2 for the other power sources in the circuit.
4. Determine the total contribution by calculating the algebraic sum of all contributions due to single sources.

(The result should be positive when the reference polarity of the unknown in the single source circuit is the same as the reference polarity of the unknown in the original circuit; otherwise it should be negative.)

Note: The superposition theorem can be applied to the linear network to determine only the unknown currents and voltages. It cannot calculate power, since power is a nonlinear variable. Power can be calculated by the voltages and currents that have been determined by the superposition theorem.

Example 5.1: Determine the branch current I_c in the circuit of Figure 5.1(a) by using the superposition theorem.

Solution:

1. Choose E_1 to apply to the circuit first and use a jump wire to replace E_2 as shown in Figure 5.1(b).

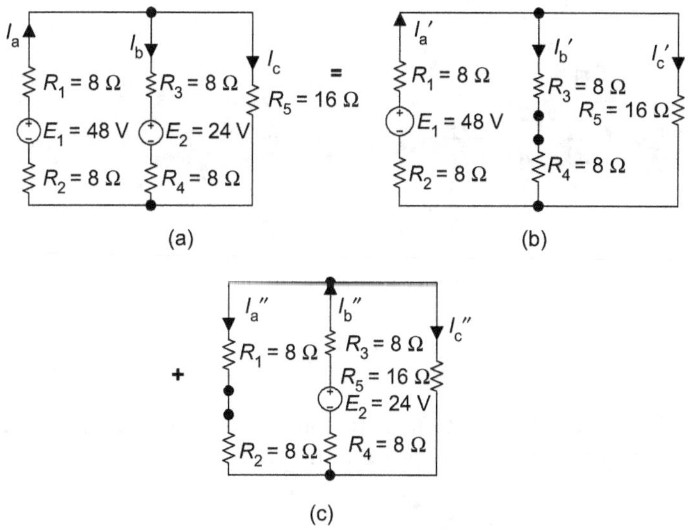

Figure 5.1 Circuit for Example 5.1

2. Determine I_c' in the circuit of Figure 5.1(b):

$$R'_{eq} = R_5//(R_3 + R_4) + (R_1 + R_2)$$
$$= \left[\frac{16 \times (8+8)}{16 + (8+8)} + (8+8)\right]\Omega = 24\,\Omega$$

(View from the E_1 branch in the circuit of Figure 5.1(b) to determine R_{eq}'.)

$$I_a' = \frac{E_1}{R_{eq}'} = \frac{48\,\text{V}}{24\,\Omega} = 2\,\text{A}$$

$$I_c' = I_a' \frac{R_3 + R_4}{R_3 + R_4 + R_5}$$
$$= 2\,\text{A} \frac{(8+8)\Omega}{(8+8+16)\Omega} = 1\,\text{A}$$

3. When E_2 is applied to the circuit, replace E_1 with a short circuit as shown in Figure 5.1(c) and calculate I_c'':

$$R_{eq}'' = R_5//(R_1 + R_2) + (R_3 + R_4)$$
$$= \left[\frac{16 \times (8+8)}{16 + (8+8)} + (8+8)\right]\Omega = 24\,\Omega$$

(View from the E_2 branch in the circuit of Figure 5.1(c) to determine R_{eq}''.)

$$I_b'' = \frac{E_2}{R_{eq}''} = \frac{24\,\text{V}}{24\,\Omega} = 1\,\text{A}$$

$$I_c = I_b \frac{R_1 + R_2}{R_1 + R_2 + R_5}$$
$$= 1\,\text{A} \frac{(8+8)\Omega}{(8+8+16)\Omega} = 0.5\,\text{A}$$

4. Calculate the sum of currents I_c' and I_c'':

$$I_c = I_c' + I_c = (1 + 0.5)\text{A} = 1.5\,\text{A}$$

Example 5.2: Determine the branch current I_2 and power P_2 of the circuit in Figure 5.2(a) by using the superposition theorem.

Solution:

1. When E is applied only to the circuit (using an open circuit to replace the current source I_1), calculate I_2' by assuming the reference direction of I_2' as shown in Figure 5.2(b).
2. Determine I_2' in the circuit of Figure 5.2(b):

$$I_2' = \frac{E}{(R_2//R_3) + R_1}$$
$$= \frac{25\,\text{V}}{[((100 \times 100)/(100 + 100)) + 50]\Omega} = 0.25\,\text{A} = 250\,\text{mA}$$

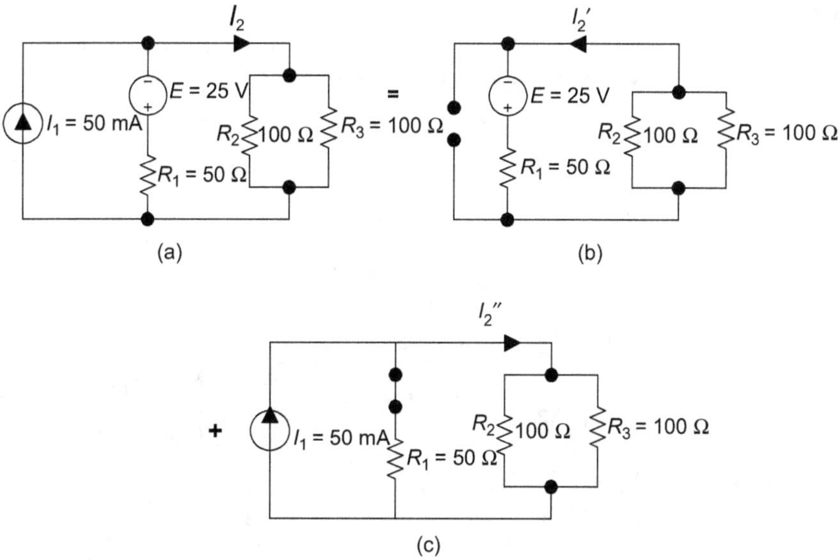

Figure 5.2 Circuit for Example 5.2

3. When the current source I_1 is applied only to the circuit (the voltage source E is replaced by a jump wire), the circuit is as shown in Figure 5.2(c). Calculate I_2' by assuming the reference direction of I_2'' as shown in the circuit of Figure 5.2(c):

$$I_2'' = I_1 \frac{R_1}{R_1 + R_2//R_3}$$

$$= 50\,\text{mA} \frac{50\,\Omega}{50 + [(100 \times 100)/(100 + 100)]\Omega} = 25\,\text{mA}$$

(Apply the current divider rule to the branches R_1 and R_2 // R_3.)

4. Calculate the sum of currents I_2' and I_2'':

$$I_2 = -I_2' + I_2'' = (-250 + 25)\text{mA} = -225\,\text{mA} = -0.225\,\text{A}$$

I_2' is negative as its reference direction in Figure 5.2(b) is opposite to that of I_2 in the original circuit of Figure 5.2(a). The negative I_2 implies that the actual direction of I_2 in Figure 5.2(a) is opposite to its reference direction.

Determine the power P_2: $P_2 = I_2^2 R_2 = (-0.225\,\text{A})^2(100\,\Omega) \approx 5.06\,\text{W}$

Example 5.3: Determine the branch current I_3 in the circuit of Figure 5.3(a) using the superposition theorem.

Solution:

1. Choose E_1 to apply to the circuit first and use a jump wire to replace E_2 and an open circuit to replace the current source I as shown in Figure 5.3(b).

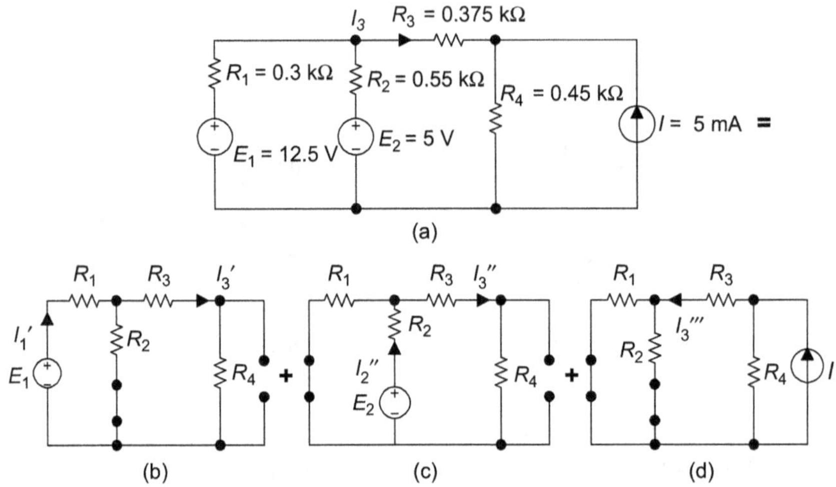

Figure 5.3 Circuit for Example 5.3

2. Use the circuit in Figure 5.3(b) to determine I_3':

$$I_3' = I_1' \frac{R_2}{R_2 + (R_3 + R_4)}$$

$$= (19.84\,\text{mA}) \frac{0.55\,\text{k}\Omega}{0.55\,\text{k}\Omega + (0.375 + 0.45)\text{k}\Omega} \approx 7.94\,\text{mA}$$

(Apply the current divider rule to the branches R_2 and $(R_3 + R_4)$.)
There

$$I_1' = \frac{E_1}{R_{eq}'} = \frac{E_1}{(R_3 + R_4)//R_2 + R_1}$$

$$= \frac{12.5\,\text{V}}{[((0.375 + 0.45) \times 0.55)/((0.375 + 0.45) + 0.55) + 0.3]\text{k}\Omega}$$

$$\approx 19.84\,\text{mA}$$

3. • Use the circuit in Figure 5.3(c) to determine I_3'':

$$I_3'' = I_2'' \frac{R_1}{R_1 + (R_3 + R_4)}$$
$$= (6.49\,\text{mA}) \frac{0.3\,\text{k}\Omega}{[0.3 + (0.375 + 0.45)]\text{k}\Omega} \approx 1.73\,\text{mA}$$

(Apply the current divider rule to the branches R_1 and $(R_3 + R_4)$.)
There

$$I_2'' = \frac{E_2}{R_{\text{eq}}''} = \frac{E_2}{(R_3 + R_4)//R_1 + R_2}$$
$$= \frac{5\,\text{V}}{[((0.375 + 0.45)(0.3))/((0.375 + 0.45) + 0.3) + 0.55]\text{k}\Omega}$$
$$\approx 6.49\,\text{mA}$$

• Use the circuit in Figure 5.3(d) to determine I_3''':

$$I_3''' = I \frac{R_4}{(R_1//R_2 + R_3) + R_4}$$
$$= (5\,\text{mA}) \frac{0.45\,\text{k}\Omega}{[((0.3 \times 0.55)/(0.3 + 0.55) + 0.375) + 0.45]\text{k}\Omega} \approx 2.21\,\text{mA}$$

(Apply the current divider rule to the branches R_4 and $(R_1 // R_2 + R_3)$.)

4. Calculate the sum of currents I_3', I_3'' and I_3''':

$$I_3 = I_3' + I_3'' - I_3''' = (7.94 + 1.73 - 2.21)\text{mA} = 7.46\,\text{mA}$$

I_3''' is negative since its reference direction is opposite to that of I_3 in the original circuit of Figure 5.3(a).

5.2 Thevenin's and Norton's theorems

5.2.1 Introduction

Thevenin's and Norton's theorems are two of the most widely used theorems to simplify the linear circuit for ease of network analysis. In 1883, French telegraph engineer M. L. Thevenin published his theorem of network analysis method. Forty-three years later, American engineer E. L. Norton in Bell Telephone laboratory published a similar theorem, but he used the current source to replace the voltage source in the equivalent circuit. These two theorems state that any complicated linear two-terminal network with power supplies can be simplified to an equivalent circuit that includes an actual voltage source (Thevenin's theorem) or an actual current source (Norton's theorem).

Here, 'the linear two-terminal network with power supplies' means:

- network: the relatively complicated circuit
- linear network: the circuits in the network are the linear circuits
- two-terminal network: the network with two terminals that can be connected to the external circuits
- network with the power supplies: network includes the power supplies

No matter how complex the inside construction of any two-terminal network with power supplies is, they can all be illustrated in Figure 5.4(a).

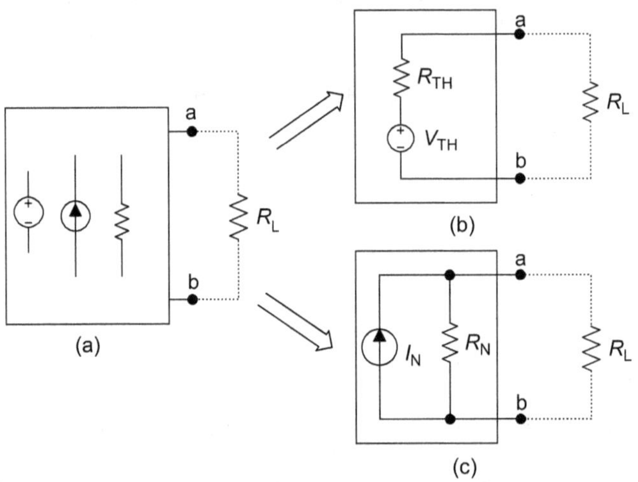

Figure 5.4 Thevenin's and Norton's theorems. (a) Linear two-terminal network with the power supply. (b) Thevenin's theorem. (c) Norton's theorem

According to Thevenin's and Norton's theorems, we can draw the following conclusion: any linear two-terminal network with power supplies can be replaced by an equivalent circuit as shown in Figure 5.4(b or c). The *equivalent* means that any load resistor branch (or unknown current or voltage branch) connected between the terminals of Thevenin's or Norton's equivalent circuit will have the same current and voltage as if it were connected to the terminals of the original circuit.

Thevenin's and Norton's theorems allow for analysis of the performance of a circuit from its terminal properties only.

Any linear two-terminal network with power supplies can be replaced by a simple equivalent circuit, which has a single power source and a single resistor.

- Thevenin's theorem: Thevenin's equivalent circuit is a *voltage source* – with an equivalent resistance R_{TH} in series with an equivalent voltage source V_{TH}.
- Norton's theorem: Norton's equivalent circuit is a *current source* – with an equivalent resistance R_N in parallel with an equivalent current source I_N.

In short we can conclude that any combination of power supplies and resistors with two terminals can be replaced by a single voltage source and a single series resistor for Thevenin's theorem, and replaced by a single current source and a single parallel resistor for Norton's theorem.

The key to applying these two theorems is to determine the equivalent resistance R_{TH} and the equivalent voltage V_{TH} for Thevenin's equivalent circuit, the equivalent resistance R_N and the equivalent current I_N for Norton's equivalent circuit. The value of R_N in Norton's equivalent circuit is the same as R_{TH} of Thevenin's equivalent circuit.

Note: The 'TH' in V_{TH} and R_{TH} means Thevenin, and the 'N' in I_N and R_N means Norton.

These two theorems are used very often to calculate the load (or a branch) current or voltage in practical applications. The load resistor can be varied sometimes (for instance, the wall plug can connect to 60 or 100 W lamps). Once the load is changed, the whole circuit has to be re-analysed or re-calculated. But if Thevenin's and Norton's theorems are used, Thevenin's and Norton's equivalent circuits will not be changed except for their external load branches. The variation of the load can be determined more conveniently by using Thevenin's or Norton's equivalent circuits.

5.2.2 Steps to apply Thevenin's and Norton's theorems

1. Open and remove the load branch (or any unknown current or voltage branch) in the network, and mark the letters a and b on the two terminals.
2. Determine the equivalent resistance R_{TH} or R_N. It equals the equivalent resistance, looking at it from the a and b terminals when all sources are turned off or equal to zero in the network. (A *voltage source* should be replaced by a *short circuit*, and a *current source* should be replaced by an *open circuit*.) That is

   $$R_{TH} = R_N = R_{ab}$$

3. Determine Thevenin's equivalent voltage V_{TH}. It equals the open-circuit voltage from the original linear two-terminal network of a and b, i.e.

$$V_{TH} = V_{ab}$$

4. Determine Norton's equivalent current I_N. It equals the short-circuit current from the original linear two-terminal network of a and b, i.e.

$$I_N = I_{sc} \text{ (where 'sc' means the short circuit)}$$

5. Plot Thevenin's or Norton's equivalent circuit, and connect the load branch (or unknown current or voltage branch) to a and b terminals of the equivalent circuit. Then the load (or unknown) voltage or current can be determined.

The above procedure for analysing circuits by using Thevenin's and Norton's theorems is illustrated in the circuits of Figure 5.5.

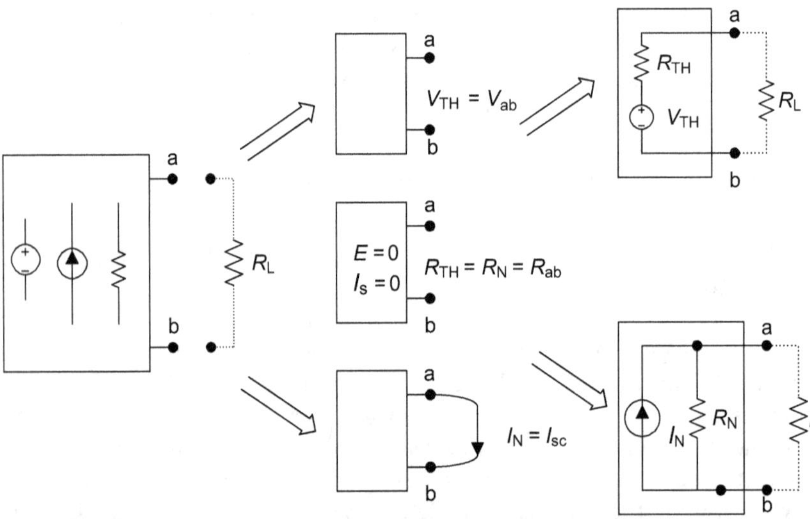

Figure 5.5 The procedure for applying Thevenin's and Norton's theorems

---

Example 5.4: Determine the load current I_L in the circuit of Figure 5.6(a) by using Thevenin's and Norton's theorems.

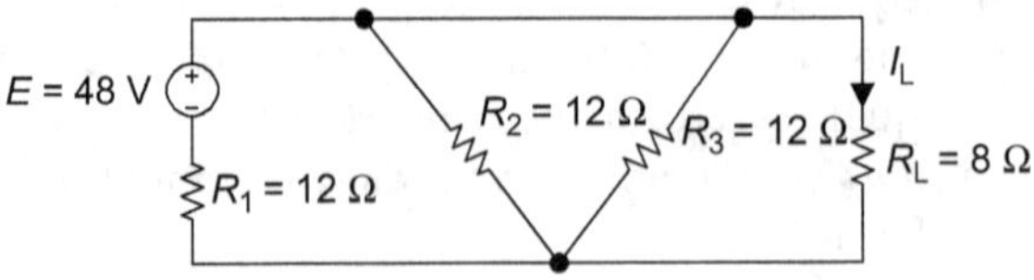

Figure 5.6(a) Circuit for Example 5.4

Solution:

1. Open and remove the load branch R_L, and mark a and b on the terminals of the load branch as shown in the circuit of Figure 5.6(b).

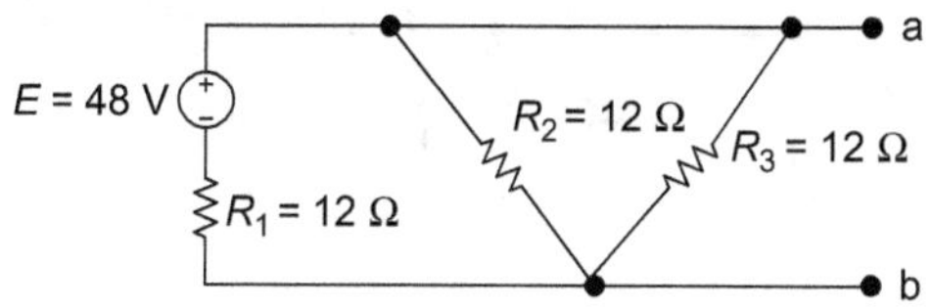

Figure 5.6(b) Circuit for Example 5.4

2. Determine Thevenin's and Norton's equivalent resistances R_{TH} and R_N (the voltage source is replaced by a short circuit) in the circuit of Figure 5.6(c).

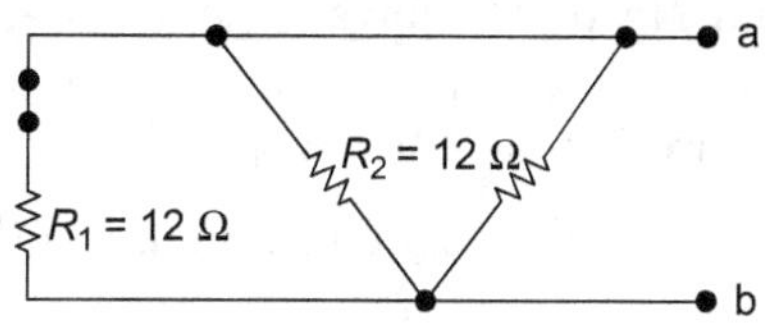

Figure 5.6(c)

$$R_{TH} = R_N = R_{ab} = R_1//R_2//R_3 = (12//12//12)\Omega = 4\,\Omega$$

3. • Determine Thevenin's equivalent voltage V_{TH}: Use the circuit in Figure 5.6(d) to calculate the open-circuit voltage across terminals a and b.

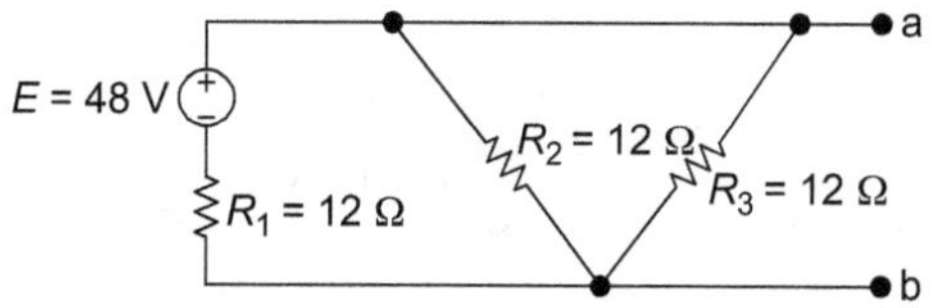

Figure 5.6(d)

$$V_{TH} = V_{ab} = E\frac{R_2//R_3}{R_1 + R_2//R_3}$$
$$= 48\,\text{V}\frac{(12//12)\Omega}{(12 + 12//12)\Omega} = 16\,\text{V}$$

(Apply the voltage divider rule to the resistors $R_2//R_3$ and R_1.)

- Determine Norton's equivalent current I_N: Use the circuit in Figure 5.6(e) to calculate the short-circuit current in terminals a and b.

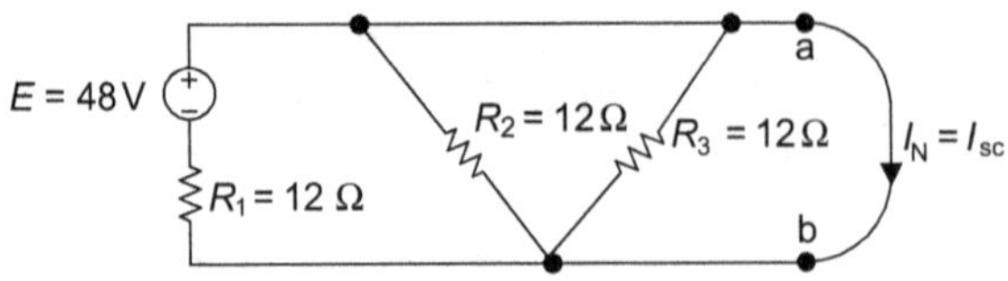

Figure 5.6(e)

$$I_N = I_{sc} = \frac{E}{R_1} = \frac{48\,\text{V}}{12\,\Omega} = 4\,\text{A}$$

Since the current in the branch E and R_1 will go through a short cut without resistance – through the branches a and b – and will not go through the branches R_2 and R_3 that have resistances, in this case $I_N = E/R_1$.

4. Plot Thevenin's and Norton's equivalent circuits as shown in Figure 5.6 (f and g). Connect the load R_L to a and b terminals of the equivalent circuits and determine the load current I_L.

- Use Thevenin's equivalent circuit in Figure 5.6(f) to determine I_L.

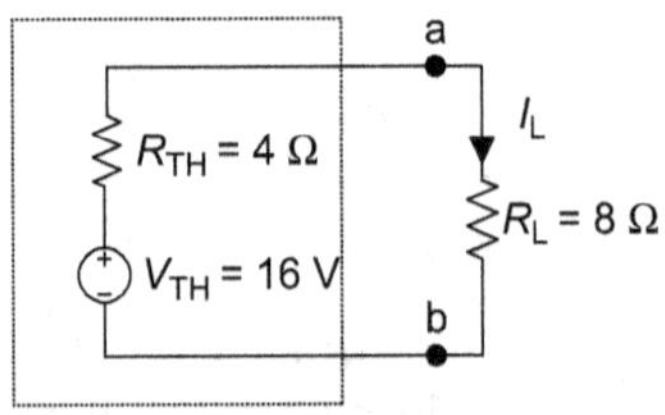

Figure 5.6(f)

$$I_L = \frac{V_{TH}}{R_{TH} + R_L} = \frac{16\,\text{V}}{(4+8)\Omega} \approx 1.33\,\text{A}$$

- Use Norton's equivalent circuit in Figure 5.6(g) to calculate I_L.

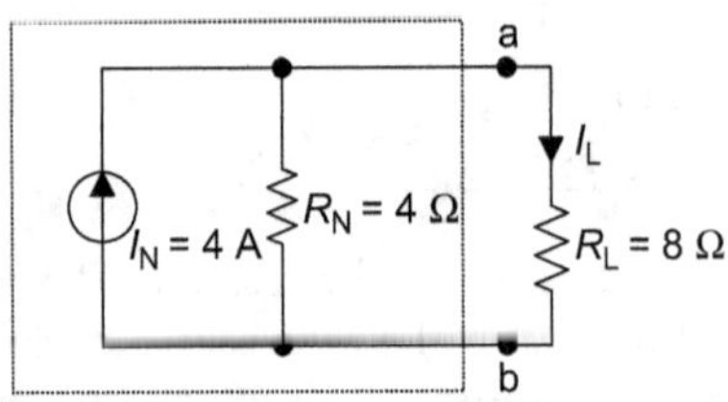

Figure 5.6(g)

$$I_L = I_N \frac{R_N}{R_N + R_L} \qquad \text{(current divider rule)}$$
$$= 4\,\text{A}\frac{4\,\Omega}{(4+8)\Omega} \approx 1.33\,\text{A}$$

5.2.3 Viewpoints of the theorems

One important way to apply Thevenin's and Norton's theorems for analysing any network is to determine the viewpoints of Thevenin's and Norton's equivalent circuits. The load branch (or any unknown current or voltage branch) belongs to the external circuit of the linear two-terminal network with power sources. The opening two terminals of the branch are the viewpoints for Thevenin's and Norton's equivalent circuits.

There could be different viewpoints for the bridge circuit as shown in Figure 5.7(a). If we want to determine the branch current I_3, we use A and B as viewpoints; if we want to determine the branch current I_2, we use D and C as viewpoints, etc. Different equivalent circuits and results will be obtained from using different viewpoints.

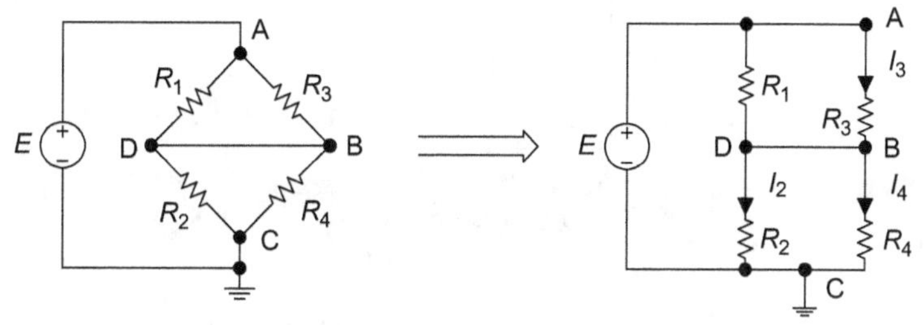

Figure 5.7(a) Viewpoints for the theorem

Example 5.5: For the circuit in Figure 5.7(a):

1. Plot Thevenin's equivalent circuit for calculating the current I_3.
2. Determine Norton's equivalent circuit for the viewpoints B–C.
3. Determine Thevenin's equivalent circuit for the viewpoints D–B.

Solution:

(a) The viewpoints for calculating I_3 should be A–B (Figure 5.7(a)).

1. Open and remove R_3 in the branch A–B of Figure 5.7(a) and mark the letters a and b, as shown in the circuit of Figure 5.7(b).

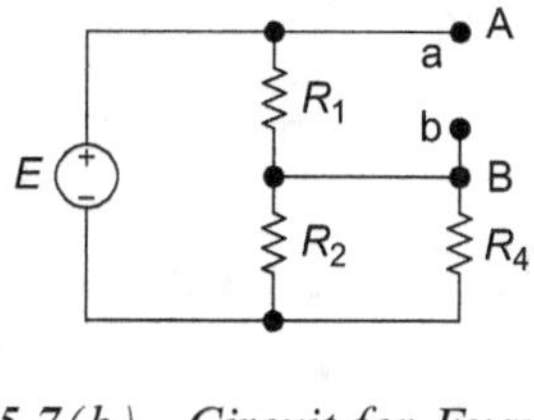

Figure 5.7(b) Circuit for Example 5.5

2. Determine R_{TH} and R_{ab}. Replace the voltage source E with a short circuit, as shown in the circuit of Figure 5.7(c).

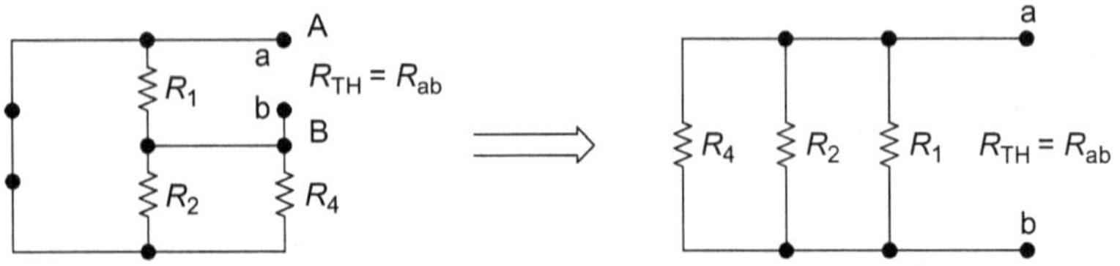

Figure 5.7(c)

$$R_{TH} = R_{ab} = (R_2//R_4)//R_1$$

3. Determine V_{TH} using the circuit in Figure 5.7(d).

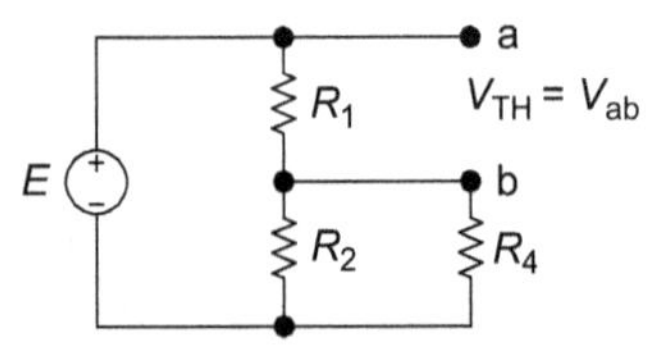

Figure 5.7(d)

$$V_{TH} = V_{ab} = E\frac{R_1}{R_1 + R_2//R_4} \quad \text{(voltage divider rule)}$$

4. Plot Thevenin's equivalent circuit as shown in the circuit of Figure 5.7(e). Connect R_3 to a and b terminals of the equivalent circuit and determine the current I_3:

$$I_3 = \frac{V_{TH}}{R_{TH} + R_3}$$

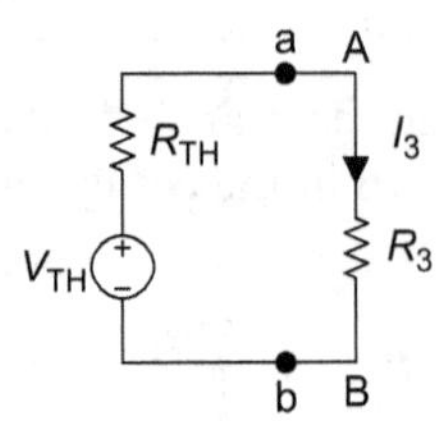

Figure 5.7(e)

(b) Norton's equivalent circuit for the viewpoints B–C:

1. Open and remove R_4 in the branch B–C of Figure 5.7(a), and mark the letters a and b on the two terminals as shown in the circuit of Figure 5.7(f).

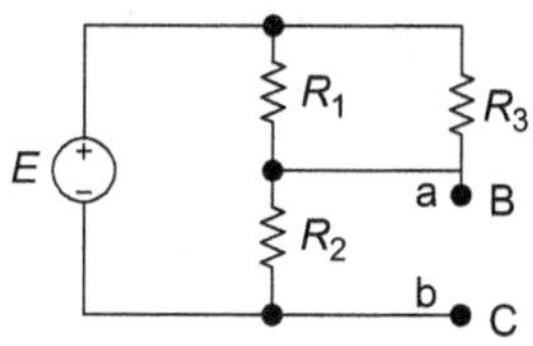

Figure 5.7(f)

2. Determine R_N. Replace the voltage source with a short circuit as shown in the circuit of Figure 5.7(g).

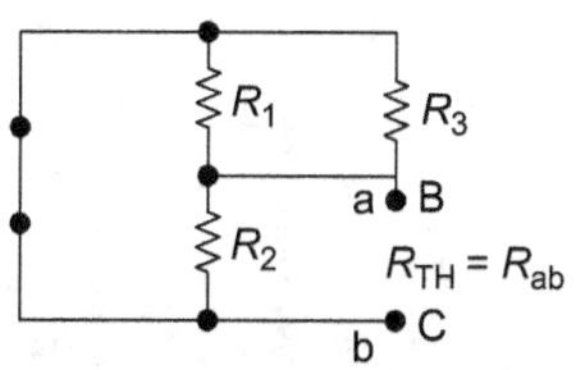

Figure 5.7(g)

3. Determine I_N using the circuit in Figure 5.7(h).

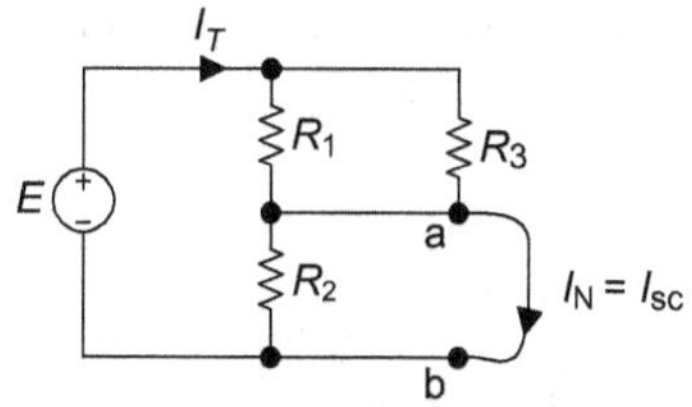

Figure 5.7(h)

Since the current will go through the short cut without resistance – the branch a and b – and will not go through the branch with resistance R_2, $I_N = I_T$:

$$I_N = I_T = \frac{E}{R_1//R_3}$$

4. Plot Norton's equivalent circuit as shown in the circuit of Figure 5.7(i).

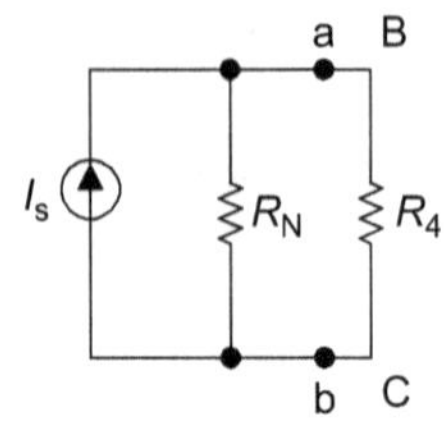

Figure 5.7(i)

(c) Thevenin's equivalent circuit for the viewpoint D–B:

1. Open branch D–B (Figure 5.7(a)) and mark the letters a and b on the two terminals as shown in the circuit of Figure 5.7(j).

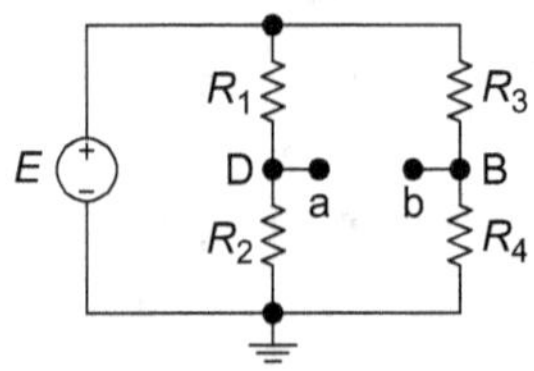

Figure 5.7(j)

2. Determine R_{TH}. Replace the voltage source with a short circuit as shown in the circuits of Figure 5.7(k).

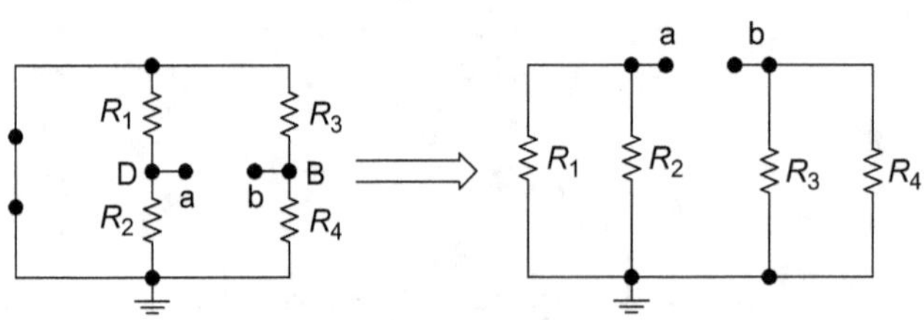

Figure 5.7(k)

$$R_{TH} = R_{ab} = (R_1//R_2) + (R_3//R_4)$$

3. Determine V_{TH} using the circuit in Figure 5.7(j):

$$V_{TH} = V_{ab} = V_a + (-V_b) = E\frac{R_2}{R_1 + R_2} - E\frac{R_4}{R_3 + R_4}$$

4. Plot Thevenin's equivalent circuit as shown in the circuit of Figure 5.7(l).

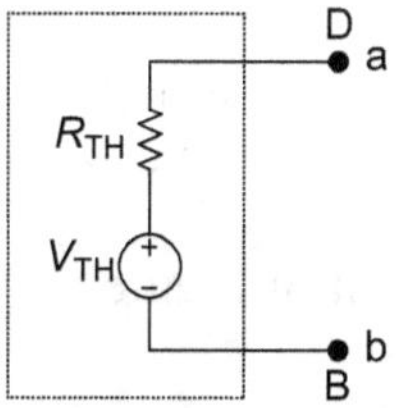

Figure 5.7(l)

Example 5.6: Determine current I_L in the circuit of Figure 5.8(a) by using Norton's theorem.

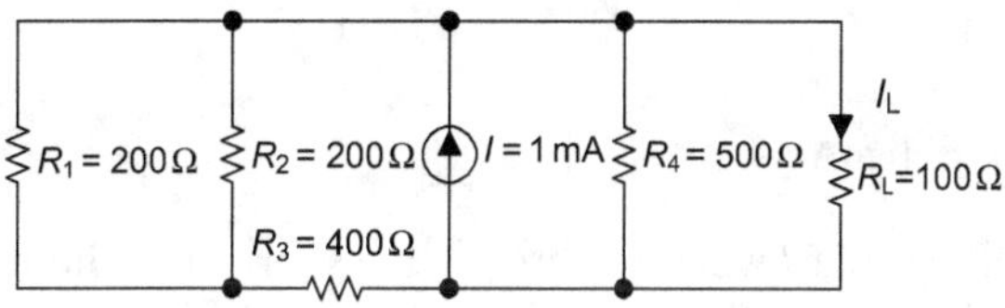

Figure 5.8(a) Circuit for Example 5.6

Solution:

1. Open and remove R_L in the load branch (Figure 5.8(b)) and mark the letters a and b on its two terminals, as shown in the circuit of Figure 5.8(b).

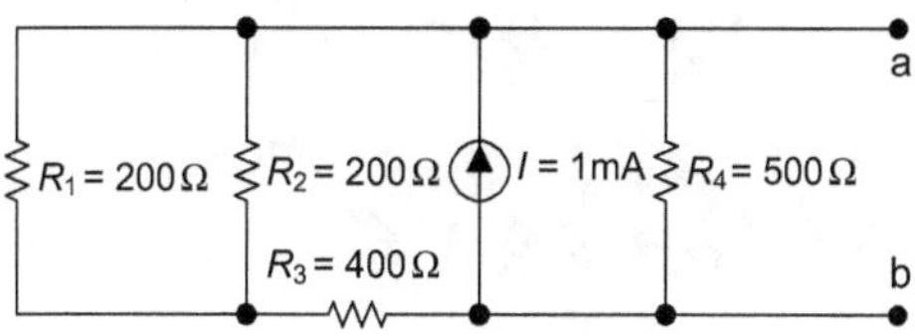

Figure 5.8(b)

2. Determine R_N. Replace the current source with an open circuit as shown in the circuit of Figure 5.8(c).

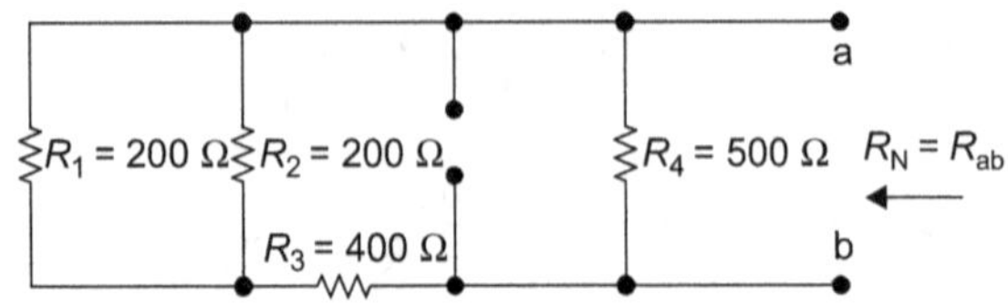

Figure 5.8(c)

$$R_N = R_{ab} = (R_1//R_2 + R_3)//R_4$$
$$= [(200//200 + 400)//500]\Omega = 250\,\Omega$$

3. Calculate I_N using the circuit of Figure 5.8(d).

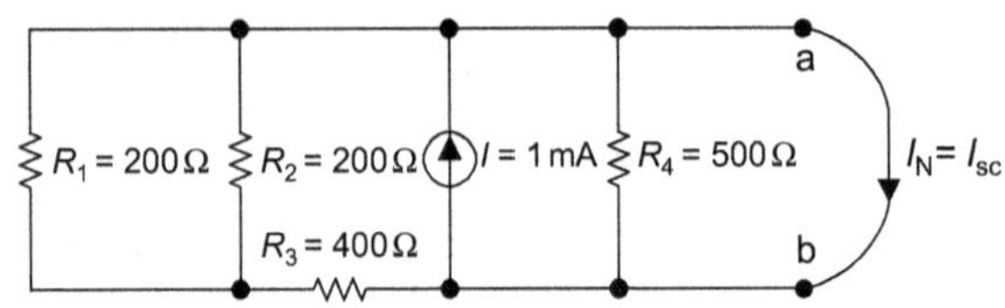

Figure 5.8(d)

$$I_N = I = 1\text{ mA}$$

Since the current I will flow through the short cut without resistance – the branch a and b – and will not go through the branch with resistance, $I_N = I$.

4. Plot Norton's equivalent circuit as shown in the circuit of Figure 5.8(e). Connect R_L to the a and b terminals of the equivalent circuit, and calculate the current I_L.

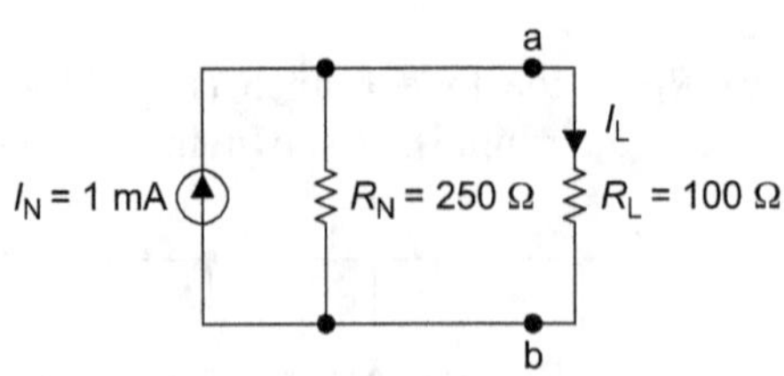

Figure 5.8(e)

$$I_L = I_N \frac{R_N}{R_L + R_N}$$
$$= 1\text{ mA}\frac{250\,\Omega}{(250 + 100)\Omega} \approx 0.71\text{ mA}$$

When applying Thevenin's and Norton's theorems to analyse networks, it is often necessary to combine theorems that we have learned in the previous chapters. This is explained in the following example.

Example 5.7: Determine Norton's equivalent circuit for the left part of the terminals a and b in the circuit of Figure 5.9(a) and determine the current I_L.

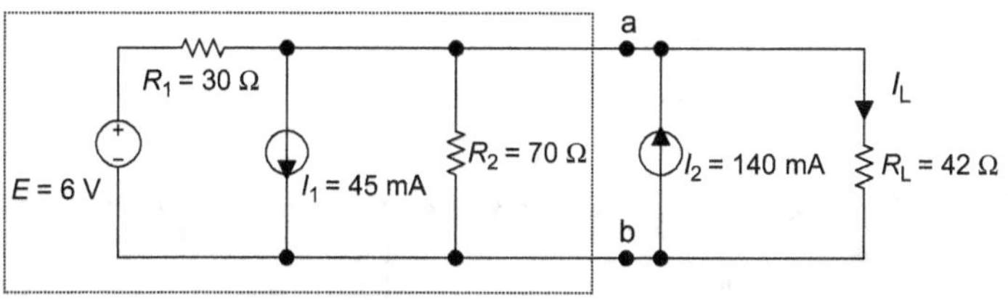

Figure 5.9(a). Circuit for Example 5.7

Solution:

1. Open and remove the current source part on the right side of the circuit from the terminals a and b (Figure 5.9(a)), as shown in the circuit of Figure 5.9(b).

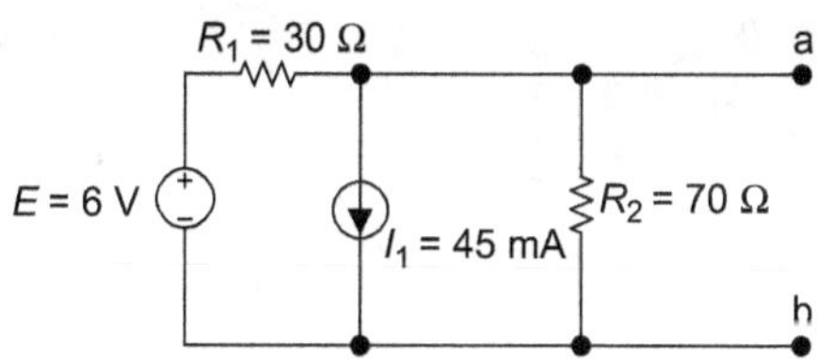

Figure 5.9(b)

2. Determine R_N. Replace the voltage source with a short circuit, and the current source with an open circuit, as shown in the circuit of Figure 5.9(c).

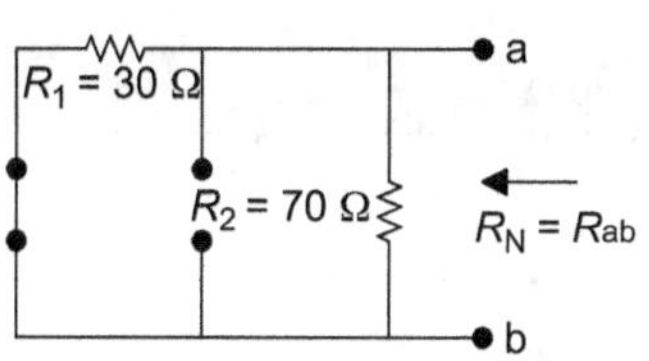

Figure 5.9(c)

$$R_N = R_{ab} = R_1//R_2 = \frac{(30 \times 70)\Omega}{(30 + 70)\Omega} = 21\,\Omega$$

3. Determine I_N using the circuit in Figure 5.9(d). Since there are two power supplies in this circuit, it is necessary to apply the network analysing method for this complex circuit. Let us try to use the superposition theorem to determine I_N.

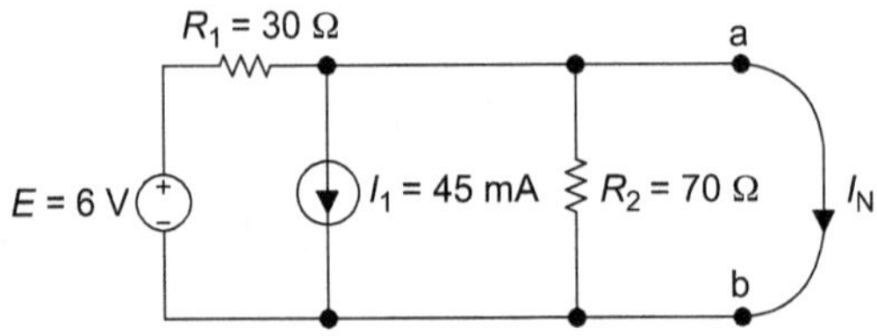

Figure 5.9(d)

- When the single voltage source E is applied only to the circuit, the circuit is shown in Figure 5.9(e).

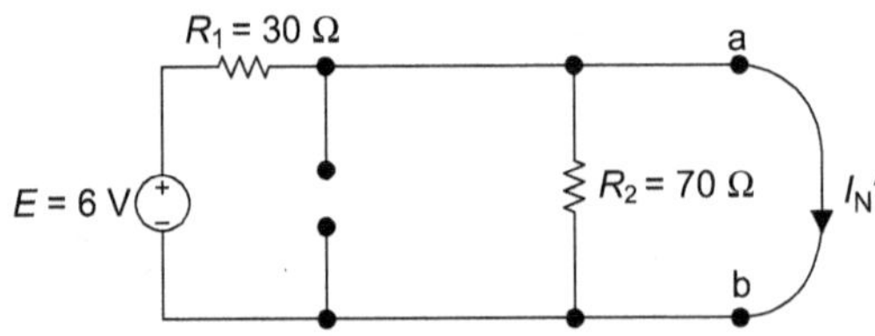

Figure 5.9(e)

Since R_2 is short circuited by the I_N' (recall that current always goes through the short cut without resistance):

$$\therefore\ I_N' = \frac{E}{R_1} = \frac{6\,\text{V}}{30\,\Omega} = 0.2\,\text{A} = 200\,\text{mA}$$

- When the single current source I_1 is applied only to the circuit, the circuit is shown in Figure 5.9(f).

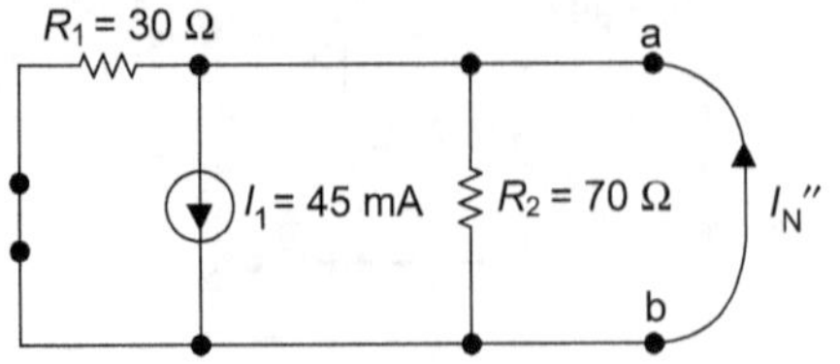

Figure 5.9(f)

Since R_1 and R_2 are short circuited by I_N'':

$$I_N'' = I_1 = 45\,\text{mA}$$

- Determine I_N:

$$I_N = I_N' - I_N'' = (200 - 45)\,\text{mA} = 155\,\text{mA}$$

4. Plot Norton's equivalent circuit. Connect the right side of the a and b terminals of the current source (Figure 5.9(a)) to the a and b terminals of Norton's equivalent circuit, as shown in the circuits of Figure 5.9(g). Determine the current I_L.

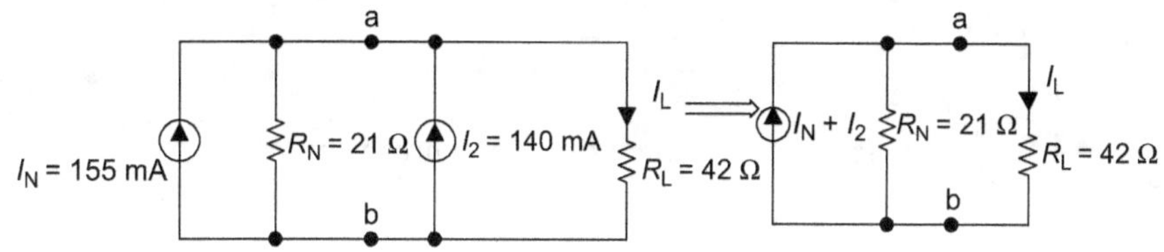

Figure 5.9(g)

$$I_L = (I_N + I_2)\frac{R_N}{R_L + R_N}$$
$$= (155 + 140)\text{mA}\frac{21\,\Omega}{(42 + 21)\Omega} \approx 98.33\,\text{mA}$$

5.3 Maximum power transfer

Practical circuits are usually designed to provide power to the load. When working in electrical or electronic engineering fields you are sometimes asked to design a circuit that will transfer the maximum power from a given source to a load. The maximum power transfer theorem can be used to solve this kind of problem. The maximum power transfer theorem states that when the load resistance is equal to the source's internal resistance, the maximum power will be transferred to the load.

From the last section, we have learned that any linear two-terminal network with power supply can be equally substituted by Thevenin's or Norton's equivalent circuits. Therefore, the maximum power transfer theorem implies that when the load resistance (R_L) of a circuit is equal to the internal resistance (R_S) of the source or the equivalent resistance of Thevenin's or Norton's equivalent circuits (R_{TH} or R_N), maximum power will be dissipated in the load. This concept is illustrated in the circuits of Figure 5.10.

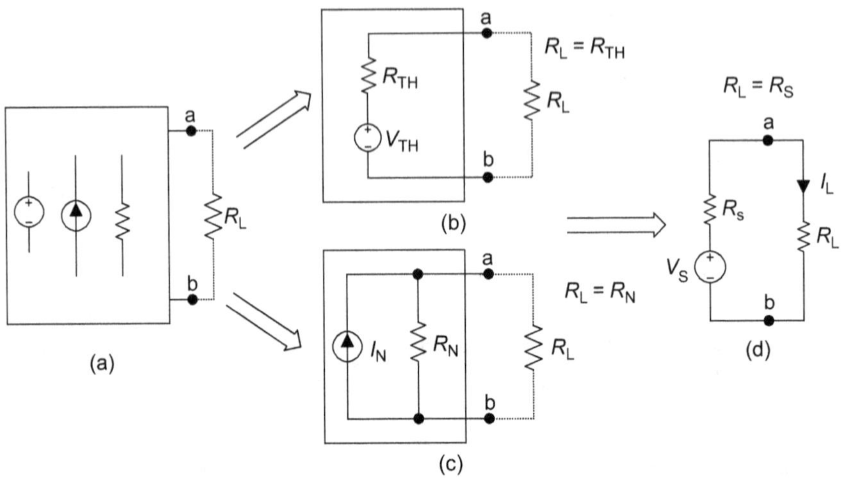

Figure 5.10 The maximum power transfer

The maximum power transfer theorem
When the load resistance is equal to the internal resistance of the source ($R_L = R_S$); or when the load resistance is equal to the Thevenin's/Norton's equivalent resistance of the network ($R_L = R_{TH} = R_N$), the maximum power can be transferred to the load.

The maximum power transfer theorem is used very often in radios, recorders, stereos, CDs, etc. If the load component is a speaker and the circuit that drives the speaker is a power amplifier, when the resistance of the speaker R_L is equal to the internal resistance R_S of the amplifier equivalent circuit, the amplifier can transfer the maximum power to the speaker, i.e. the maximum volume can be delivered by the speaker.

Using the equivalent circuit in Figure 5.10(d) to calculate the power consumed by the load resistor R_L gives

$$P_L = I_L{}^2 R_L = \left(\frac{V_S}{R_S + R_L}\right)^2 R_L \tag{5.1}$$

When $R_L = R_S$, the maximum power that can be transferred to the load is

$$P_L = \frac{V_S{}^2}{(2R_S)^2} R_S = \frac{V_S{}^2}{4R_S}$$

The maximum load power

$$P_L = \frac{V_S^2}{4R_S}$$

If $V_S = 10\,\text{V}$, $R_S = 30\Omega$, and $R_L = 30\Omega$

Then

$$P_L = \frac{V_S^2}{4R_S} = \frac{10^2\,\text{V}}{4(30\,\Omega)} \approx 833\,\text{mW}$$

The maximum power transfer theorem can be proved by using an experiment circuit as shown in Figure 5.11. When the variable resistor R_L is adjusted, it will change the value of the load resistor. Replacing the load resistance R_L with different values in (5.1) gives different load power P_L, as shown in Table 5.1.

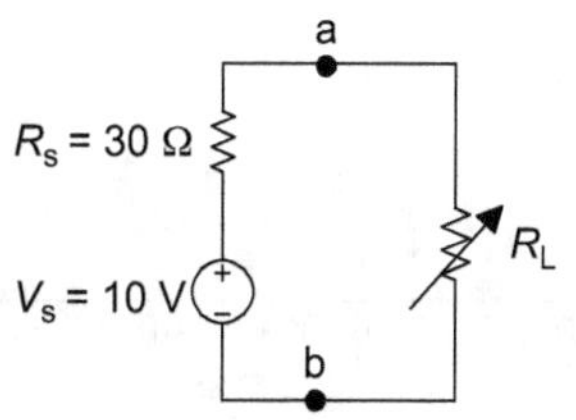

Figure 5.11 The experiment circuit

Table 5.1 The load power

R_L (Ω)	P_L (W)
10	0.625
20	0.8
30	0.833
40	0.816
50	0.781

Such as when $R_L = 10\ \Omega$:

$$P_L = I^2 R_L = \left(\frac{V_S}{R_S + R_L}\right)^2 R_L$$

$$= \left(\frac{10\,\text{V}}{(30+10)\Omega}\right)^2 10\,\Omega = 0.625\,\text{W}$$

When $R_L = 20\ \Omega$:

$$P_L = I^2 R_L = \left(\frac{V_S}{R_S + R_L}\right)^2 R_L$$

$$= \left(\frac{10\,\text{V}}{(30+20)\Omega}\right)^2 20\,\Omega = 0.8\,\text{W}$$

The R_L and P_L curves can be plotted from Table 5.1 as shown in Figure 5.12.

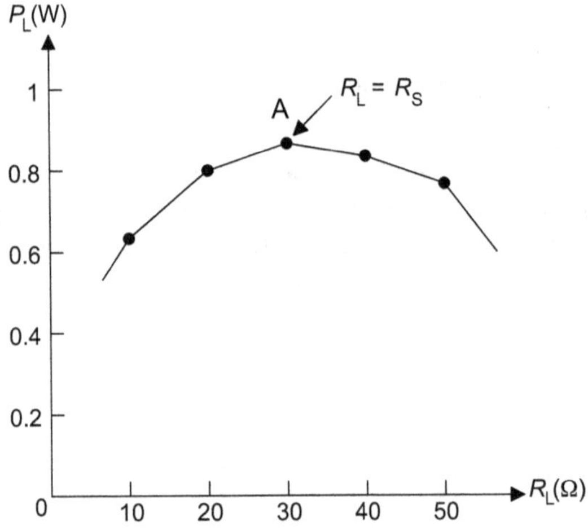

Figure 5.12 R_L–P_L curve

Table 5.1 and Figure 5.12 shows that only when $R_L = R_S$ (30 Ω), the power for the resistor R_L reaches the maximum point A (0.833 W).

Proof of the maximum power transfer equation: The maximum power transfer equation $R_L = R_S$ can be derived by the method of getting the maximum value in calculus (skip this part if you haven't learned calculus yet). Take the derivative for R_L in (5.1), and let its derivative equal to zero, giving the following:

$$\frac{dP}{dR_L} = \frac{d\left[(V_S/(R_S + R_L))^2 R_L\right]}{dR_L} = 0$$

$$V_S^2 \frac{(R_S + R_L)^2 - 2R_L(R_S + R_L)}{(R_S + R_L)^4} = 0$$

$$V_S^2 \frac{R_S^2 + 2R_S R_L + R_L^2 - 2R_L R_S - 2R_L^2}{(R_S + R_L)^4} = 0$$

$$V_S{}^2 \frac{R_S{}^2 - R_L{}^2}{(R_S + R_L)^4} = 0$$

$$V_S{}^2 \frac{(R_S + R_L)(R_S - R_L)}{(R_S + R_L)^4} = 0$$

$$V_S{}^2 \frac{R_S - R_L}{(R_S + R_L)^3} = 0$$

i.e. $R_S - R_L = 0$ or $R_L = R_S$ hence proved.

5.4 Millman's and substitution theorems

5.4.1 Millman's theorem

Millman's theorem is named after the Russian electrical engineering professor Jacob Millman (1911–1991) who proved this theorem. A similar method, known as Tank's method, had already been used before Millman's proof.

The method using series–parallel power sources was stated in chapter 4. However, the series–parallel method can only be used in power sources that have the same polarities and values. Millman's theorem in this chapter can be used to analyse circuits of parallel voltage sources that have different polarities and values. This can be shown in the circuits of Figure 5.13.

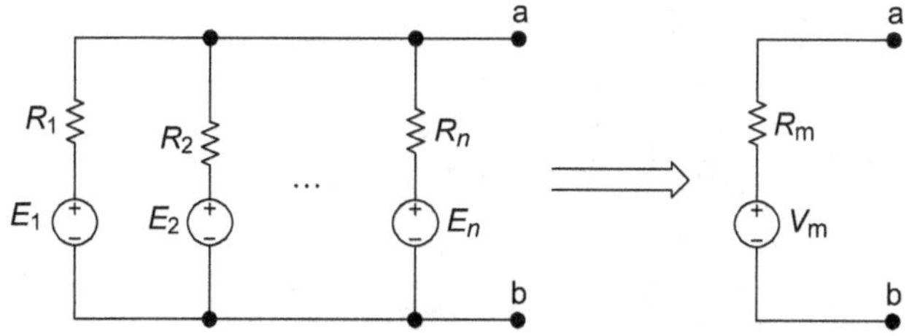

Figure 5.13 Millman's theorem

Millman's theorem states that for a circuit of parallel branches, with each branch consisting of a resistor or a voltage/current source, this circuit can be replaced by a single voltage source with voltage V_m in series with a resistor R_m as shown in Figure 5.13.

Millman's theorem, therefore, can determine the voltage across the parallel branches of a circuit.

Millman's theorem

When several voltage sources or branches consisting of a resistor are in parallel, they can be replaced by a single voltage source.

$$V_m = R_m I_m = R_m \left(\frac{E_1}{R_1} + \frac{E_2}{R_2} + \cdots + \frac{E_n}{R_n} \right) R_m = R_1 // R_2 // \cdots // R_n$$

Note: V_m is the algebraic sum for all the individual terms in the equation. It will be positive if E_n and V_m have the same polarities, otherwise it will be negative. The letter m in V_m and R_m means Millman.

Example 5.8: Determine the load voltage V_L in the circuit of Figure 5.14 using Millman's theorem.

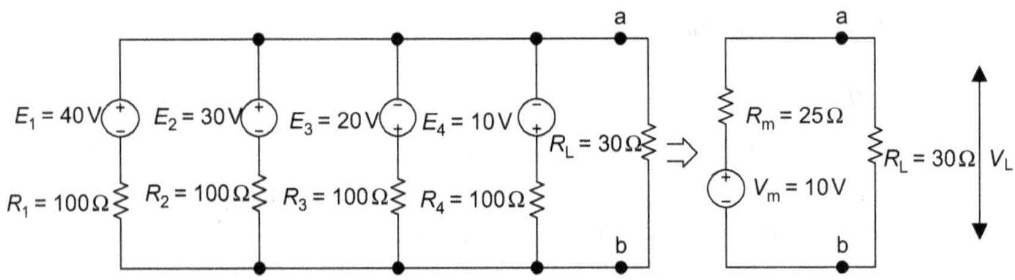

Figure 5.14 Circuit for Example 5.8

Solution:

$$R_m = R_1 // R_2 // R_3 // R_4 = (100//100//100//100)\Omega = 25\,\Omega$$

$$V_m = R_m I_m = R_m \left(\frac{E_1}{R_1} + \frac{E_2}{R_2} - \frac{E_3}{R_3} - \frac{E_4}{R_4} \right)$$

$$= (25\,\Omega) \left(\frac{40\,\text{V}}{100\,\Omega} + \frac{30\,\text{V}}{100\,\Omega} - \frac{20\,\text{V}}{100\,\Omega} - \frac{10\,\text{V}}{100\,\Omega} \right) = 10\,\text{V}$$

$$V_L = V_m \frac{R_L}{R_L + R_m} = (10\,\text{V}) \frac{30\,\Omega}{(30 + 25)\Omega} \approx 5.455\,\text{V}$$

5.4.2 Substitution theorem

Substitution theorem

A branch in a network that consists of any component can be replaced by an equivalent branch that consists of any combination of components, as long as the currents and voltages on that branch do not change after the substitution.

This theorem can be illustrated in the circuits of Figures 5.15 and 5.16.

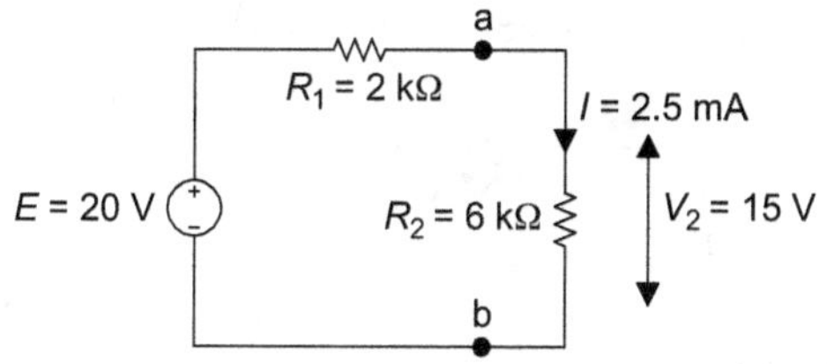

Figure 5.15 Circuit 1 of the substitution theorem

The current and voltage of branch a–b in the circuit of Figure 5.15 can be determined as follows:

The voltage across branch a–b: $V_2 = E\dfrac{R_2}{R_1 + R_2} = 20\,\text{V}\dfrac{6\text{k}\Omega}{(2+6)\text{k}\Omega} = 15\,\text{V}$

The current in branch a–b: $I = \dfrac{E}{R_1 + R_2} = \dfrac{20\text{V}}{(2+6)\text{k}\Omega} = 2.5\,\text{mA}$

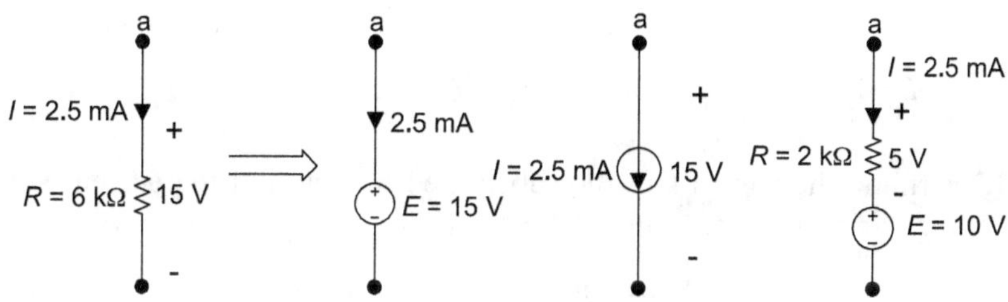

Figure 5.16 Circuit 2 of the substitution theorem

According to the definition of the substitution theorem, any branch in the circuit of Figure 5.16 can replace the a–b branch in the circuit of Figure 5.15, since their voltages and currents are the same as the voltages and currents in branch a–b in the circuit of Figure 5.15.

Example 5.9: Use a current source with a 30 Ω internal resistor to replace the a–b branch in the circuit of Figure 5.17(a).

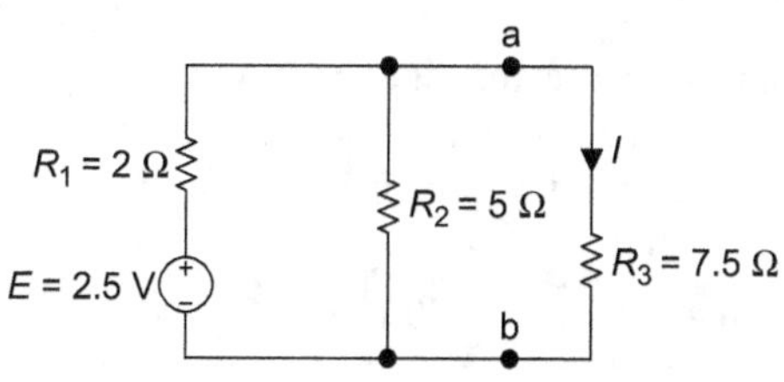

Figure 5.17(a) Circuit for Example 5.9

Solution:

- Figure 5.17(b) shows the resultant circuit after the current source with a 30 Ω internal resistor replaced the a–b branch in the circuit of Figure 5.17(a).

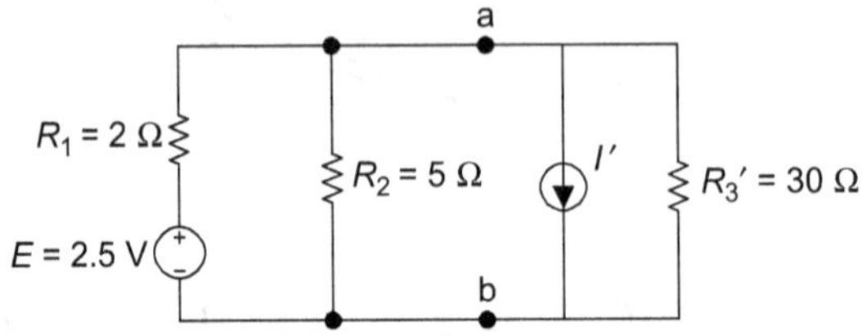

Figure 5.17(b)

- Determine the voltage and current in the a–b branch of the circuit in Figure 5.17(a).

$$V_{ab} = E\frac{R_2//R_3}{R_1 + R_2//R_3} = 2.5\,\text{V}\frac{((5 \times 7.5)/(5 + 7.5))\Omega}{(2 + (5 \times 7.5)/(5 + 7.5))\Omega} = 1.5\,\text{V}$$

(voltage divider rule)

$$I = \frac{V_{ab}}{R_3} = \frac{1.5\,\text{V}}{7.5\,\Omega} = 0.2\,\text{A} = 200\,\text{mA}$$

- Determine the currents in the substituted branch and the current source branch using the circuit in Figure 5.17(c).

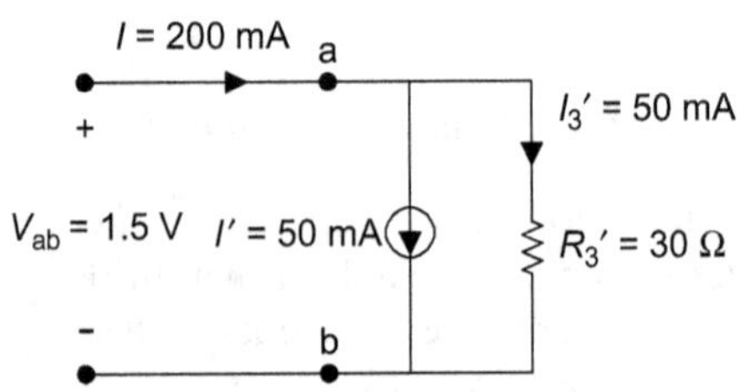

Figure 5.17(c)

$$I_3' = \frac{V_{ab}}{R_3'} = \frac{1.5\,\text{V}}{30\,\Omega} = 50\,\text{mA}$$

Therefore, to maintain the terminal voltage $V_{ab} = 1.5$ V in the original branch, the current I_3' in the R_3' branch should be 50 mA. Using KCL we can get the current I' in the current source branch:

$$I' = I - I_3' = (200 - 50)\text{mA} = 150\,\text{mA}$$

Summary

Linear circuit: includes the linear components (such as resistors).

Superposition theorem

- The unknown voltages or currents in any linear network are the sum of the voltages or currents of the individual contributions from each single power supply, by setting the other inactive sources to zero.
- Steps to apply superposition theorem:

  1. Turn off all power sources except one, i.e. replace the voltage source with the short circuit, and replace the current source with an open circuit. Redraw the original circuit with a single source.
  2. Analyse and calculate this circuit by using the single source series–parallel analysis method, and repeat steps 1 and 2 for the other power sources in the circuit.
  3. Determine the total contribution by calculating the algebraic sum of all contributions due to single sources.

The result should be positive when the reference polarity of the unknown in the single source circuit is the same as the reference polarity of the unknown in the original circuit; otherwise it should be negative.

The linear two-terminal network with the sources

It is a linear complex circuit that has power sources and two terminals.

Thevenin's and Norton's theorems

- Any linear two-terminal network with power supplies can be replaced by a simple equivalent circuit that has a single power source and a single resistor.
- Thevenin's theorem: The equivalent circuit is a voltage source (with an equivalent resistance R_{TH} in series with an equivalent voltage source V_{TH}).
- Norton's theorem: The equivalent circuit is a current source (with an equivalent resistance R_N in parallel with an equivalent current source I_N).
- Steps to apply Thevenin's and Norton's theorems:

  1. Open and remove the load branch (or any unknown current or voltage branch) in the network, and mark the letter a and b on the two terminals.
  2. Determine the equivalent resistance R_{TH} or R_N. It equals the equivalent resistance, looking at it from the a and b terminals when all sources are turned off or equal to zero in the network. (A *voltage source* should be replaced by a *short circuit*, and a *current source* should be replaced by an *open circuit*.) That is

$$R_{TH} = R_N = R_{ab}$$

3. Determine Thevenin's equivalent voltage V_{TH}. It equals the open-circuit voltage from the original linear two-terminal network of a and b, i.e. $V_{TH} = V_{ab}$
4. Determine Norton's equivalent current I_N. It equals to the short-circuit current from the original linear two-terminal network of a and b, i.e. $I_N = I_{sc}$.
5. Plot Thevenin's or Norton's equivalent circuit, and connect the load branch (or unknown current or voltage branch) to a and b terminals of the equivalent circuit. Then the load (or unknown) voltage or current can be determined.

Maximum power transfer theorem

When the load resistance is equal to the internal resistance of the source ($R_L = R_S$); or when the load resistance is equal to the Thevenin's/Norton's equivalent resistance of the network ($R_L = R_{TH} = R_N$), maximum power will be transferred to the load.

Millman's theorem

When several voltage sources or branches consisting of a resistor are in parallel, they can be replaced by a branch with a voltage source.

$$V_m = R_m I_m = R_m\left(\frac{E_1}{R_1} + \frac{E_2}{R_2} + \cdots + \frac{E_n}{R_n}\right),\ R_m = R_1 // R_2 // \ldots // R_n$$

Substitution theorem

A branch in a network that consists of any component can be replaced by an equivalent branch that consists of any combination of components, as long as the currents and voltages on that branch do not change after the substitution.

Experiment 5A: Superposition theorem

Objectives

- Understand the superposition theorem through experiment.
- Construct a circuit with two voltage sources, and collect and evaluate experimental data to verify applications of the superposition theorem.
- Analyse experimental data, circuit behaviour and performance, and compare them to theoretical equivalents.

Equipment and components

- Breadboard
- Multimeter
- Dual-output DC power supply
- Resistors: 5.1, 7.5, and 11 kΩ

Background information

- Superposition theorem: The unknown voltages or currents in any linear network are the sum of the voltages or currents of the individual contributions from each single power supply, by setting the other inactive sources to zero.
- Steps to apply superposition theorem:

  1. Turn off all power sources except one, i.e. replace the voltage source with a short circuit (by placing a jump wire), and replace the current source with an open circuit. Redraw the original circuit with a single source.
  2. Analyse and calculate this circuit using the single source series–parallel analysis method, and repeat steps 1 and 2.
  3. Determine the total contribution by calculating the algebraic sum of all contributions due to single sources.

(The result should be positive when the reference polarity of the unknown in the single source circuit is the same with the polarity of the unknown in the original circuit; otherwise it should be negative.)

Procedure

1. Measure the values of resistors listed in Table L5.1 using a multimeter (ohmmeter function) and record in Table L5.1.

Table L5.1

Resistance	R_1	R_2	R_3
Nominal value	5.1 kΩ	7.5 kΩ	11 kΩ
Measured value			

2. Construct a circuit on the breadboard as shown in Figure L5.1(b).
3. Calculate the equivalent resistance R_{eq}' from the terminals a and b in the circuit of Figure L5.1(b). Record the value in Table L5.2.
4. Calculate currents I_1' and I_3' in the circuit of Figure L5.1(b). Record the values in Table L5.2.
5. Measure the equivalent resistance R_{eq}' and currents I_1' and I_3' using the multimeter (ohmmeter and ammeter functions) in Figure L5.1(b). Record the values in Table L5.2.
6. Reconstruct a circuit as shown in Figure L5.1(c) on the breadboard.
7. Calculate the equivalent resistance R_{eq}'' from the terminals c and d in the circuit of Figure L5.1(c). Record the values in Table L5.2.
8. Calculate currents I_1'' and I_3'' in the circuit of Figure L5.1(c). Record the values in Table L5.2.

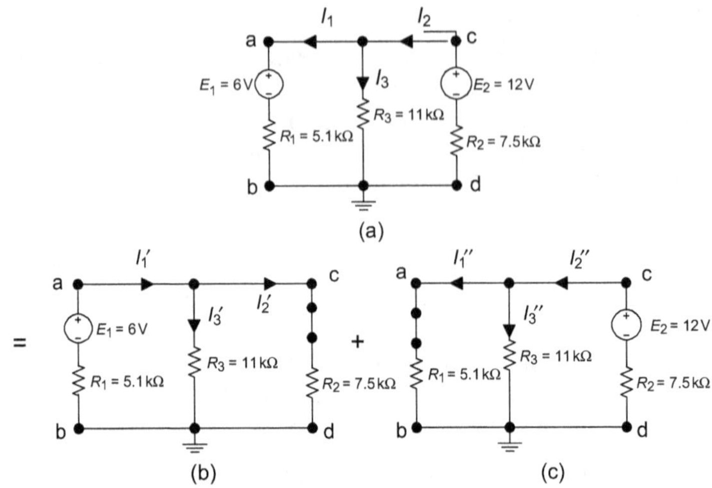

Figure L5.1 Superposition theorem

Table L5.2

Resistance	R_{eq}'	I_1'	I_3'	R_{eq}''	I_1''	I_3''	I_3
Formula for calculations							
Calculated value							
Measured value							

9. Measure the equivalent resistance R_{eq}'' and currents I_1'' and I_3'' in Figure L5.1(c) using a multimeter (ohmmeter and ammeter functions). Record the values in Table L5.2.
10. Calculate $I_3 = I_3' + I_3''$ from Table L5.2 using the calculated values. Record the value in Table L5.2.
11. Construct a circuit as shown in Figure L5.1(a) on the breadboard and measure current I_3 in the circuit of Figure L5.1(a). Record the value in Table L5.2.
12. Compare the measured values and calculated values; are there any significant differences? If so, explain the reasons.

Conclusion

Write your conclusions below:

Experiment 5B: Thevenin's and Norton's theorems

Objectives

- Understand Thevenin's and Norton's theorems and the maximum power transfer theorem through this experiment.
- Construct electric circuits, and collect and evaluate experimental data to verify the applications of Thevenin's and Norton's theorems.
- Construct electric circuits, and collect and evaluate experimental data to verify the applications of the maximum power transfer theorem.
- Analyse experimental data, circuit behaviour and performance, and compare them to the theoretical equivalents.

Equipment and components

- Breadboard
- Multimeter
- DC power supply
- Resistors: 300 Ω, 750 Ω, 620 Ω, 180 Ω, 1 kΩ, and one 10 kΩ variable resistor

Background information

- Any linear two-terminal network (complex circuit) with power supplies can be replaced by a simple equivalent circuit that has a single power source and a single resistor.
- Thevenin's theorem: Thevenin's equivalent circuit is an actual voltage source that has an equivalent resistance R_{TH} in series with an equivalent voltage source V_{TH}.
- Norton's theorem: Norton's equivalent circuit is an actual current source that has an equivalent resistance R_N in parallel with an equivalent current source I_N.
- The maximum power transfer theorem: When the load resistance is equal to the internal resistance of the source ($R_L = R_S$); or when the load resistance is equal to the Thevenin/Norton equivalent resistance of the circuit ($R_L = R_{TH} = R_N$), maximum power will be dissipated in the load.

Procedure

Part I: Thevenin's and Norton's theorems

1. Measure the values of the resistors listed in Table L5.3 using a multimeter (ohmmeter function) and record in Table L5.3.

Table L5.3

Resistor	R_1	R_2	R_3	R_L
Colour code value	300 Ω	620 Ω	750 Ω	180 Ω
Measured Value				

2. Calculate R_{TH} and V_{TH} of Thevenin's equivalent circuit in Figure L5.2(a). Record the values in Table L5.4.

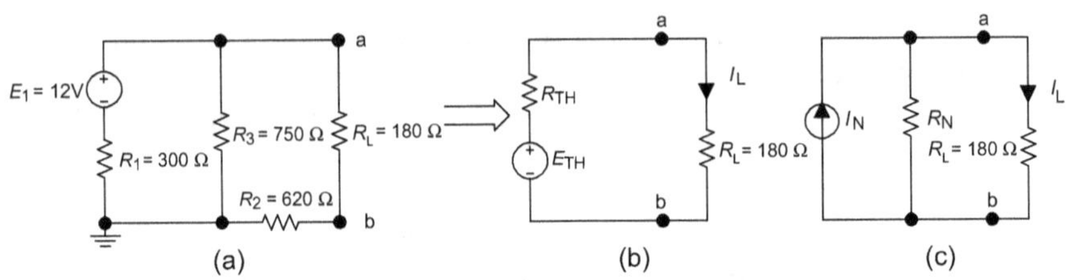

Figure L5.2 Thevenin's and Norton's equivalent circuits

Table L5.4

	R_{TH}	V_{TH} (V_{ab})	R_N	I_N (I_{SC})	I_L (Figure L5.2(b))	I_L (Figure L5.2(c))
Formula for calculations						
Calculated value						
Measured value						

3. Calculate R_N and I_N of Norton's equivalent circuit in Figure L5.2(a). Record the values in Table L5.4.
4. Calculate the load current I_L in Thevenin's and Norton's equivalent circuits of Figure L5.2(b and c). Record the values in Table L5.4.
5. Construct a circuit as shown in Figure L5.3(a) on the breadboard.

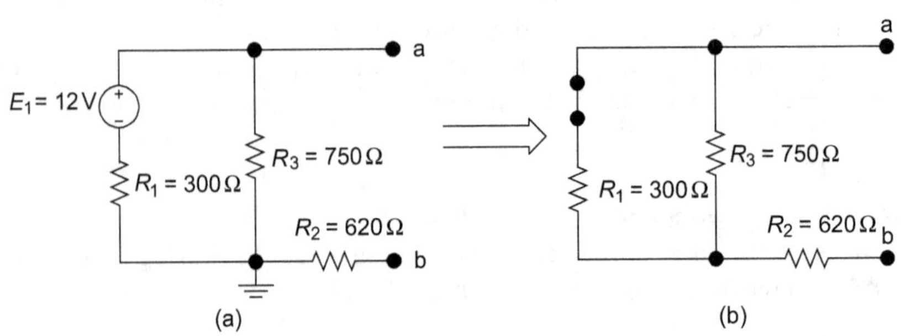

Figure L5.3 Thevenin's and Norton's circuits

6. Measure the open-circuit voltage V_{ab} ($V_{ab} = V_{TH}$) on the two terminals a and b in the circuit of Figure L5.3(a) using a multimeter (voltmeter function). Record the value in Table L5.4.

7. Measure short-circuit current I_{sc} ($I_{sc} = I_N$) from a to b in Figure L5.3(a) using a multimeter (ammeter function). Record the value in Table L5.4.
8. Disconnect the power supply E in Figure L5.3(a), and use a jump wire connecting the two terminals of E as shown in Figure L5.3(b). Measure the equivalent resistance at terminals of a to b ($R_{ab} = R_{TH} = R_N$) in Figure L5.3 (b) using a multimeter (ohmmeter function). Record the value in Table L5.4.
9. Construct a circuit as shown in Figure L5.2(b) (with measured V_{TH} and R_{TH}) on the breadboard, and measure the load current I_L using a multimeter (ammeter function). Record the value in Table L5.4.
10. Compare the measured values and the calculated values. Are there any significant differences? If so, state the reasons.

Part II: Maximum power transfer

1. Construct a circuit as shown in Figure L5.4 on the breadboard. R_S represents the internal resistance for power supply or the Thevenin's or Norton's equivalent resistances.

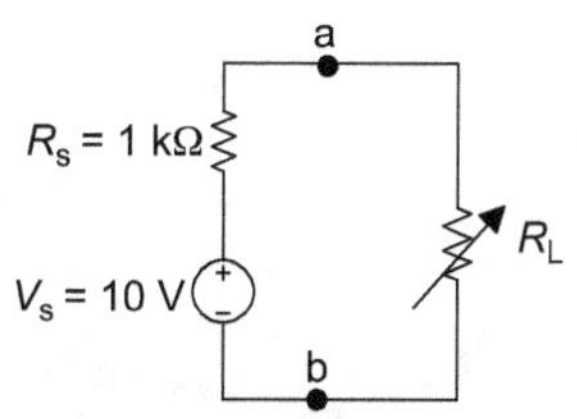

Figure L5.4 Maximum power transfer circuit

2. Measure the open-circuit voltage V_{ab} in the circuit of Figure L5.4 (without connecting the load R_L) using a multimeter (voltmeter function). Record the value in Table L5.5.

Table L5.5

Load Resistor R_L	**Measured R_L Value**	**Measured V_{R_L} Value**	**Calculated P_L Value**
R_{L1}			
R_{L2}			
R_{L3}			
R_{L4}			
R_{L5}	1 kΩ		
R_{L6}			
R_{L7}			
R_{L8}			
R_{L9}			
Open-circuit voltage V_{ab} =			

3. Connect the 10-kΩ variable resistor R_L to the circuit in Figure L5.4 and change the value of the variable resistor nine times from lower to higher values (one of the values should be 1 kΩ). Measure each R_L and V_{R_L} using a multimeter (ohmmeter and voltmeter functions). Record the values in Table L5.5.
4. Calculate power dissipated in each load resistor and record the values in Table L5.5.
5. Sketch the R_L–P_L curve (use P_L as vertical axis and R_L as horizontal axis).
6. When the load resistance is equal to the internal resistance of the source ($R_L = R_S = 1$ kΩ), is P_L at the maximum point on the curve? Why?

Conclusion

Write your conclusions below:

Chapter 6

Capacitors and inductors

Objectives

After completing this chapter, you will be able to:

- describe the basic structure of the capacitor and inductor
- explain the charging and discharging behaviours of a capacitor
- understand the storing and releasing energy of an inductor
- define capacitance and inductance
- list the factors affecting capacitance and inductance
- understand the relationship between voltage and current in capacitive and inductive circuits
- calculate energy stored in capacitors and inductors
- determine the equivalent capacitance and inductance in series, parallel and series–parallel configurations

There are three important fundamental circuit elements: the resistor, capacitor and inductor. The resistor (R) has appeared in circuit analysis in the previous chapters. The other two elements – the capacitor (C) and inductor (L) will be introduced in this chapter. Both of these electric elements can store energy that has been absorbed from the power supply, and release it to the circuit. A capacitor can store energy in the electric field, and an inductor can store energy in the magnetic field. This is different with a resistor that consumes or dissipates electric energy.

Three basic circuit components

- Resistor (R)
- Capacitor (C)
- Inductor (L)

A circuit containing only resistors has limited applications. Practical electric circuits usually combine the above three basic elements and possibilities along with other devices.

6.1 Capacitor

6.1.1 The construction of a capacitor

A capacitor has applications in many areas of electrical and electronic circuits, and it extends from households to industry and the business world. For instance, it is used in flash lamps (for flash camera), power systems (power supply smoothing, surge protections), electronic engineering, communications, computers, etc. There are many different types of capacitors, but no matter how differently their shapes and sizes, they all have the same basic construction.

A capacitor has two parallel conductive metal plates separated by an isolating material (the dielectric). The dielectric can be of insulating material, such as paper, vacuum, air, glass, plastic film, oil, mica, ceramics, etc. The basic construction of a capacitor is shown in Figure 6.1.

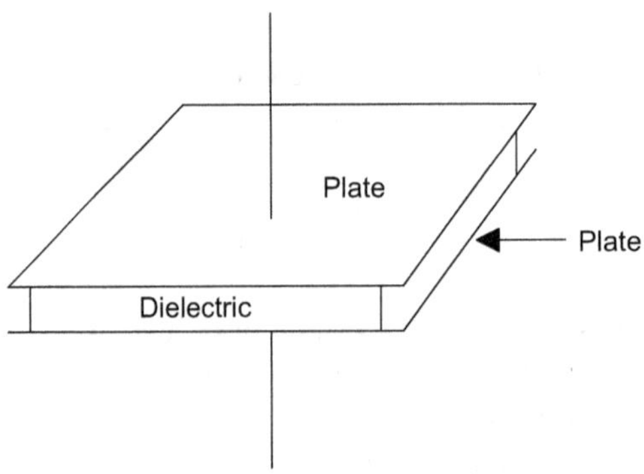

Figure 6.1 The basic construction of a capacitor

A capacitor can be represented by a capacitor schematic symbol as its circuit model. Similar to resistors, there are two basic types of capacitors, variable and fixed, and their schematic symbols are shown in Figure 6.2 (a and b).

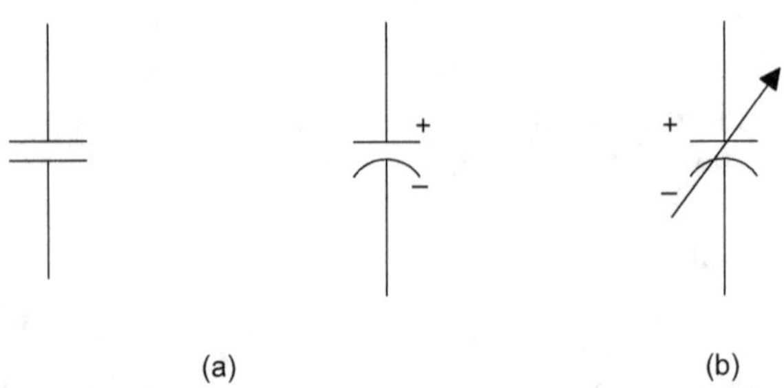

Figure 6.2 Symbols of capacitor. (a) Fixed: unpolarized and polarized and (b) variable

A variable capacitor is a capacitor that possesses a value that may be changed manually or automatically. A fixed capacitor is a capacitor that possesses a fixed value and cannot be adjusted. For a fixed *polarized* capacitor, connect its positive (+) lead to the higher voltage point in the circuit, and negative (−) lead to the lower voltage point. For an *unpolarized* capacitor, it does not matter which lead connects to where.

Electrolytic capacitors are usually polarized, and *non-electrolytic* capacitors are unpolarized. Electrolytic capacitors can have higher working voltages and store more charges than non-electrolytic capacitors.

Capacitor C
An energy storage element that has two parallel conductive metal plates separated by an isolating material (the dielectric).

6.1.2 Charging a capacitor

A purely capacitive circuit with an uncharged capacitor ($V_C = 0$), a three-position switch, and a DC (direct current) voltage source (E) is shown in Figure 6.3(a). With the switch at position 0, the circuit is open, and the potential difference between the two metal plates of the capacitor is zero ($V_C = 0$). Two plates of the capacitor have the same size and are made by the same conducting material, so they should have the same number of charges at the initial condition.

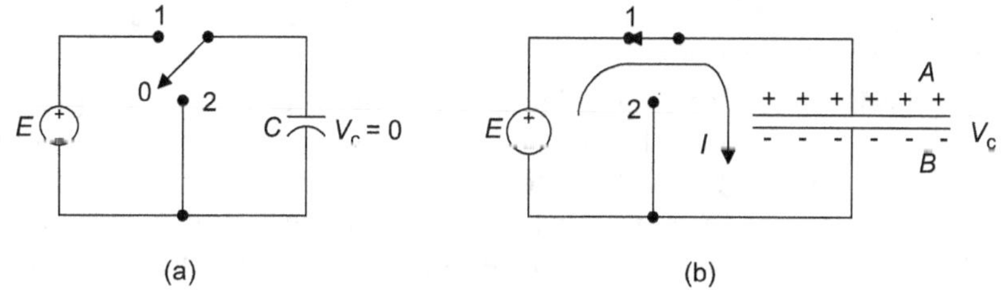

Figure 6.3 Charging a capacitor

Once the three-position switch is turned on to position 1 as shown in Figure 6.3(b), the DC voltage source is connected to the two leads of the capacitor. From the rule 'opposites attract and likes repel', we know that the positive pole of the voltage source will attract electrons from the positive plate of the capacitor, and the negative pole of the voltage source will attract positive charges from the negative plate of the capacitor; this causes current I to flow in the circuit.

Plate A loses electrons and shows positive; plate B loses positive charges and thus shows negative. Thus, the electric field is built up between the two metal plates of the capacitor, and the potential difference (V_C) appears on the capacitor with positive (+) on plate A and negative (−) on plate B, as shown in Figure 6.3(b).

Once voltage across the capacitor V_C has reached the source voltage V_S, i.e. $V_C = E$, there is no more potential difference between the source and capacitor, the charging current cesses to flow ($I = 0$), and the process of charging the capacitor is completed. This is the process of charging a capacitor.

6.1.3 Energy storage element

When the switch is turned off to position 0 in the circuit shown in Figure 6.3(a), the capacitor and power supply will disconnect. If the voltage across the capacitor V_C is measured at this time using a multimeter (voltmeter function), V_C should still be the same with the source voltage ($V_C = E$) even without a power supply connected to it. This is why a capacitor is called an energy storage element, as it can store charges absorbed from the power supply and store electric energy obtained from charging. Once a capacitor has transferred some charges through charging, an electric field is built up between the two plates of the capacitor, and it can maintain the potential difference across it.

The isolating material (dielectric) between the two metal plates isolates the charges between the two plates. Charges will not be able to cross the insulating material from one plate to another. So the energy storage element capacitor will keep its charged voltage V_C for a long time (duration will depend on the quality and type of the capacitor). Since the insulating material will not be perfect and a small leakage current may flow through the dielectric, this may eventually slowly dissipate the charges.

6.1.4 Discharging a capacitor

When the switch is closed to position 2 as shown in the circuit of Figure 6.4, the capacitor and wires in the circuit forms a closed path. At this time, the capacitor is equivalent to a voltage source, as voltage across the capacitor V_C will cause the current to flow in the circuit. Since there is no resistor in this circuit, it is a short circuit, and a high current causes the capacitor to release its charges or stored energy in a short time. This is known as discharging a capacitor. After the capacitor has released all its stored energy, the voltage across the capacitor will be zero ($V_C = 0$), the current in the circuit ceases to flow ($I = 0$) and the discharge process is completed.

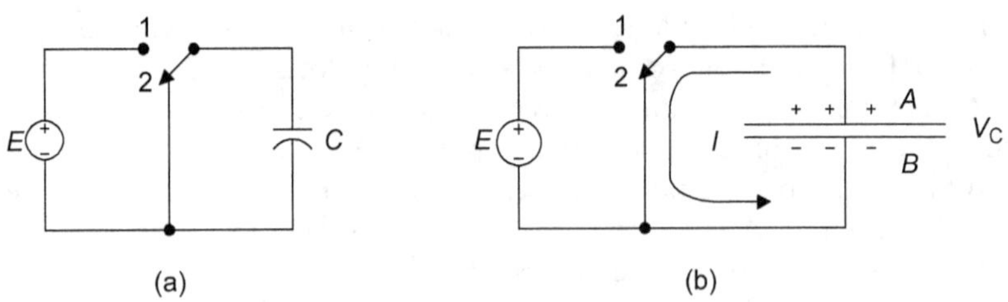

Figure 6.4 Discharging a capacitor

The capacitor cannot release energy that is more than it has absorbed and stored, therefore it is a passive component. A passive component is a component that absorbs (but not produce) energy.

The concept of a capacitor may be analogous to a small reservoir. It acts as a reservoir that stores and releases water. The process of charging a capacitor from the power supply is similar to a reservoir storing water. The process of discharging a capacitor is similar to a reservoir releasing water.

There is an important characteristic that implies in the charge and discharge of a capacitor. That is, the voltage on the capacitor won't be able to change instantly; it will always take time, i.e. gradually increase (charging) or decrease (discharge).

Charging/Discharging a capacitor

An electric element that can store and release charges that it absorbed from the power supply.

- Charging: The process of storing energy.
- Discharging: The process of releasing energy.

6.1.5 Capacitance

As previously mentioned, once the source voltage is applied to two leads of a capacitor, the capacitor starts to store energy or charges. The charges (Q) that are stored are proportional to the voltage (V) across it. This can be expressed by the following formula:

$$Q = CV \quad \text{or} \quad C = \frac{Q}{V}$$

This is analogous to a pump pumping water to a reservoir. The higher the pressure, the more water will be pumped into the reservoir. The higher the voltage, the more charges a capacitor can store.

The voltage and charge (V–Q) characteristic of a capacitor is shown in Figure 6.5, demonstrating that the capacitor voltage is proportional to the amount of charges a capacitor can store.

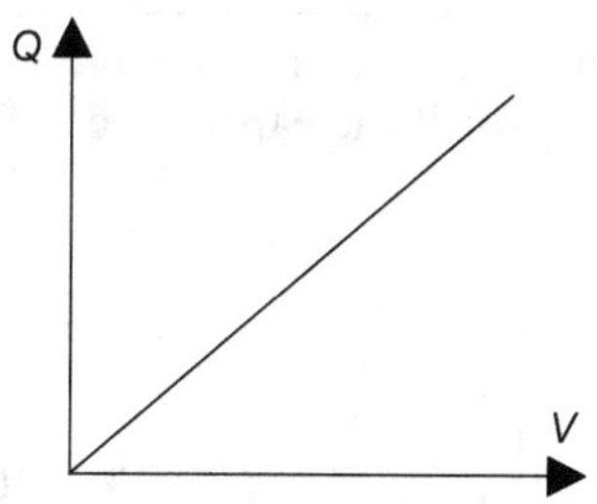

Figure 6.5 Q–V characteristic of a capacitor

C is the capacitance, which is the value of the capacitor and describes the amount of charges stored in the capacitor. Just as a resistor is a component and resistance is the value of a resistor, capacitor is a component and capacitance is the value of a capacitor. Resistor is symbolized by R while resistance is *R*: capacitor is symbolized by C while capacitance is *C*.

Capacitance *C*

C, the value of the capacitor, is directly proportional to its stored charges, and inversely proportional to the voltage (*V*) across it.

$$C = \frac{Q}{V}$$

Quantity	Quantity symbol	Unit	Unit symbol
Capacitance	C	Farad	F
Charge	Q	Coulomb	C
Voltage	V	Volt	V

A capacitor can store 1 C charge when 1 V of voltage is applied to it. That is,

$$1\,\text{F} = \frac{1\,\text{C}}{1\,\text{V}}$$

Farad is a very large unit of measure for most practical capacitors. Microfarad (μF) or picofarad (pF) are more commonly used units for capacitors.

Recall

$$1\,\mu\text{F} = 10^{-6}\,\text{F} \quad \text{and}$$

$$1\,\text{pF} = 10^{-12}\,\text{F}$$

Note: μ is a Greek letter called 'mu' (see Appendix A for a list of Greek letters).

Example 6.1: If a 50 μC charge is stored on the plates of a capacitor, determine the voltage across the capacitor if the capacitance of the capacitor is 1000 pF.

Solution:

$$Q = 50\,\mu\text{C}, \quad C = 1000\,\text{pF}, \quad V = ?$$

$$V = \frac{Q}{C} = \frac{50\,\mu\text{C}}{1000\,\text{pF}} = \frac{50 \times 10^{-6}\,\text{C}}{1000 \times 10^{-12}\,\text{F}} = 0.05 \times 10^{6}\,\text{V} = 50\,\text{KV}$$

6.1.6 Factors affecting capacitance

There are three basic factors affecting the capacitance of a capacitor, and they are determined by the construction of a capacitor as shown below:

- The area of plates (A): A is directly proportional to the charge Q; the larger the plate area, the more electric charges that can be stored.
- The distance between the two plates (d): The shorter the distance between two plates, the stronger the produced electric field that will increase the ability to store charges. Therefore, the distance (d) between the two plates is inversely proportional to the capacitance (C).
- The dielectric constant (k): Different insulating materials (dielectrics) will have a different impact on the capacitance. The dielectric constant (k) is directly proportional to the capacitance (C).

The factors affecting the capacitance of a capacitor are illustrated in Figure 6.6.

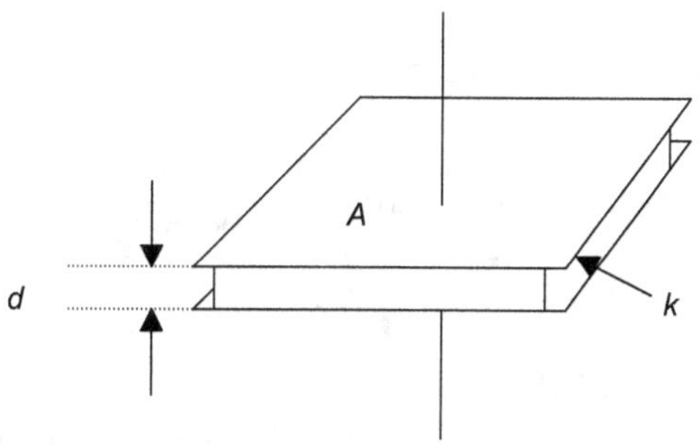

Figure 6.6 Factors affecting capacitance

Factors affecting capacitance

$$C = 8.85 \times 10^{-12}\frac{kA}{d}$$

Quantity	Quantity symbol	Unit	Unit symbol
Plates area	A	Square meter	m^2
Distance	d	Meter	m
Dielectric constant	k	No unit	
Capacitance	C	Farad	F

Dielectric constants for some commonly used capacitor materials are listed in Table 6.1.

Table 6.1 Dielectric constants of some insulating materials

Material	Dielectric constant
Vacuum	1
Air	1.0006
Paper (dry)	2.5
Glass (photographic)	7.5
Mica	5
Oil	4
Polystyrene	2.6
Teflon	2.1

Example 6.2: Determine the capacitance if the area of plates for a capacitor is 0.004 m^2, the distance between the plates is 0.006 m and the dielectric for this capacitor is mica.

Solution:

$$A = 0.004\,\text{m}^2, d = 0.006\,\text{m and } k = 5$$

$$C = 8.85 \times 10^{-12}\frac{kA}{d} = 8.85 \times 10^{-12}\frac{5 \times 0.004\text{m}^2}{0.006\text{m}} = 29.5\,\text{pF}$$

6.1.7 *Leakage current*

The dielectric between two plates of the capacitor is insulating material, and practically no insulating material is perfect (i.e. 100 per cent of the insulation). Once voltage is applied across the capacitor, there may be a very small current through the dielectric, and this is called the leakage current in the capacitor. Although the leakage current is very small, it is always there. That is why the charges or the energy stored on the capacitor plates will eventually leak off. But the leakage current is so small that it can be ignored for the application. (Electrolytic capacitors have higher leakage current.)

Leakage current
A very small current through the dielectric.

6.1.8 *Breakdown voltage*

As mentioned earlier, a capacitor charging acts as a pump pumping water into a reservoir, or a water tank. The higher the pressure, the more water will be pumped into the tank. If the tank is full and still continues to increase pressure, the tank may break down or become damaged by such high pressure.

This is similar to a capacitor. If the voltage across a capacitor is too high and exceeds the capacitor's working or breakdown voltage, the capacitor's dielectric will break down, causing current to flow through it. As a result, this may explode or permanently damage the capacitor. Therefore, when using a capacitor, pay attention to the maximum working voltage, which is the maximum voltage a capacitor can have. The applied voltage of the capacitor can never exceed the capacitor's breakdown voltage.

Breakdown voltage
The voltage that causes a capacitor's dielectric to become electrically conductive. It may explode or permanently damage the capacitor.

6.1.9 Relationship between the current and voltage of a capacitor

The relationship between the current and voltage for a resistor is Ohm's law for a resistor. The relationship between the current and voltage for a capacitor is Ohm's law for a capacitor. It can be obtained mathematically as follows.

A quantity that varies with time (such as a capacitor that takes time to charge/discharge) is called instantaneous quantity, which is the quantity at a specific time. Usually the lowercase letters symbolize instantaneous quantities, and the uppercase letters symbolize the constants or average quantities. The equation $Q = CV$ in terms of instantaneous quantity is $q = Cv$.

Note: If you haven't learned calculus, just keep in mind that $i = C(\Delta v/\Delta t)$ or $i_C = C(\mathrm{d}v_C/\mathrm{d}t)$ is Ohm's law for a capacitor, and skip the following mathematic derivation process, where Δv and Δt or $\mathrm{d}v$ and $\mathrm{d}t$ are very small changes in voltage and time.

Differentiating the equation $q = Cv$ yields

$$\frac{\mathrm{d}q}{\mathrm{d}t} = C\frac{\mathrm{d}v}{\mathrm{d}t}$$

Recall that current is the rate of movement of charges, and has the $i = \mathrm{d}q/\mathrm{d}t$ notation in calculus.

Substitute $i = \mathrm{d}q/\mathrm{d}t$ into the equation of $\mathrm{d}q/\mathrm{d}t = C(\mathrm{d}v/\mathrm{d}t)$ yields

$$i = C\frac{\mathrm{d}v}{\mathrm{d}t} \quad \text{or} \quad i = C\frac{\Delta v}{\Delta t}$$

This is Ohm's law for a capacitor. The relationship between voltage and current of a capacitor can be expressed by Figure 6.7(b).

The relationship of voltage and current for a capacitor shows that when the applied voltage at two leads of the capacitor changes, the charges (q) stored on the plates of the capacitor will also change. This will cause current to flow in

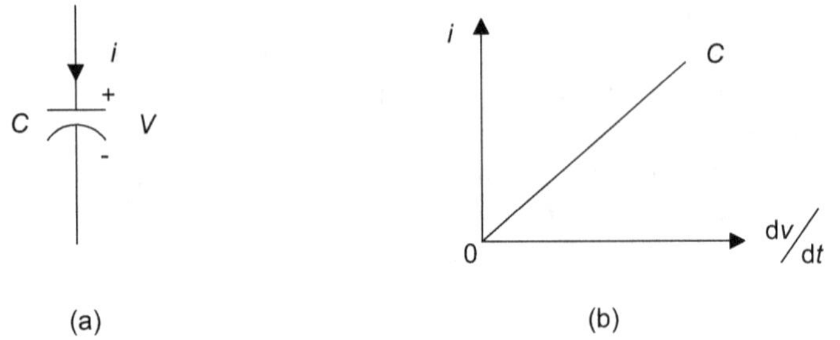

Figure 6.7 Relationship between v and i of a capacitor

the capacitor circuit. Current and the rate of change of voltage are directly proportional to each other.

The reference polarities of capacitor voltage and current should be mutually related. That is, the reference polarities of voltage and current of a capacitor should be consistent, as shown in Figure 6.7(a).

Ohm's law for a capacitor

The current of a capacitor i_C is directly proportional to the ratio of capacitor voltage $\mathrm{d}v_C/\mathrm{d}t$ (or $\Delta v_C/\Delta t$) and capacitance C.

$$i_C = C\frac{\mathrm{d}v_C}{\mathrm{d}t} \quad \text{or} \quad i = C\frac{\Delta v_C}{\Delta t}$$

where $\mathrm{d}v_C$ and $\mathrm{d}t$ or Δv_C, and Δt are very small changes in voltage and time.

The relationship of voltage and current in a capacitive circuit shows that the faster the voltage changes with time, the greater the amount of capacitive current flows through the circuit. Similarly, the slower the voltage changes with time, the smaller the amount of current, and if voltage does not change with time, the current will be zero. Zero current means that the capacitor acts like an open circuit for DC voltage at this time. Voltage that does not change with time is DC voltage, meaning that current is zero when DC voltage is applied to a capacitor. Therefore, the capacitor may play an important role for blocking the DC current. This is a very important characteristic of a capacitor.

DC blocking

Current through a capacitor is zero when DC voltage applied to it (open-circuit equivalent). A capacitor can block DC current.

Note: Although there is a DC voltage source applied to the capacitive circuit in Figures 6.3 and 6.4, the capacitor charging/discharging happened at the moment when the switch turned to different locations, i.e. when the voltage across the capacitor changes within a moment. When the capacitor charging/dischaging has finished, the capacitor is equivalent to an open circuit for that circuit.

6.1.10 Energy stored by a capacitor

As mentioned earlier, a capacitor is an energy storage element. It can store energy that it absorbed from charging and maintain voltage across it. Energy stored by a capacitor in the electric field can be derived as follows.

The instantaneous electric power of a capacitor is given by $p = vi$. Substituting this into the capacitor's current $i = C(\mathrm{d}v/\mathrm{d}t)$ yields

$$p = Cv\frac{\mathrm{d}v}{\mathrm{d}t}$$

Since the relationship between power and work is $P = W/t$ (energy is the ability to do work), and, instantaneous power for this expression is $p = \mathrm{d}w/\mathrm{d}t$, substituting it into $p = Cv(\mathrm{d}v/\mathrm{d}t)$ yields

$$\frac{\mathrm{d}w}{\mathrm{d}t} = Cv\frac{\mathrm{d}v}{\mathrm{d}t}$$

Integrating the above expression:

$$\int_0^t \frac{\mathrm{d}w}{\mathrm{d}t}\mathrm{d}t = C\int_0^V v\frac{\mathrm{d}v}{\mathrm{d}t}\mathrm{d}t$$

gives

$$W = \frac{1}{2}Cv^2$$

Note: If you haven't learned calculus, just keep in mind that $W = ½(Cv^2)$, and skip the above mathematic derivation process.

Energy stored by a capacitor

$$W_C = \frac{1}{2}Cv^2$$

Quantity	**Quantity symbol**	**Unit**	**Unit symbol**
Energy	W	Joule	J
Capacitor	C	Farad	F
Voltage	V	Volt	V

The expression for energy stored by a capacitor demonstrates that the capacitor's energy depends on the values of the capacitor and voltage across the capacitor.

Example 6.3: A 15 V voltage is applied to a 2.2 μF capacitor. Determine the energy this capacitor has stored.

Solution:

$$\begin{aligned} W_C &= \frac{1}{2}Cv^2 \\ &= \frac{1}{2}(2.2\,\mu\text{F}) \times (15\,\text{V})^2 \\ &= 247.5\,\mu\text{J} \end{aligned}$$

6.2 Capacitors in series and parallel

Same as resistors, capacitors may also be connected in series or parallel to obtain a suitable resultant value that may be either higher or lower than a single capacitor value.

The total or equivalent capacitance C_{eq} will decrease for a series capacitive circuit and it will increase for a parallel capacitive circuit. The total or equivalent capacitance has the opposite form with the total or equivalent resistance R_{eq}.

6.2.1 Capacitors in series

A circuit of *n* capacitors is connected in series as shown in Figure 6.8.

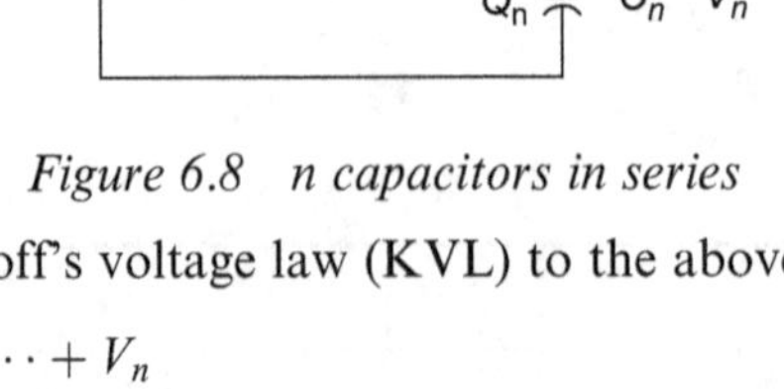

Figure 6.8 n capacitors in series

Applying Kirchhoff's voltage law (KVL) to the above circuit gives

$$E = V_1 + V_2 + \cdots + V_n$$

and since $V = Q/C$, substituting it into the above expression yields

$$\frac{Q_{eq}}{C_{eq}} = \frac{Q_1}{C_1} + \frac{Q_2}{C_2} + \cdots + \frac{Q_n}{C_n}$$

where $E = Q_{eq}/C_{eq}$, Q_{eq} is the equivalent (or total) charges and C_{eq} is the equivalent (or total) capacitance for a series capacitive circuit respectively. Since only one current flows in a series circuit, each capacitor will store the same amount of charges, i.e. $Q_{eq} = Q_1 = Q_2 = \cdots = Q_n = Q$ therefore,

$$\frac{Q}{C_{eq}} = \frac{Q}{C_1} + \frac{Q}{C_2} + \cdots + \frac{Q}{C_n}$$

Dividing by Q on both sides of the above expression gives

$$\frac{1}{C_{eq}} = \frac{1}{C_1} + \frac{1}{C_2} + \cdots + \frac{1}{C_n} \quad \text{or} \quad C_{eq} = \frac{1}{(1/C_1) + (1/C_2) + \cdots + (1/C_n)}$$

This is the equation for calculating the series equivalent (total) capacitance. This formula has the same form with the formula for calculating equivalent parallel resistance $(1/R_{eq}) = (1/R_1) + (1/R_2) + \cdots + (1/R_n)$. When there are two capacitors in series, it also has the same form with the formula for calculating two resistors in parallel, i.e. $C_{eq} = C_1C_2/(C_1 + C_2)$ and $R_{eq} = R_1R_2/(R_1 + R_2)$.

Equivalent (total) series capacitance

- n capacitors in series: $C_{eq} = \frac{1}{(1/C_1)+(1/C_2)+\cdots+(1/C_n)}$
- Two capacitors in series: $C_{eq} = \frac{C_1C_2}{C_1+C_2}$

Example 6.4: Determine the charges Q stored by each capacitor in the circuit of Figure 6.9.

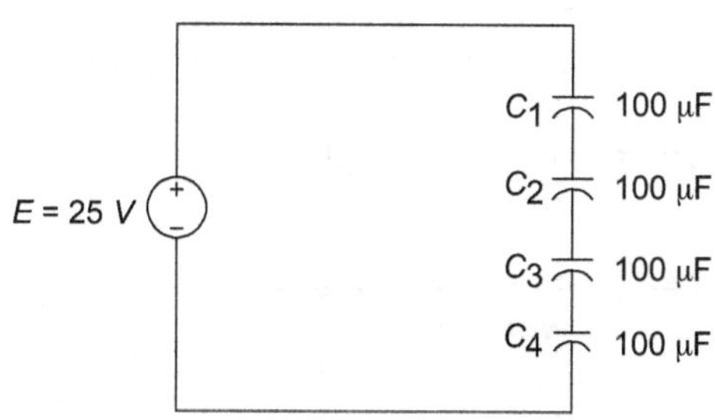

Figure 6.9 Circuit for Example 6.4

Solution: Since $Q = CV$, or $Q = C_{eq}E$, solve for C_{eq} first.

$$C_{eq} = \frac{1}{(1/C_1) + (1/C_2) + (1/C_3) + (1/C_4)}$$

$$= \frac{1}{[(1/100) + (1/100) + (1/100) + (1/100)\mu F]} = 25\ \mu F$$

Therefore,

$$Q = C_{eq}E = (25\ \mu F)\,(25\ V) = 625\ \mu C$$

Example 6.4 shows that when capacitors are connected in series, the total or equivalent capacitance C_{eq} (25 μF) is less than any one of the individual capacitances (100 μF).

The physical characteristic of the series equivalent capacitance is that the single series equivalent capacitance C_{eq} has the total *dielectric* (or total *distance* between the plates) of all the individual capacitors. The formula for factors affecting the capacitance ($C = 8.85 \times 10^{-12}kA/d$) shows that if the distance between the plates of a capacitor (d) increases, the capacitance (C) will decrease. This is shown in Figure 6.10.

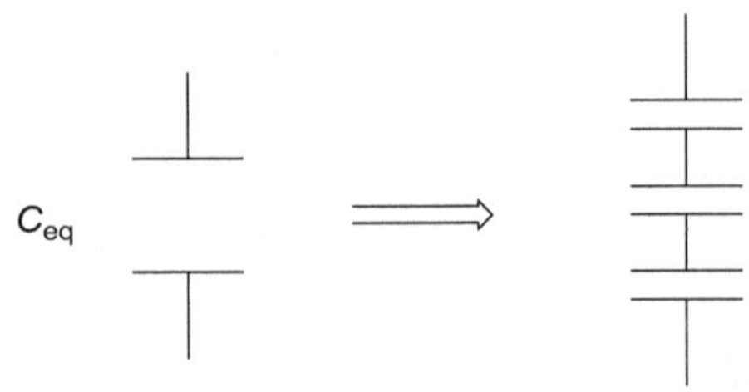

Figure 6.10 The physical characteristic of series C_{eq}

6.2.2 Capacitors in parallel

A circuit of n capacitors connected in parallel is shown in Figure 6.11.

The charge stored on the individual capacitor in this circuit is

$$Q_1 = C_1V, \quad Q_2 = C_2V, \quad Q_n = C_nV, \quad (\text{where } V = E)$$

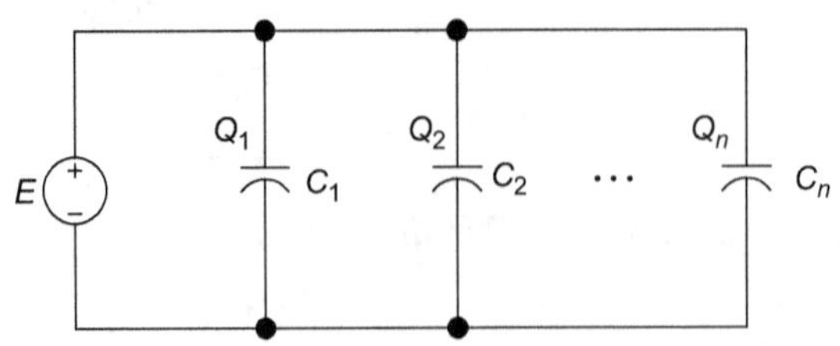

Figure 6.11 n capacitors in parallel

The total charge Q_{eq} in this circuit should be the sum of all stored charges on the individual capacitor, i.e.

$$Q_{eq} = Q_1 + Q_2 + \cdots + Q_n$$

therefore,

$$C_{eq}V = C_1V + C_2V + \cdots + C_nV$$

dividing both sides by V yields

$$C_{eq} = C_1 + C_2 + \cdots + C_n$$

This is the equation for calculating the parallel equivalent (total) capacitance. As you may have noticed, this equation has the same form with the equation for calculating series resistances ($R_{eq} = R_1 + R_2 + \cdots + R_n$).

Equivalent (total) parallel capacitance

$$C_{eq} = C_1 + C_2 + \cdots + C_n$$

Equation for calculating capacitance is exactly opposite the equations for calculating resistance. Capacitors in series result in parallel form as resistances, and capacitors in parallel result in series form as resistances.

Example 6.5: Determine the total charge in all the capacitors in the circuit of Figure 6.12.

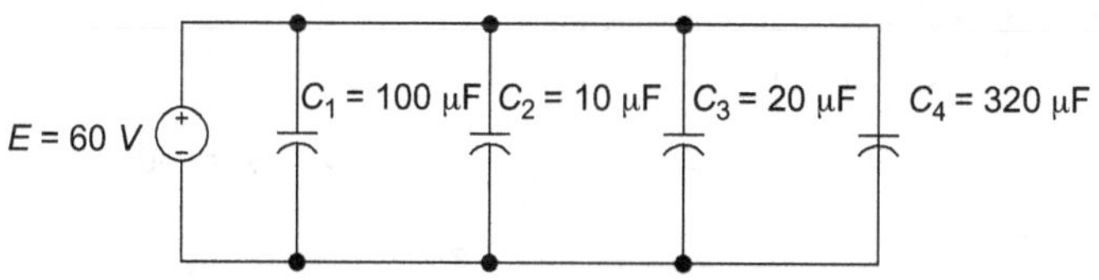

Figure 6.12 Circuit for Example 6.5

Solution: Since $Q = CV$, i.e. $Q_{eq} = C_{eq}E$ and $C_{eq} = C_1 + C_2 + \cdots + C_n = (100 + 10 + 20 + 320)\mu\text{F} = 450\,\mu\text{F}$ therefore,

$$\begin{aligned} Q_{eq} &= C_{eq}E \\ &= (450\,\mu\text{F})(60\,\text{V}) \\ &= 27\,000\,\mu\text{C} \end{aligned}$$

From Example 6.5, we can see that when capacitors are connected in parallel, the total or equivalent capacitance C_{eq} (450 μF) is greater than any one of the individual capacitances ($C_1 = 100\ \mu F$, $C_2 = 10\ \mu F$, $C_3 = 20\ \mu F$ and $C_4 = 320\ \mu F$).

The physical characteristic of the equation for calculating the parallel equivalent capacitance is that a single parallel equivalent capacitor C_{eq} has the total *area* of plates of the individual capacitors. From the formula of factors affecting the capacitance ($C = 8.85 \times 10^{-12} kA/d$), we can see that if the area of plates (A) of a capacitor increases, the capacitance will increase. This is shown in Figure 6.13.

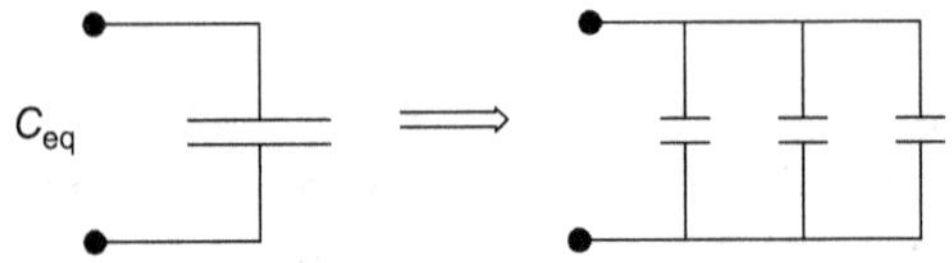

Figure 6.13 The physical characteristic of parallel C_{eq}

6.2.3 Capacitors in series–parallel

Similar to resistors, capacitors may also be connected in various combinations. When serial and parallel capacitors are combined together, series–parallel capacitor circuits result and an example is shown in the following.

Example 6.6: Determine the equivalent capacitance through two terminals a and b in the circuit of Figure 6.14.

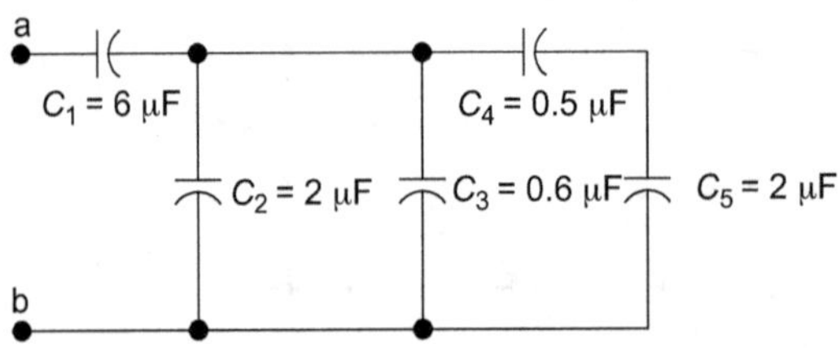

Figure 6.14 Circuit for Example 6.6

Solution:

$$C_{4,5} = \frac{C_4 C_5}{C_4 + C_5} = \frac{(0.5\mu F)(2\mu F)}{(0.5 + 2)\mu F} = 0.4\ \mu F$$

$$C_{2,3,4,5} = C_2 + C_3 + C_{4,5} = (2 + 0.6 + 0.4)\mu F = 3\ \mu F$$

$$C_{eq} = \frac{C_1 C_{2,3,4,5}}{C_1 + C_{2,3,4,5}} = \frac{(6\mu F)(3\mu F)}{(6 + 3)\mu F} = 2\ \mu F$$

6.3 Inductor

We have learned about two of the three important fundamental passive circuit elements (components that absorb but not produce energy), the resistor and the capacitor. The third element is the inductor (or coil). Inductors have many applications in electrical and electronic devices, including electrical generators, transformers, radios, TVs, radars, motors, etc. As previously mentioned, both capacitors and inductors are energy storage elements. The difference between the two is that a capacitor stores transferred energy in the *electric field*, and an inductor stores transferred energy in the *magnetic field.* Since inductors are based on the theory of electromagnetism induction, let us review some concepts of electromagnetism induction you may have learned in physics that will be used in the following section.

6.3.1 Electromagnetism induction

6.3.1.1 Electromagnetic field

All stationary electrical charges are surrounded by electric fields, and the movement of a charge will produce a magnetic field. When the charge changes its velocity of motion (or when the charge is accelerated), an electromagnetic field is generated. Therefore, whenever a changing current flows through a conductor, the area surrounding the conductor will produce an electromagnetic field.

The electromagnetic field can be visualized by inserting a current-carrying conductor (wire) through a hole in a cardboard and sprinkling some iron filings on it. As the changing current flows through the conductor, the iron fillings will align themselves with the circles surrounding the conductor; these are magnetic lines of force. The direction of these lines of force can be determined by the right-hand spiral rule, as shown in Figure 6.15. The area shows that the magnetic characteristics are called the magnetic field, as it is produced by the changing current-carrying conductor, and therefore, it is also called the electromagnetic field. This is the principle of electricity producing magnetism.

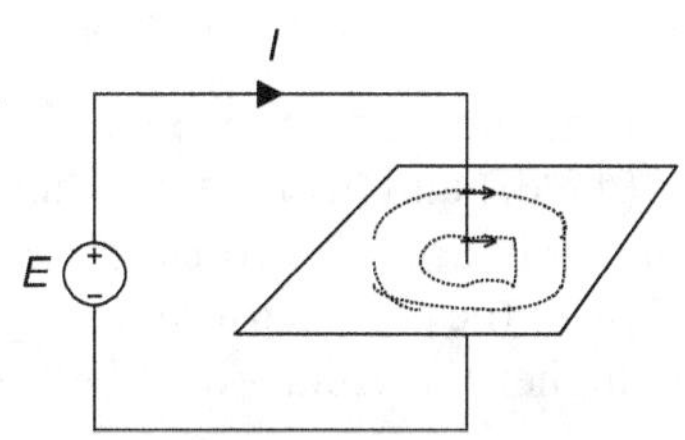

Figure 6.15 Electricity produces magnetism

- Right-hand spiral rule:
  Thumb = the direction of current.
  Four fingers = the direction of magnetic lines of force or direction of the flux (the total magnetic lines of force).

Electromagnetic field
The surrounding area of a conductor with a changing current can generate an electromagnetic field.

6.3.1.2 Faraday's law

In 1831, British physicist and chemist Michael Faraday discovered how an electromagnetic field can be induced by a changing magnetic flux. When there is a relative movement between a conductor and a magnetic field (or a changing current through the conductor), it will induce a changing magnetic flux Φ (the total number of magnetic lines of force) surrounding the conductor, hence an electromagnetic field is generated. This electromagnetic field will produce an induced voltage and current.

For example, in Figure 6.16, if a magnet bar is moved back and forth in a coil of wire (conductor), or if the coil is moved back and forth close to the magnet and through the magnetic field, the magnetic lines of flux will be cut and a voltage v_L across the coil will be induced (v_L can be measured by using a voltmeter.) Or, an electromotive force (emf, e_L) that has an opposite polarity with v_L will be induced, and this will result in an induced current in the coil. This is the principle of a magnet producing electricity.

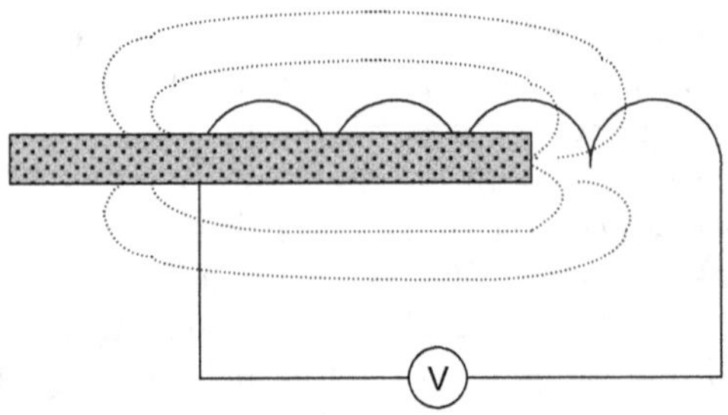

Figure 6.16 Magnet produces electricity

Faraday observed that the induced voltage (v_L) is directly proportional to the rate of change of flux ($d\phi/dt$) and also the number of turns (N) in the coil, and is expressed mathematically as $v_L = N(d\phi/dt)$. In other words, the faster the relative movement between the conductor and magnetic fields, or the more the turns the coil has, the higher the voltage will be produced.

Faraday's law

- When there is a relative movement between a conductor and magnetic field, the changing magnetic flux will induce an electromagnetic field and produce an induced voltage (v_L).
- v_L is directly proportional to the rate of change of flux ($d\phi/dt$) and the number of turns (N) in the coil, $v_L = N(d\phi/dt)$.

6.3.1.3 Lenz's law

In 1834, Russian physicist Heinrich Lenz developed a companion result with the Faraday's law. Lenz defined the polarity of induced effect and stated that an induced effect is always opposed to the cause producing it. When there is a relative movement between a conductor and a magnetic field (or a changing current through the conductor), an induced voltage (v_L) or induced emf (e_L) and also an induced current (i) will be produced. The polarity of the induced emf is always opposite to the change of the original current.

When the switch is turned on in the circuit of Figure 6.17, the current (cause) in the circuit will increase, but the induced emf (effect) will try to stop it from increasing. When the switch is turned off, the current i will decrease, but the polarity of induced emf (e_L) changes and will try to stop it from decreasing. This is because an induced current in the circuit flows in a direction that can create a magnetic field that will counteract the change in the original magnetic flux.

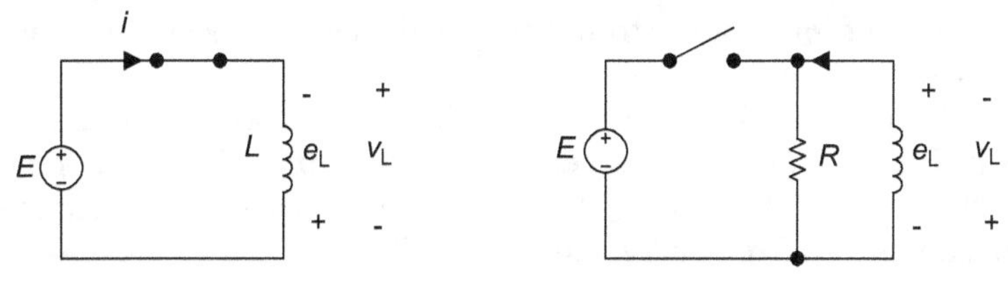

Figure 6.17 Lenz's law

Mathematically, Lenz's law can be expressed as follows:

$$\text{If } i > 0, \ \frac{di}{dt} > 0, \quad \text{then } e_L = -L\frac{di}{dt}, \quad \left(\text{or } v_L = L\frac{di}{dt}\right)$$

where di/dt is the rate of change of current, and the minus sign for e_L is to remind us that the induced emf always acts to oppose the change in magnetic flux that generates the emf and current.

The induced voltage (v_L) and induced emf (e_L) have opposite polarities ($E = -V$); this emf is also called the counter emf. However, the induced voltage (v_L) has the same polarity with the direction of induced current (i). This is similar to the concept of the mutually related reference polarity of voltage and current.

Lenz's law

- When there is a changing current through the conductor, an induced voltage (v_L) or induced emf (e_L) and also an induced current (i) will be produced.

- The polarity of the induced emf (e_L) is always opposite to the change of the original current. $e_L = -L\frac{di}{dt}$ or $v_L = L\frac{di}{dt}$

The letter L in the above equation is called inductance (or self-inductance), which is discussed below.

6.3.2 *Inductor*

An inductor (L) is made by winding a given length of wire into a loop or coil around a core (centre of the coil). Inductors may be classified as air-core inductors or iron-core inductors. An air-core inductor is simply a coil of wire. But this coil turns out to be a very important electric/electronic element because of its magnetic properties. Iron-core provides a better path for the magnetic lines of force and a stronger magnetic field for the iron-core inductor as compared to the air-core inductor.

The schematic symbol for an air-core inductor looks like a coil of wire as shown in Figure 6.18(a). The schematic symbol for an iron-core inductor is shown in Figure 6.18(b). Similar to resistors and capacitors, the inductor can be also classified as fixed and variable.

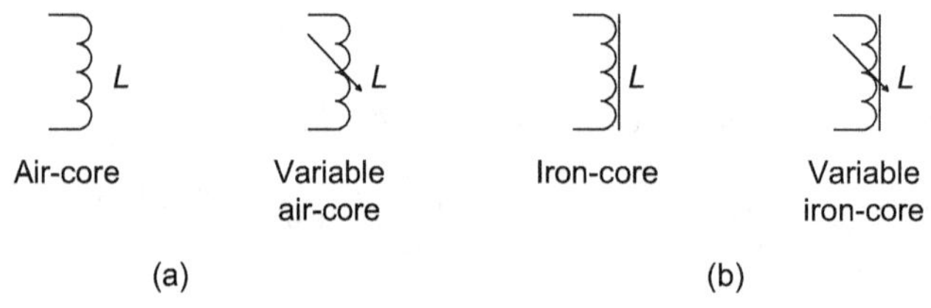

Figure 6.18 Schematic symbols for inductors

Inductor L
An inductor is an energy storage element that is made by winding a given length of wire into a loop or coil around a core.

6.3.3 *Self-inductance*

When current flows through an inductor (coil) that is the same as a current-carrying conductor, a magnetic field will be induced around the inductor. According to the principle of electromagnetic induction, Faraday's law and Lenz's law, when there is a relative movement between an inductor and magnetic field or when current changes in the inductor, the changing magnetic flux will induce an electromagnetic field resulting in an induced voltage (v_L), or induced emf (e_L), and also an induced current (i).

The measurement of the changing current in an inductor that is able to generate induced voltage is called inductance. The inductor is symbolized by L while inductance is symbolized by L, and the unit of inductance is henry (H). The resistor, capacitor and inductor are circuit components, and the resistance, capacitance and inductance are the value or capacity of these components. So inductance is the capacity to store energy in the magnetic field of an inductor.

Inductance L (or self-inductance)

The measurement of the changing current in an inductor that is able to generate induced voltage is called inductance that is measured in henries (H).

Quantity	Quantity symbol	Unit	Unit symbol
Inductance (or Self-Inductance)	L	Henry	H

6.3.4 *Relationship between inductor voltage and current*

Lenz's law $v_L = L(di/dt)$ shows the relationship between current and voltage for an inductor, and it is Ohm's law for an inductor. There, the inductance (L) and the current rate of change (di/dt) determine the induced voltage (v_L). The induced voltage v_L is directly proportional to the inductance L and the current rate of change di/dt. This relationship can be illustrated as in Figure 6.19.

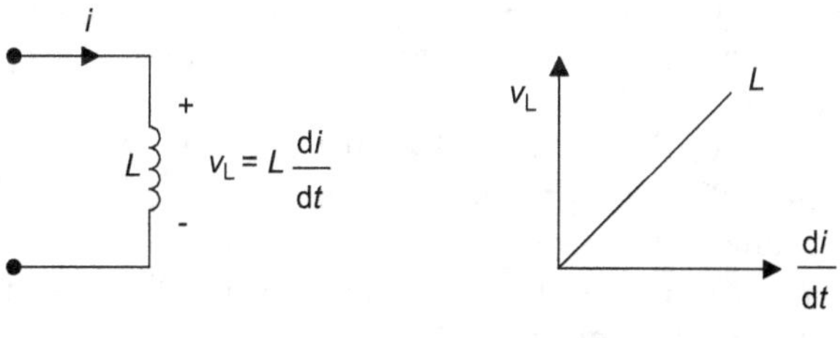

Figure 6.19 Characteristics of an inductor's voltage and current

Ohm's law for an inductor

An inductor's voltage v_L is directly proportional to the inductance L and the rate of change current di/dt: $v_L = L(di/dt)$.

Ohm's law for an inductor $v_L = L(di/dt)$ has a similar form as Ohm's law for a capacitor $i_C = C(dv_C/dt)$. These two are very important formulas that will be used in future circuits.

The larger the inductance, or the greater the change of current, the higher the induced voltage in the coil. When the current does not change with time (DC current), i.e. $di/dt = 0$, the inductor voltage (v_L) is also zero. Zero voltage means that an inductor acts like a short circuit for DC current. Therefore, the inductor may play an important role for passing the DC current. This is a very important characteristic of an inductor and is opposite to that of a capacitor. Recall that a capacitor can block DC and acts like an open circuit for DC.

Passing DC

- Voltage across an inductor is zero when a DC current flows through it (short-circuit equivalent).
- An inductor can pass DC.

6.3.5 *Factors affecting inductance*

There are some basic factors affecting the inductance of an inductor (iron-core). These parameters are determined by the construction of an inductor as shown in the following (if all other factors are equal):

- The number of turns (N) for the coil: More turns for a coil will produce a stronger magnetic field resulting in a higher induced voltage and inductance.
- The length of the core (l): A longer core will make a loosely spaced coil and a longer distance between each turn, and therefore producing a weaker magnetic field resulting in a smaller inductance.
- The cross-section area of the core (A): A larger core area requires more wire to construct a coil, and therefore it can produce a stronger magnetic field resulting in a higher inductance.
- The permeability of the material of the core (μ): A core material with higher permeability will produce a stronger magnetic field resulting in a higher inductance. (Permeability of the material of the core determines the ability of *material* to produce a magnetic field. Different materials have different degrees of permeability.)

Factors affecting the inductance of an inductor are illustrated in Figure 6.20.

Factors affecting inductance

$$L = \frac{N^2 A\mu}{l}$$

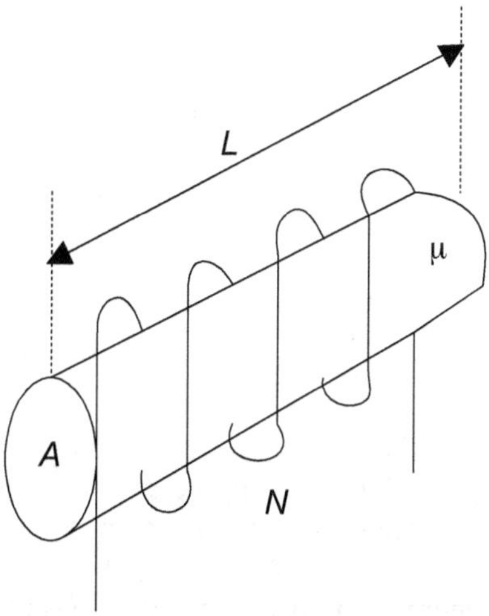

Figure 6.20 Factors affecting inductance

Where inductance is symbolized by L, measured in henries (H); area of the core is symbolized by A, measured in m^2; permeability is symbolized by μ; number of turns is symbolized by N.

From the expression of the factor affecting inductance, we can see that either when the number of turns of a coil increases, or when the cross-section area of the core increases, or when core material with higher permeability is chosen, or when the length of core is reduced, the inductance of an inductor will increase.

6.3.6 The energy stored by an inductor

Same as a capacitor, an inductor is also an energy storage element. When voltage is applied to two leads of an inductor, the current flows through the inductor and will generate energy, and this energy is then absorbed by the inductor and stored in the magnetic field as electromagnetic field builds up. The energy stored by an inductor can be derived as follows:

The instantaneous electric power of an inductor is given by

$$p = iv_{\mathrm{L}}$$

Since the relationship between power and work is $P = W/t$ (energy is the ability to do work), and the instantaneous power for this expression is $p = \mathrm{d}w/\mathrm{d}t$.

Substituting $p = \mathrm{d}w/\mathrm{d}t$ and $v_{\mathrm{L}} = L(\mathrm{d}i/\mathrm{d}t)$ into the instantaneous power expression $p = iv_{\mathrm{L}}$ gives

$$\frac{\mathrm{d}w}{\mathrm{d}t} = Li\frac{\mathrm{d}i}{\mathrm{d}t}$$

Integrating both sides:

$$\int_0^t \frac{dw}{dt}dt = \int_0^t Li\frac{di}{dt}dt$$

Therefore

$$w = L\int_0^t i di, \quad \text{i.e.} \quad w = \frac{1}{2}Li^2$$

Note: If you haven't learned calculus, just keep in mind that $W_L = (½)Li^2$, and skip the above mathematic derivation process.

This equation has a similar form with the energy equation of a capacitor ($W_C = (½)Cv^2$).

The equation for energy stored by an inductor shows that the inductor's energy depends on the inductance and the inductor's current.

When current increases, an inductor absorbs energy and stores it in the magnetic field of the inductor. When current decreases, an inductor releases the stored energy to the circuit. Same as a capacitor, an inductor will not be able to release more energy than it has stored, so it is also called a passive element.

Energy stored by an inductor

$$W_L = \frac{1}{2}Li^2$$

where inductance L is measured in H, energy W is measured in J and current i is measured in A.

Example 6.7: Current in a 0.01 H inductor is $i(t) = 5e^{-2t}$A, determine the energy stored by the inductor and induced voltage v_L.

Solution:

$$W_L = \frac{1}{2}Li^2 = \frac{1}{2}(0.01\,\text{H})(5e^{-2t}\text{A})^2 = \frac{1}{2}(0.01\,\text{H})(25e^{-4t}\text{A}) = 0.125e^{-4t}\text{J}$$

$$v_L = L\frac{di}{dt} = 0.01\frac{d}{dt}(5e^{-2t}) = 0.01\,\text{H}(-2)(5)(e^{-2t}\text{A}) = -0.1e^{-2t}\text{V}$$

Note: If you haven't learned calculus, skip the v_L part.

6.3.7 Winding resistor of an inductor

When winding a given length of wire into a loop or coil around a core, an inductor is formed. A coil or inductor always has resistance. This is because there is always a certain internal resistance distributed in the wire, and the longer the wire, the more turns of coils there are, and thus the wire will have a significantly higher internal resistance. This is called the winding resistance of a coil (R_w). An inductor circuit with winding resistance is shown in Figure 6.21.

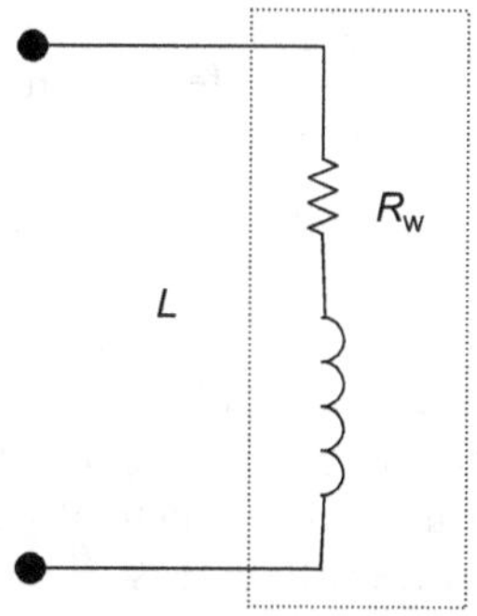

Figure 6.21 Winding resistance

Winding resistance R_w
The internal resistance in the wire of an inductor.

Example 6.8: The winding resistance for an inductor in the circuit of Figure 6.22 is 5 Ω. When the current approaches a steady state (does not change any more), the energy stored by the inductor is 4 J. What is the inductance of the inductor?

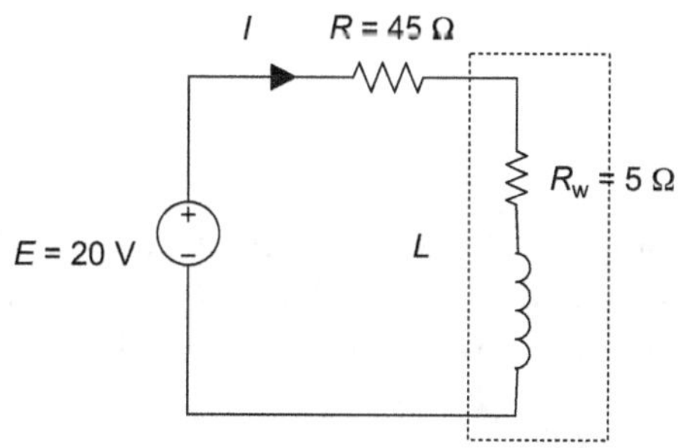

Figure 6.22 Circuit for Example 6.8

Solution:

$$E = 20\,\text{V},\ R = 45\,\Omega,\ R_W = 5\,\Omega,\ \text{and}\ W_L = 4\,\text{J}.\quad L = ?$$

$$I = \frac{E}{R + R_W} = \frac{20\,\text{V}}{(45 + 5)\Omega} = 0.4\,\text{A}$$

From $W_L = \frac{1}{2}(Li^2)$
solving for L:

$$L = \frac{2W_L}{I^2} = \frac{2 \times 4\,\text{J}}{(0.4\,\text{A})^2} = 50\,\text{H}$$

($i = I$ since the current approaches to steady state)

6.4 Inductors in series and parallel

Similar to resistors and capacitors, inductors may also be connected in series or in parallel to obtain a suitable resultant value that may be either higher or lower than a single inductor value. The equivalent (total) series or parallel inductance has the same form as the equivalent (total) series or parallel resistance. The equivalent inductance will increase if inductors are in series, and the equivalent (total) inductance will decrease if inductors are in parallel.

6.4.1 Inductors in series

A circuit of n inductors connected in series is shown in Figure 6.23.

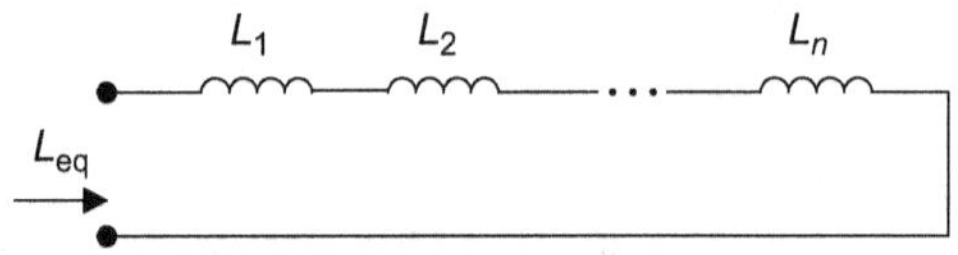

Figure 6.23 Inductors in series

Equivalent series inductance

$$L_{eq} = L_1 + L_2 + \cdots + L_n$$

This is the equation for calculating the equivalent (total) series inductance. As you may have noticed, this formula has the same form as the formula for calculating series resistances ($R_{eq} = R_1 + R_2 + \cdots + R_n$).

6.4.2 Inductors in parallel

A circuit of n inductors connected in parallel is shown in Figure 6.24.

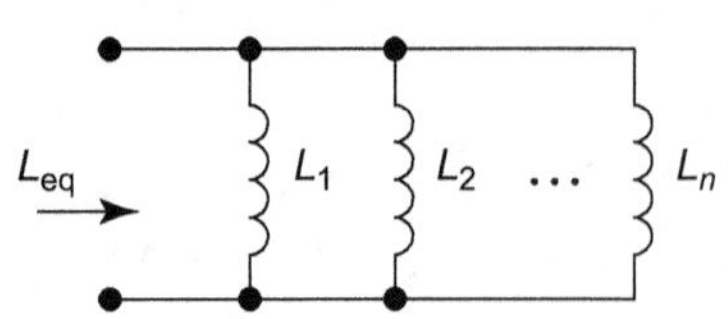

Figure 6.24 Inductors in parallel

Equivalent parallel inductance

- n inductors in parallel: $L_{eq} = \frac{1}{(1/L_1)+(1/L_2)+\cdots+(1/L_n)}$
- Two inductors in parallel: $L_{eq} = \frac{L_1 L_2}{L_1+L_2}$

These are the equations for calculating the equivalent parallel inductance. As you may have noticed, these equations have the same forms as the equations for calculating parallel resistance $R_{eq} = 1/(1/R_1) + (1/R_2) + \cdots + (1/R_n)$, also $R_{eq} = R_1R_2/(R_1 + R_2)$.

6.4.3 *Inductors in series–parallel*

Similar to resistors and capacitors, inductors may also be connected in various combinations of series and parallel. An example of a series–parallel inductive circuit is shown in the following.

Example 6.9: Determine the equivalent inductance for the series–parallel inductive circuit shown in Figure 6.25.

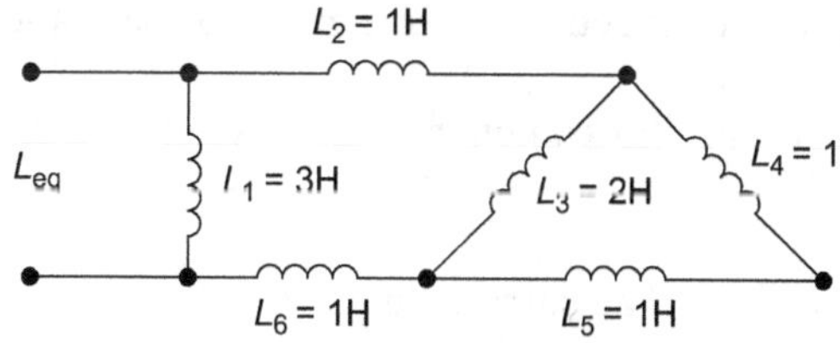

Figure 6.25 Circuit for Example 6.9

Solution:

$$L_{eq} = [(L_4 + L_5)//L_3 + (L_2 + L_6)]//L_1$$

$$L_{eq} = \frac{[((1+1)\times 2)/((1+1)+2)+1+1]\times 3}{[((1+1)\times 2)/((1+1)+2)+1+1]+3}\text{H} = 1.5\,\text{H}$$

Example 6.10: There are three inductors in a series–parallel inductive circuit: 40, 40 and 50 H. If $L_{eq} = 70$ H, how are these inductors connected?

Solution:

$$L_{eq} = 50\,\text{H} + (40\,\text{H})//(40\,\text{H}) = 70\,\text{H}$$

$$\text{or} \quad L_{eq} = 50\,\text{H} + \frac{40 \times 40}{40 + 40}\,\text{H} = 70\,\text{H}$$

So two 40 H inductors are in parallel, and then in series with a 50 H inductor.

Summary

Capacitor

- Capacitor (C): An energy storage element that has two conductive plates separated by an isolating material (the dielectric).
- Capacitor charging: Capacitor stores absorbed energy.
- Capacitor discharging: Capacitor releases energy to the circuit.
- Capacitance (C): The value of the capacitor, $C = Q/U$.
- Factors affecting capacitance:

$$C = 8.85 \times 10^{-12} \frac{kA}{d}$$

- Leakage current: A very small current through the dielectric.
- Breakdown voltage: The voltage that causes a capacitor's dielectric to become electrically conductive, it can explode or permanently damage the capacitor.
- Blocking DC: A capacitor can block DC current (open-circuit equivalent).

Inductor

- Electromagnetic field: The surrounding area of a conductor with a changing current can generate an electromagnetic field.
- Faraday's law:

$$v_{\text{L}} = N\frac{\text{d}\phi}{\text{d}t}$$

- Lenz's law :

$$e_{\text{L}} = -L\frac{\text{d}i}{\text{d}t} \quad \text{or} \quad v_{\text{L}} = L\frac{\text{d}i}{\text{d}t}$$

- Inductor (L): An energy storage element that is made by winding a given length of wire into a loop or coil around a core.
- Inductance (L): The measurement of the changing current in an inductor that produces the ability to generate induced voltage is called inductance.

- Factors affecting inductance:

$$L = \frac{N^2 A\mu}{l}$$

- Winding resistance (R_w): The internal resistance in the wire of an inductor.

The characteristics of the resistor, capacitor and inductor

Characteristic	Resistor	Capacitor	Inductor
Ohm's law	$V = IR$	$i_C = C(dv_C/dt)$	$v_L = L(di/dt)$
Energy	$W = pt$ or $dw = pdt$	$W_C = ½(Cv^2)$	$W_L = ½(Li^2)$
Series	$R_{eq} = R_1 + R_2 + \cdots + R_n$	$C_{eq} = 1/[(1/C_1) + (1/C_2) + \cdots + (1/C_n)]$ Two capacitors: $C_{eq} = C_1C_2/(C_1 + C_2)$	$L_{eq} = L_1 + L_2 + \cdots + L_n$
Parallel	$R_{eq} = 1/[(1/R_1) + (1/R_2) + \cdots + (1/R_n)]$ Two resistors: $R_{eq} = R_1R_2/(R_1 + R_2)$	$C_{eq} = C_1 + C_2 + \cdots + C_n$	$L_{eq} = 1/[(1/L_1) + (1/L_2) + \cdots + (1/L_n)]$ Two inductors: $R_{eq} = L_1L_2/(L_1 + L_2)$
Elements in DC		Open-circuit equivalent	Short-circuit equivalent

Experiment 6: Capacitors

Objectives

- Understand the characteristics of a capacitor through experiment.
- Verify the equations of capacitors in series and parallel through experiment.
- Apply the voltage divider rule in a capacitive circuit.
- Analyse experimental data, circuit behaviour and performance, and compare them to the theoretical equivalents.

Background information

- Parallel equivalent capacitance:

$$C_{eq} = C_1 + C_2 + \cdots + C_n$$

- Series equivalent capacitance:

$$C_{eq} = \frac{1}{(1/C_1) + (1/C_2) + \cdots + (1/C_n)}$$

When $n = 2$:

$$C_{eq} = \frac{C_1 C_2}{C_1 + C_2}$$

- Voltage divider rule for capacitive circuit:

$$V_{C_1} = E\frac{C_{eq}}{C_1}, \quad V_{C_2} = E\frac{C_{eq}}{C_2}$$

Recall the voltage divider rule for resistive circuit:

$$V_{R_1} = E\frac{R_1}{R_{eq}} \quad V_{R_2} = E\frac{R_2}{R_{eq}}$$

Equipment and components

- Multimeter
- Breadboard
- DC power supply
- Z meter or LCZ meter (or any other measuring instruments that can be used to measure capacitance):
    - Z meter: A measuring instrument that can be used to measure the values of capacitors.
    - LCZ meter: A measuring instrument that can be used to measure the values of capacitors and inductors.
- Capacitors:
    - 30 and 470 μF electrolytic capacitors each
    - Four non-electrolytic capacitors with any values

Note: Electrolytic capacitors are polarized. Connect the positive lead of the capacitor to the positive terminal of the DC power supply, and negative lead to the negative terminal of the DC power supply. Non-electrolytic capacitors are non-polarized, so they can be connected either way in a circuit.

Procedure

Part I: Capacitors in series and parallel

1. Take four non-electrolytic capacitors, short circuit each capacitor with a bit of wire to release or discharge the stored charges on the capacitor. Record their nominal values in Table L6.1. (A capacitor can hold its stored charges for days or weeks and can shock you even when it is not connected to a circuit.)

Table L6.1

Capacitor	C_1	C_2	C_3	C_4
Nominal value				
Measured value				

2. Measure the value of each capacitor using a Z meter or LCZ meter and record in Table L6.1.
3. Connect each circuit as shown in Figure L6.1(a, b and c) on the breadboard. Calculate the equivalent capacitance for each circuit. Record the values in Table L6.2.

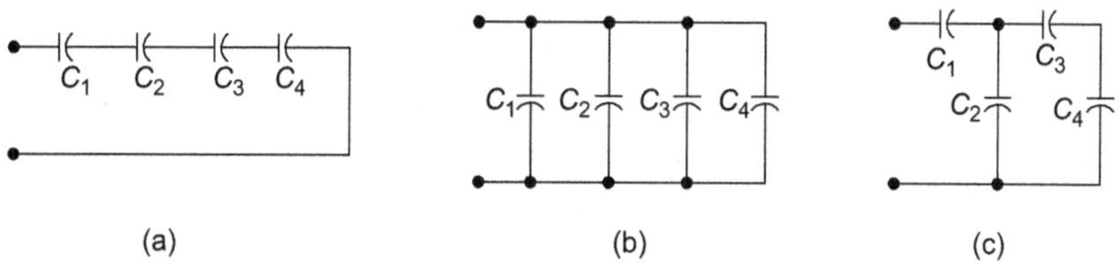

Figure L6.1 Capacitors in series and parallel

Table L6.2

Equivalent capacitance	**C_{eq} for Figure L6.1(a)**	**C_{eq} for Figure L6.1(b)**	**C_{eq} for Figure L6.1(c)**
Formula for calculations			
Calculated value			
Measured value			

4. Measure the equivalent capacitance for each circuit in Figure L6.1. Record the values in Table L6.2.
5. Compare the measured values and calculated values, are there any significant differences? If so, explain the reasons.

Part II: Apply voltage divider rule in the capacitive circuit

1. Take two capacitors with the value shown in Table L6.3, short circuit each capacitor with a bit of wire to release or discharge the stored charges on the capacitor.

Table L6.3

Capacitor	C_1	C_2
Nominal value	30 μF	470 μF
Measured value		

2. Connect a series capacitive circuit as shown in Figure L6.2 on the breadboard.

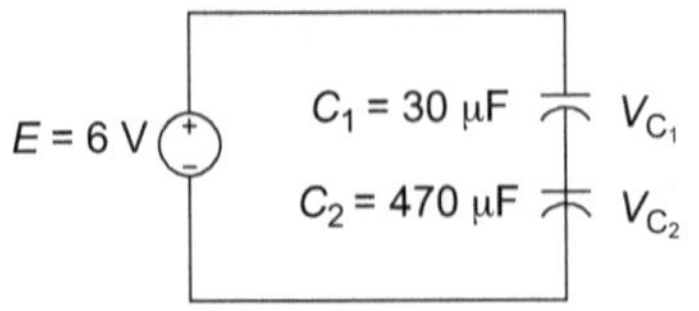

Figure L6.2 Capacitor in series

3. Calculate V_{C_1} and V_{C_2} in the circuit of Figure L6.2 using the voltage divider rule. Record the values in Table L6.4.

Table L6.4

	V_{C_1}	V_{C_2}
Formula for calculations		
Calculated value		
Measured value		

4. Measure the voltage across each capacitor from Figure L6.2 using the multimeter (voltmeter function). Record the values in Table L6.4.
5. Compare the measured values and calculated values, are there any significant differences? If so, explain the reasons.

Note: Take down the measurement quickly, otherwise the capacitor will discharge gradually.

Conclusion

Write your conclusions below:

Chapter 7
Transient analysis of circuits

Objectives

After completing this chapter, you will be able to:

- understand the first-order circuits and concepts of the step response and source-free response of the circuits
- understand the initial conditions in the switching circuit
- understand the concepts of the transient and steady states of RL and RC circuits
- determine the charging/discharging process in an RC circuit
- determine the energy storing/releasing process in an RL circuit
- understand the concepts of time constants for RL and RC circuits
- plot the voltages and currents verse time curves for RL and RC circuits
- understand the relationship between the time constant and the charging/discharging in an RC circuit
- understand the relationship between the time constant and the energy storing/releasing in an RL circuit

7.1 Transient response

7.1.1 The first-order circuit and its transient response

There are three basic elements in an electric circuit, the resistor R, capacitor C and inductor L. The circuits in this chapter will combine the resistor(s) R with an energy storage element capacitor C or an inductor L to form an RL (resistor–inductor) or RC (resistor–capacitor) circuit. These circuits exhibit the important behaviours that are fundamental to much of analogue electronics, and they are used very often in electric and electronic circuits.

Analysis RL or RC circuits still use Kirchhoff's current law (KCL) and Kirchhoff's voltage law (KVL). The main difference between these types of circuits and pure resistor circuits is that the pure resistor circuits can be analysed by algebraic methods. Since the relationship of voltages and currents in the capacitor and inductor circuits is expressed by the derivative, differential equations (the equations with the derivative) will be used to describe RC and RL circuits.

RL or RC circuits that are described by the first-order differential equations, or the circuits that include resistor(s), and only one single energy storage element (inductor or capacitor), are called the first-order circuits.

First-order circuit

- The circuit that contains resistor(s), and a single energy storage element (L or C).
- RL or RC circuits that are described by the first-order differential equations.

We have discussed the concept of charging/discharging behaviour of the energy storage element capacitor C in chapter 6. Another energy storage element inductor L also has the similarly energy storing/releasing behaviour. The difference is that charging/discharging of a capacitor is in the electric filed, and the energy storing/releasing of an inductor is in the magnetic field.

There are two types of circuit states in RL or RC circuit, the transient and steady states.

The *transient* state is the dynamic state that occurs by a sudden change of voltage, current, etc. in a circuit. That means the dynamic state of the circuit has been changed, such as the process of charging/discharging a capacitor or energy storing/releasing for an inductor as the result of the operation of a switch.

The *steady-state* is an equilibrium condition that occurs in a circuit when all transients have finished. It is the stable-circuit state when all the physical quantities in the circuit have stopped changing. For the process of charging/discharging a capacitor or energy storing/releasing for an inductor, it is the result of the operation of a switch in the circuit after a certain time interval.

Transient state: The dynamic state that occurs when the physical quantities have been changed suddenly.
Steady state: An equilibrium condition that occurs when all physical quantities have stopped changing and all transients have finished.

7.1.2 Circuit responses

A response is the effect of an output resulting from an input. The first-order RL or RC circuit has two responses, one is called the step response, and the other is the source-free response.

The *step response* for a general system states that the time behaviour of the outputs when its inputs change from zero to unity value (1) in a very short time. And the step response for an RC or RL circuit is the circuit responses (outputs) when the initial state of the energy store elements L or C is zero and the input (DC power source) is not zero in a very short time.

It is when a DC source voltage is instantly applied to the circuit, the energy store elements L or C hasn't stored energy yet and the output current or voltage generated in this first-order circuit. Or the charging process of the energy storing process of the capacitor or inductor. The step response can be analogized as a process to fill up water in a reservoir or a water bottle.

Following are some of the basic terms for a step response:

- The initial state: the state when an energy storage element hasn't stored energy yet.
- Input (excitation): the power supply.
- Output (response): the resultant current and voltage.

Step response

The circuit response when the initial condition of the energy store elements (L or C) is zero, and the input (DC power source) is not zero in a very short time, i.e. the charging/storing process of the C or L.

The *source-free response* or natural response is opposite to the step response. It is the circuit response when the input is zero, and the initial condition of the capacitor or inductor is not zero (the energy has been stored to the capacitor or inductor). It is the discharging or energy releasing process of the capacitor or inductor in an RC or RL circuit. The source-free response can also be analogized as the process to release water in a reservoir or a water bottle.

Source-free (or natural) response

The circuit response when the input (DC power source) is zero, and the initial condition of the energy store elements (L or C) is not zero, i.e. the discharging/releasing process of the C or L.

When an RC or RL circuit that is initially at 'rest' with zero initial condition and a DC voltage source is switched on to this circuit instantly, this DC voltage source can be analogized as a unit-step function, since it 'steps' from zero to a unit constant value (1). So the step response can be also called the unit-step response.

The unit-step response is defined as follows: All initial conditions of the circuit are zero at time less than zero ($t < 0$), i.e. at the moment of time before the power turns on. And the response $v(t)$ or output voltage for this condition is obviously also zero. After the power turns on ($t > 0$), the response $v(t)$ will be a constant unit value 1, as shown in the following mathematical expression and also can be illustrated in Figure 7.1(b).

$$v(t) = \begin{cases} 0, & t < 0 \\ 1, & t > 0 \end{cases}$$

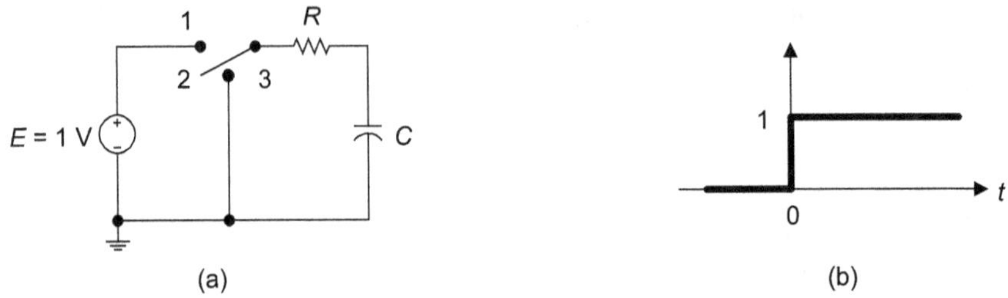

Figure 7.1 Step function

This unit-step function can be expressed as the switch in the circuit of Figure 7.1(a), when $t = 0$, the switch turns to position 1, a DC power source is connected to the RC circuit, and produces a unit-step response to the circuit.

7.1.3 The initial condition of the dynamic circuit

The process of charging and discharging of a capacitor needs a switch to connect or disconnect to the DC source in the RC circuit, as shown in the circuit of Figure 7.1(a). The instantly turned on or turned off the switch, or the source input that is switched 'on' or 'off' in an RC or RL circuit is called the switching circuit. At the moment when the circuit is suddenly switched, the capacitor voltage and inductor current will not change instantly, this concept can be described as $t = 0-$ and $t = 0+$.

- $t = 0-$ is the instant time interval *before* switching the circuit (turn *off* the switch).
- $t = 0+$ is the instant time interval *after* switching the circuit (turn *on* the switch).

At this switching moment, the non-zero initial capacitor voltage and inductor current can be expressed as follows:

$v_C(0+) = v_C(0-)$ and $i_L(0+) = i_L(0-)$

And

- v_C $(0-)$ is the capacitor voltage at the instant time *before* the switch is closed.
- v_C $(0+)$ is the capacitor voltage at the instant time *after* the switch is closed.
- i_L $(0-)$ is the inductor current at the instant time *before* the switch is closed.
- i_L $(0+)$ is the inductor current at the instant time *after* the switch is closed.

Initial conditions

- Switching circuit: the instantly turned on or turned off switch in the circuit.
- $t = 0-$: the instant time interval before the switch is closed.
- $t = 0+$: the instant time interval after the switch is closed.
- At the instant time before/after the switch is closed, v_C and i_L do not change instantly: $v_C(0+) = v_C(0-)$ and $i_L(0+) = i_L(0-)$.

7.2 The step response of an RC circuit

Chapter 6 has discussed the charging and discharging process of a capacitor. When there are no resistors in the circuit, a pure capacitive circuit will fill with electric charges instantly, or release the stored electric charges instantly. But there is always a small amount of resistance in the practical capacitive circuits. Sometimes a resistor will be connected to a capacitive circuit that is used very often in the different applications of the electronic circuits.

Figure 7.2 is a resistor–capacitor series circuit that has a switch connecting to the DC power supply. Such a circuit is generally referred to as an *RC* circuit. All important concepts of step response (charging) or source-free response (discharging) and transient and steady state of an RC circuit can be analysed by this simple circuit.

7.2.1 The charging process of an RC circuit

Assuming the capacitor has not been charged yet in the circuit of Figure 7.2, the switch is in position 2 (middle).

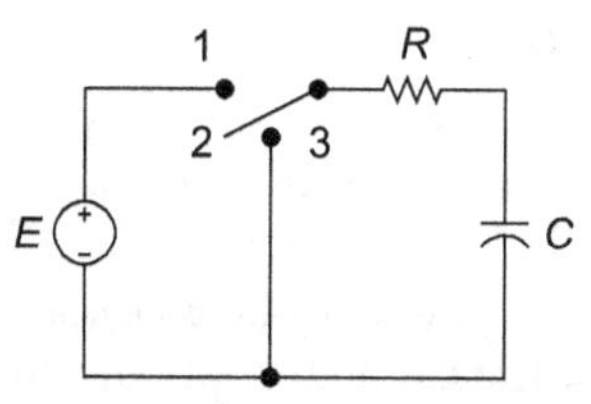

Figure 7.2 An RC circuit

What will happen when the switch is turned to position 1, and the DC power source (E) is connected to the RC series circuit as shown in the circuit of Figure 7.3?

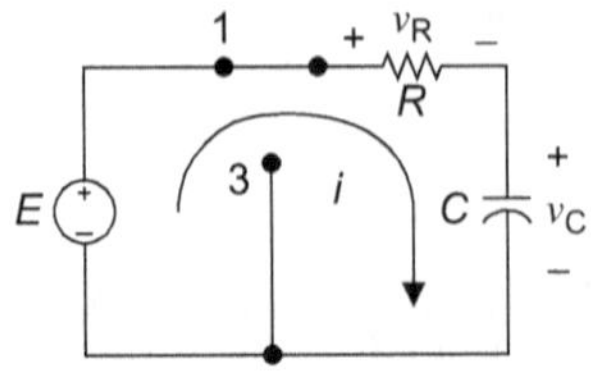

Figure 7.3 RC charging circuit

The energy storing element capacitor C will start charging. Since there is a resistor in this circuit, the process of the capacitor's charging will not finish instantly, the capacitor will gradually store the electric charges.

This RC circuit is similar to a reservoir (or a water bottle) filling with water to capacity. If the door of the reservoir opened only to a certain width, the reservoir will need more time to fill up with the water (or the water bottle will need more time to fill up with water if the tab didn't fully open).

The voltage across the capacitor v_C is not instantaneously equal to the source voltage E when the switch is closed to 1. The capacitor voltage v is zero at the beginning. It needs time to overcome the resistance R of the circuit to gradually charge to the source voltage E. After this charging time interval or the transient state of the RC circuit, the capacitor can be fully charged; this is shown Figure 7.4.

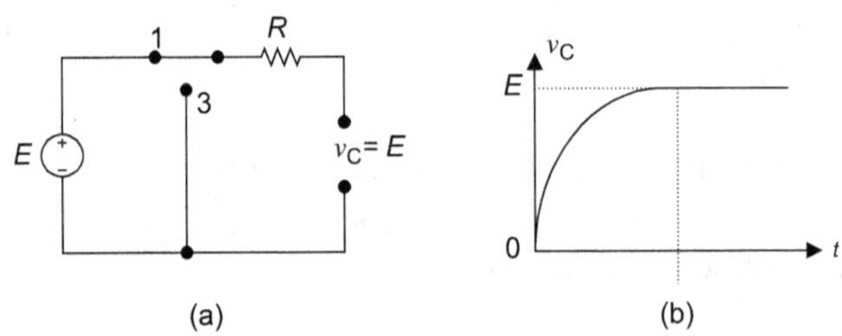

Figure 7.4 The charging process of an RC circuit

Figure 7.4(b) indicates that capacitor voltage v_C increases exponentially from zero to its final value (E). The voltage across the capacitor will be increased until it reaches the source voltage (E), at that time no more charges will flow onto the plates of the capacitor, i.e. the circuit current stops flowing. And the capacitor will reach a state of dynamic equilibrium (steady state).

The state of the circuit voltage or current *after* charging is called the steady state. Once they reach the steady state, the current and voltage in the circuit will not change any more, and at this time, the capacitor voltage is equal to the source voltage, i.e. $v_C = E$. The circuit current is zero, and the capacitor is equivalent to an open circuit as shown in the circuit of Figure 7.4(a). For this open circuit, the current stops to flow. Therefore, there is no voltage drop across the resistor.

The phenomena of the capacitor voltage v_C increases *dexponentially* from zero to its final value E (or a charging process) in an RC circuit can be also analysed by the quantity analysis method as follows.

7.2.2 Quantity analysis for the charging process of the RC circuit

The polarities of the capacitor and resistor voltages of the RC circuit are shown in Figure 7.3. Applying KVL to this circuit will result in

$$v_R + v_C = E \tag{7.1}$$

The voltage drop across the resistor is Ri (Ohm's law) while the current through this circuit is $i = C(dv_C/dt)$ (from chapter 6, section 1.9), i.e.

$$v_R = Ri \quad i = C\frac{dv_C}{dt}$$

Therefore,

$$v_R = RC\frac{dv_C}{dt} \tag{7.2}$$

Substituting (7.2) into (7.1) yields

$$RC\frac{dv_C}{dt} + v_C = E \tag{7.3}$$

- Determine the capacitor voltage v_C

Note: If you haven't learned calculus, then just keep in mind that (7.4) is the equation for the capacitor voltage v_C during the discharging process in an RC circuit, and skip the following mathematical derivation process.

The first-order differentia (7.3) can be rearranged as

$$v_C - E = -RC\frac{dv_C}{dt}$$

Divide both sides by $-RC$

$$-\frac{1}{RC}(v_C - E) = \frac{dv_C}{dt}$$

rearrange

$$-\frac{dt}{RC} = \frac{dv_C}{v_C - E}$$

Integrating the above equation on both sides yields

$$-\frac{1}{RC}\int_0^t dt = \int_0^{v_C} \frac{dv_C}{v_C - E}$$

$$-\frac{t}{RC}\Big|_0^t = \ln|v_C - E|_0^{v_C}$$

rearrange

$$-\frac{t}{RC} = \ln|v_C - E| - \ln|-E|$$

$$\ln\left|\frac{v_C - E}{-E}\right| = -\frac{t}{RC}$$

Taking the natural exponent (e) on both sides results in

$$e^{\ln|(v_c - E)/(-E)|} = e^{-t/RC}$$

$$\frac{v_C - E}{-E} = e^{-t/RC}$$

Solve for v_C

$$v_C = E(1 - e^{-t/RC}) \qquad (7.4)$$

The above equation is the capacitor voltage during the charging process in an RC circuit.

- Determine the resistor voltage v_R

Applying KVL in the circuit of Figure 7.3

$$v_R + v_C = E$$

rearrange

$$v_R = E - v_C \qquad (7.5)$$

Substituting the capacitor voltage $v_C = E(1 - e^{-t/RC})$ into (7.5) yields

$$v_R = E - E(1 - e^{-t/RC})$$

Therefore, the resistor voltage is

$$v_R = Ee^{-t/RC}$$

- Determine the charging current i

  Dividing both sides of the equation $v_R = Ee^{-t/RC}$ by R yields

$$\frac{v_R}{R} = \frac{E}{R}\,e^{-t/RC}$$

  Applying Ohm's law to the left side of the above equation will result in the charging current i

$$i = \frac{E}{R}e^{-t/RC}$$

Charging equations for an RC circuit

- Capacitor voltage: $v_C = E(1 - e^{-t/RC})$
- Resistor voltage: $v_R = Ee^{-t/RC}$
- Charging current: $i = \frac{E}{R}e^{-t/RC}$

Mathematically, these three equations indicate that capacitor voltage increases exponentially from initial value *zero* to the final value E; the resistor voltage and the charging current decay exponentially from initial value E and E/R (or I_{max}) to zero, respectively. And t is the charging time in the equations.

According to the above mathematical equations, the curves of v_C, v_R and i versus time can be plotted as in Figure 7.5.

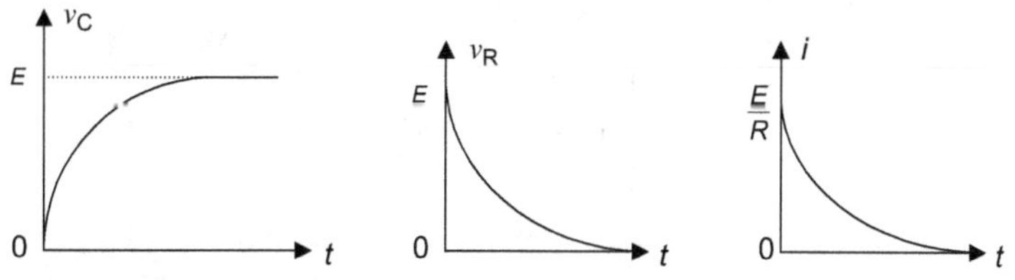

Figure 7.5 v_C, v_R *and i versus t*

Example 7.1: For the circuit shown in Figure 7.3, if $E = 25$ V, $R = 2.5$ kΩ and $C = 2.5$ μF, the charging time $t = 37.5$ ms. Determine the resistor voltage v_R and capacitor voltage v_C.

Solution:

$RC = (2.5\text{ k}\Omega)(2.5\ \mu\text{F}) = 6.25$ ms

- $v_R = Ee^{-t/RC}$

$$= (25\text{ V})(e^{(-37.5/6.25)\text{ms}})$$

$$= (25\text{ V})(e^{-6})$$

$$\approx 0.062\text{ V}$$

- $v_C = E(1 - e^{-t/RC})$
  $= (25\ V)(1 - e^{(-37.5/6.25)ms})$
  $= (25\ V)(1 - e^{-6})$
  $= 24.938\ V$

These results can be checked by using KVL: $v_R + v_C = E$.
Substituting the values into KVL yields

$$v_R + v_C = E$$
$$(0.062 + 24.938)V = 25\ V \quad \text{(checked)}$$

Thus, the sum of the capacitor voltage and resistor voltage must be equal to the source voltage in the *RC* circuit.

7.3 The source-free response of the RC circuit

7.3.1 The discharging process of the RC circuit

Consider a capacitor C that has initially charged to a certain voltage value v_0 (such as the DC source voltage *E*) through the charging process of the last section in the circuit of Figure 7.4(a). The voltage across the capacitor is $v_C = E$, whose function will be the same as a voltage source in the right loop of the RC circuit in Figure 7.6.

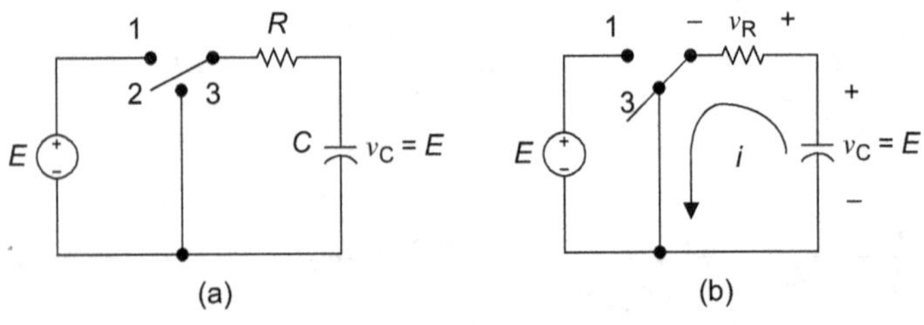

Figure 7.6 Discharging process of the RC circuit

Once the switch turns to position 3 as shown in the circuit of Figure 7.6(b), the capacitor will start discharging, but now it will be different than a pure capacitive circuit that can discharge instantly. The discharging time will increase since there is a resistor in the circuit. It needs some time to overcome the resistance and eventually release all the charges from the capacitor. Once the capacitor has finished the discharge, the capacitor voltage v_C will be 0, the discharging curve is shown in Figure 7.7.

This is similar to a reservoir that has an opened door to release the water (or the water bottle has an opened lid to pour water). But the releasing door of a reservoir is not open wide enough, so it will need some time to release all the water.

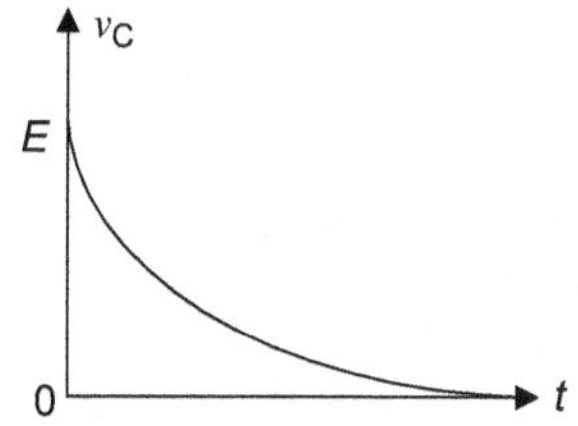

Figure 7.7 Discharge curve of the RC circuit

7.3.2 Quantity analysis of the RC discharging process

The equations used to calculate the capacitor voltage v_C, resistor voltage v_R and discharging current i of the capacitor discharging circuit can be determined by the following mathematical analysis method.

Applying KVL to the circuit in Figure 7.6(b) will result in

$$v_R - v_C = 0 \quad \text{or} \quad v_R = v_C \tag{7.6}$$

Since

$$v_R = iR \quad \text{and} \quad i = -C\frac{dv_C}{dt}$$

Substitute i into the equation of v_R

$$v_R = -R_C\frac{dv_C}{dt} \tag{7.7}$$

The negative sign (−) in the above equation is because the current i and voltage v_C in the circuit of Figure 7.6(b) have opposite polarities.

Substitute (7.7) into the left-hand side of (7.6)

$$-RC\frac{dv_C}{dt} = v_C$$

Divide both sides of the above equation by $-RC$

$$\frac{dv_C}{dt} = -\frac{1}{RC}v_C \tag{7.8}$$

- Determine the capacitor voltage v_C

Note: If you haven't learned calculus, then just keep in mind that (7.10) is the equation for the capacitor voltage v_C during the charging process in an RC circuit, and skip the following mathematical derivation process.

Integrating (7.8) on both sides yields

$$\int\frac{dv_C}{v_C} = -\frac{1}{RC}\int dt$$

$$\ln |v_C| = -\frac{1}{RC}t + \ln A$$

(ln A – the constant of the integration)

or

$$\ln |v_C| - \ln A = -\frac{1}{RC}t$$

$$\text{Rearrange } \ln \left|\frac{v_C}{A}\right| = -\frac{t}{(RC)}$$

Taking the natural exponent (e) on both sides of the above equation:

$$e^{\ln |v_C/A|} = e^{-t/RC}$$

Therefore,

$$\frac{v_C}{A} = e^{-t/RC} \quad \text{or} \quad v_C = Ae^{-t/RC} \tag{7.9}$$

As the capacitor has been charged to an initial voltage value v_0 before being connected to the circuit in Figure 7.6(b), the initial condition (initial value) of the capacitor voltage should be

$$v_C(0-) = V_0$$

v_0 can be any initial voltage value for the capacitor, such as the source voltage E.

Immediately before/after the switch is closed to the position 3 in the circuit of Figure 7.6(b), v_C does not change instantly (from the section 7.1.3), therefore,

$$v_C(0+) = v_C(0-) \quad \text{or} \quad v_C = V_0$$

When $t = 0$, substituting $v_C = v_0$ in (7.9) yields

$$V_0 = Ae^{-0/RC}$$

That is

$$V_0 = A$$

Substitute $v_0 = \mathrm{A}$ into (7.9)

$$v_C = V_0 e^{-t/RC} \tag{7.10}$$

This is the equation of the capacitor voltage for the RC discharging circuit.

- Determine the resistor voltage v_R

According to (7.6)

$$v_R = v_C$$

Substitute (7.10) into (7.6) yields

$$v_R = V_0 e^{-t/RC} \tag{7.11}$$

- Determine the discharge current i
Since

$$i = \frac{v_R}{R} \quad \text{(Ohm's law)}$$

Substitute (7.11) into the above Ohm's law will result in

$$i = \frac{V_0}{R} e^{-t/RC}$$

Discharging equations for an RC circuit

- Capacitor voltage: $v_C = V_0 e^{-t/RC}$
- Resistor voltage: $v_R = V_0 e^{-t/RC}$
- Discharging current: $i = \frac{V_0}{R} e^{-t/RC}$

In the above equations, t is the discharging time. V_0 is the initial capacitor voltage.

These three equations mathematically indicate that capacitor voltage v_C and the resistor voltage v_R decay exponentially from initial value V_0 to the final value zero; and the discharging current i decays exponentially from the initial value v_0/R (or I_{max}) to the final value zero. The curves of v_C, v_R and i versus time t can be illustrated as shown in Figure 7.8.

The discharging voltage and current decay exponentially from the initial value to zero, this means that the capacitor gradually releases the stored energy, and eventually the energy stored in the capacitor will be released to the circuit completely, and it will be received by the resistor and convert to heat energy.

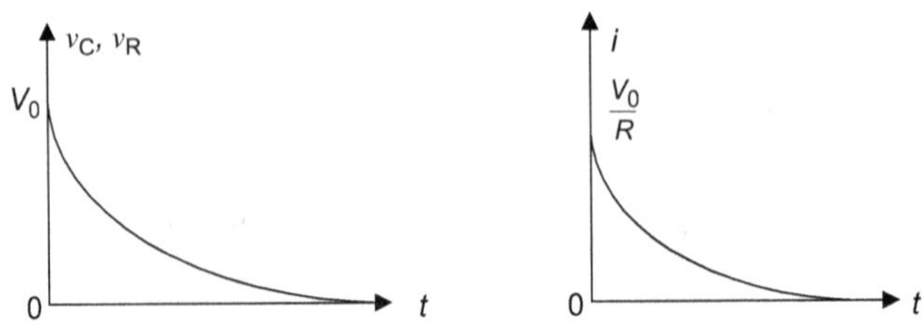

Figure 7.8 The curves of v_C, v_R and i versus t

7.3.3 RC time constant τ

In an RC circuit, the charging and discharging is a gradual process that needs some time. The time rate of this process depends on the values of circuit capacitance C and resistance R. The variation of the R and C will affect rate of the charging and discharging. The product of the R and C is called the RC time constant and it can be expressed as a Greek letter τ (tau), i.e. $\tau = RC$.

Generally speaking, the time constant is the time interval required for a system or circuit to change from one state to another, i.e. the time required from the transient to the steady state or to charge or discharge in an RC circuit.

RC time constant

$\tau = RC$

Quantity	Quantity symbol	Unit	Unit symbol
Resistance	R	Ohm	Ω
Capacitance	C	Farad	F
Time constant	τ	Second	s

The time constant τ represents the time the capacitor voltage reaches (increases) to 63.2% of its final value (steady state) or the time the capacitor voltage decays (decreases) below to 36.8% of its initial value. The higher the R and C values (or when the time constant τ increases), the longer the charging or discharging time; lesser the v_C variation, longer the time to reach the final or initial values. This can be shown in Figure 7.9.

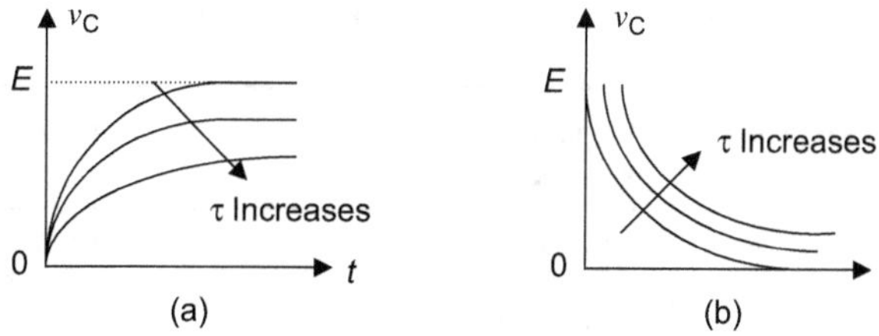

Figure 7.9 The effect of the time constant τ to v_C. (a) Charging and (b) discharging

7.3.4 The RC time constant and charging/discharging

The capacitor charging/discharging voltages when the time is 1 τ and 2 τ can be determined from the equations of the capacitor voltage in the RC charging/discharging circuit.

That is

$$v_C = E(1 - e^{-t/\tau}) \quad \text{and} \quad v_C = V_0 e^{-t/\tau}$$

For example, when $v_0 = E = 100$ V,

- at $t = 1\,\tau$
  - Capacitor charging voltage:

$$\begin{aligned} v_C &= E(1 - e^{-t/\tau}) \\ &= 100\,\text{V}(1 - e^{-1\tau/\tau}) \\ &\approx 63.2\,\text{V} \end{aligned}$$

  - Capacitor discharging voltage:

$$\begin{aligned} v_C &= V_0 e^{-t/\tau} \\ &= 100\,\text{V}\,e^{-1\tau/\tau} \\ &\approx 36.8\,\text{V} \end{aligned}$$

- at $t = 2\,\tau$
  - Capacitor charging voltage:

$$\begin{aligned} v_C &= E(1 - e^{-t/\tau}) \\ &= 100\,\text{V}(1 - e^{-2\tau/\tau}) \\ &\approx 86.5\,\text{V} \end{aligned}$$

  - Capacitor discharging voltage:

$$\begin{aligned} v_C &= V_0 e^{-t/\tau} \\ &= 100\,\text{V}\,e^{-2\tau/\tau} \\ &\approx 13.5\,\text{V} \end{aligned}$$

Table 7.1 The capacitor charging/discharging voltages

Charging/discharging time	Capacitor charging voltage: $v_C = E(1 - e^{-t/\tau})$	Capacitor discharging voltage: $v_C = V_0 e^{-t/\tau}$
1 τ	63.2% of E	36.8% of E
2 τ	86.5% of E	13.5% of E
3 τ	95.0% of E	5% of E
4 τ	98.2% of E	1.8% of E
5 τ	99.3% of E	0.67% of E

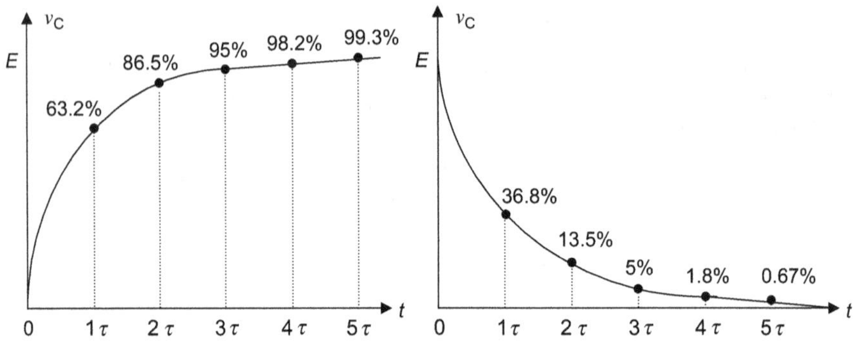

Figure 7.10 The charging/discharging curves of the capacitor voltage

Using the same method as above, the capacitor charging/discharging voltages can be determined when the time is 3, 4 and 5 τ. These results are summarized in Table 7.1 and Figure 7.10.

The data in Table 7.1 and graphs in Figure 7.10 mean that when the time constant is 1 τ, the capacitor will charge to 63.2% of the final value (source voltage), and discharge to 36.8% of the initial value (the initial capacitor voltage). If the final and initial value are 100 V, it will charge to 63.2 V and discharge to 36.8 V.

When the time constant is 2 τ, the capacitor will charge to 86.5% of the final value, and discharge to 13.5% of the initial value.

According to this sequence, if the time constant is 5 τ, the capacitor will charge to 99.3% of the final value, and discharge to 0.67% of the initial value.

When the time is 5 τ, the circuit will reach the steady state, which means that the capacitor will charge approaching to the source voltage E, or discharge approaching to zero. Therefore, when time has passed 4 to 5 τ, charging/discharging of the capacitor will be almost finished. After 5 τ, the transient state of RC circuit will be finished and enter the steady state of the circuit.

Time constant τ for and charging/discharging

- When $t = 1\ \tau$: the capacitor charges to 63.2% of the final value and discharges to 36.8% of the initial value.

- When $t = 5\,\tau$: the capacitor charges to 99.3% of the final value and discharges to 0.67% of the initial value.

Example 7.2: In the circuit of Figure 7.6(a), the source voltage is 100 V, the resistance is 10 kΩ, and the capacitance is 0.005 μF. In how much time can the capacitor voltage be discharged to 5 V after the switch is turned to position 3?

Solution:

$$E = 100\,\text{V}, \quad R = 10\,\text{k}\Omega, \quad C = 0.005\,\mu\text{F}, \quad t = ?$$

The time constant τ for discharging is: $\tau = RC = (10\ \text{k}\Omega)(0.005\ \mu\text{F}) = 50\ \mu\text{s}$.

The capacitor voltage discharging to 5 V is 5% of the initial value E (100 V). Table 7.1 and Figure 7.10 indicate that the time capacitor discharges to 5% of the initial value is 3 τ. Therefore, the capacitor discharging time is

$$t = 3\,\tau = 3(50\,\mu\text{s}) = 150\,\mu\text{s}$$

Example 7.3: In an RC circuit, $R = 5$ kΩ, the transient state has last 1 s in this circuit. Determine the capacitance C.

Solution: The transient state in the RC circuit will last 5 τ, therefore,

$$5\,\tau = 1\,\text{s} \quad \text{or} \quad \tau = \frac{1}{5} = 0.2\,\text{s}$$

$$\because \tau = RC \quad \text{therefore } C = \frac{\tau}{R} = \frac{(0.2\,\text{s})}{(5\text{k}\,\Omega)} = 40\,\mu\text{s}$$

7.4 The step response of an RL circuit

Figure 7.11(a) is a resistor and inductor series circuit, it runs through a switch connecting to the DC power supply. Such a circuit is generally referred to as an *RL* circuit. An RC circuit stores the charges in the *electric* field, and an RL circuit stores the energy in the *magnetic* field. We will use the term charging/discharging in an RC circuit, and the term energy storing/releasing in an RL circuit. All important concepts of the *magnetic* storing/releasing or transient and steady state of RL circuit can be analysed by a simple circuit as shown in Figure 7.11.

The step response (storing) and source-free response (releasing) of an RL circuit is similar to the step response and source-free response of an RC circuit. After understanding the RC circuit, its method of analysis can be used to analyse the RL circuit in a similar fashion.

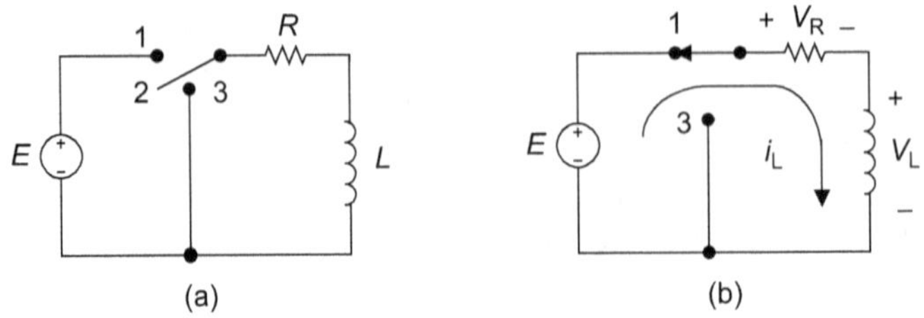

Figure 7.11 RL circuit

Figure 7.11(a) is a circuit that can be used to analyse RL step response and source-free response.

7.4.1 Energy storing process of the RL circuit

In the circuit of Figure 7.11(a), assuming the energy has not been stored in the inductor yet, the switch is in position 2. What will happen when the switch is turned to position 1, and the DC power source is connected to the RL series circuit as shown in the circuit of Figure 7.11(b)?

As it has been mentioned in chapter 6, when the switch in Figure 7.11(a) turns to position 1, the current will flow through this RL circuit, the electromagnetic field will be built up in the inductor L, and will produce the induced voltage V_L. The inductor L absorbs the electric energy from the DC source and converts it to magnetic energy. This energy storing process of the inductor in an RL circuit is similar to the electron charging process of the capacitor in an RC circuit.

Since there is a resistor R in the circuit of Figure 7.11(b), it will be different as a pure inductor circuit that can store energy instantly. After the switch is turned to position 1, the current needs time to overcome the resistance in this RL circuit. Therefore, the process of the inductor's energy storing will not finish instantly. The current i_L in the RL circuit will reach the final value (maximum value) after a time interval, as shown in Figure 7.12.

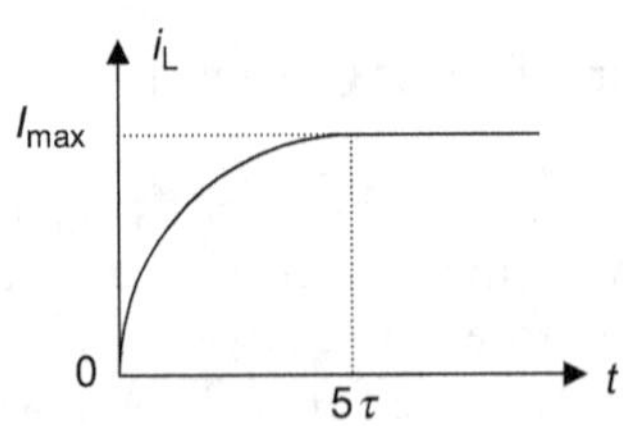

Figure 7.12 Current versus time curve in the RL circuit

The phenomenon of the inductor current i_L in an RL circuit increases exponentially from zero to its final value (I_{max}) or from the transient to the steady state can also be analysed by the quantitative analysis method below.

7.4.2 Quantitative analysis of the energy storing process in an RL circuit

The polarities of the inductor and resistor voltages of an RL circuit are shown in the circuit of Figure 7.11(b). Applying KVL to this circuit will result in

$$v_L + v_R = E \tag{7.12}$$

Substituting $v_L = L(\mathrm{d}i/\mathrm{d}t)$ (from chapter 6, section 6.3.4) and $v_R = Ri$, where $i = i_L$, with (7.12) yields

$$L\frac{\mathrm{d}i_L}{\mathrm{d}t} + Ri_L = E$$

Applying a similar analysis method for the RC charging circuit in section 7.2 will yield the equation of the current in RL circuit during the process of energy storing as given in the following sections.

- Determine the current i_L

$$\begin{aligned} i_L &= \frac{E}{R}(1 - \mathrm{e}^{-t/(L/R)}) \\ &= \frac{E}{R}(1 - \mathrm{e}^{-t/\tau}) \\ &= I_{max}(1 - \mathrm{e}^{-t/\tau}) \end{aligned} \tag{7.13}$$

The time constant of RL circuit is

$$\tau = \frac{L}{R}$$

The final value for the current is

$$I_{max} = \frac{E}{R}$$

- Determine the resistor voltage v_R

Applying Ohm's law

$$v_R = Ri \tag{7.14}$$

Keeping in mind that $i = i_L$ and substituting i by the current i_L in (7.14) yields

$$\begin{aligned} v_R &= R\frac{E}{R}(1 - \mathrm{e}^{-t/\tau}) \\ &= E(1 - \mathrm{e}^{-t/\tau}) \end{aligned}$$

The final value for the resistor voltage is

$$E = I_{\max} R$$

- Determine the inductor voltage v_L

According to (7.12)

$$v_L + v_R = E$$

Substitute v_R and solving for v_L

$$\begin{aligned} v_L &= E - v_R \\ &= E - E(1 - e^{-t/\tau}) \\ &= Ee^{-t/\tau} \end{aligned}$$

Energy storing equations for an RL circuit

- Circuit current: $i_L = \frac{E}{R}(1 - e^{-t/\tau})$
- Resistor voltage: $v_R = E(1 - e^{-t/\tau})$
- Inductor voltage: $v_L = Ee^{-t/\tau}$

In the above equations, t is the energy storing time, and $\tau = L/R$ is the time constant of the RL circuit.

These three equations mathematically indicate that circuit current and resistor voltage increase exponentially from initial value zero to the final value E/R and E, respectively; the inductor voltage decays exponentially from initial value E to zero.

According to the above mathematical equations, the curves of i_L, v_R and v_L versus time can be illustrated in Figure 7.13.

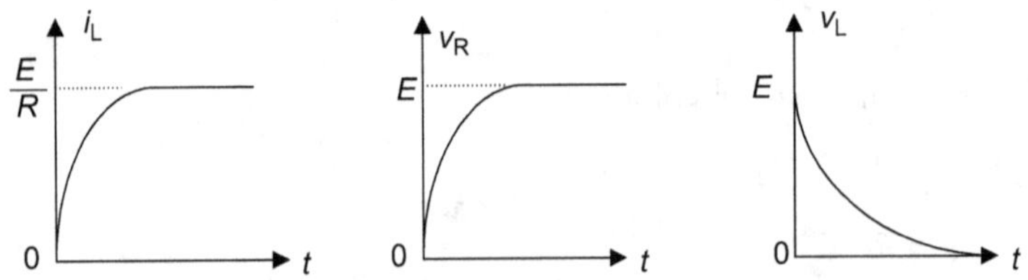

Figure 7.13 Curves of i_L, v_R and v_L versus time

Example 7.4: The resistor voltage $v_R = 10(1 - e^{-2t})$V and circuit current $i_L - 2(1-e^{-2t})$ A in an RL circuit is shown in the circuit of Figure 7.11(b). Determine the time constant τ and inductance L in this circuit.

Solution: The given resistor voltage

$$v_R = E(1 - e^{-t/\tau})$$
$$= 10(1 - e^{-2t}) \text{V}$$

with

$$E = 10\,\text{V} \quad \text{and} \quad -\frac{t}{\tau} = -2t$$

or

$$\tau = \frac{1}{2}\,\text{s} = 0.5\,\text{s}$$

The given current

$$i_L = \frac{E}{R}(1 - e^{-t/\tau}) = 2(1 - e^{-2t})\,\text{A}$$

with

$$\frac{E}{R} = 2\,\text{A} \quad E = 10\,\text{V}, \; R = \frac{E}{I} = \frac{10\,\text{V}}{2\,\text{A}} = 5\,\Omega$$

The time constant

$$\tau = \frac{L}{R}$$

Solve for L

$$L = R\tau = (5\,\Omega)(0.5\,\text{s}) = 2.5\,\text{H}$$

7.5 Source-free response of an RL circuit

7.5.1 Energy releasing process of an RL circuit

Consider an inductor L that has initially stored energy and has the induced voltage v_L through the energy storing process of the last section. If the switch turns to position 3 at this moment (Figure 7.14(b)), the inductor voltage v_L has a function just like a voltage source in the right loop of this RL circuit.

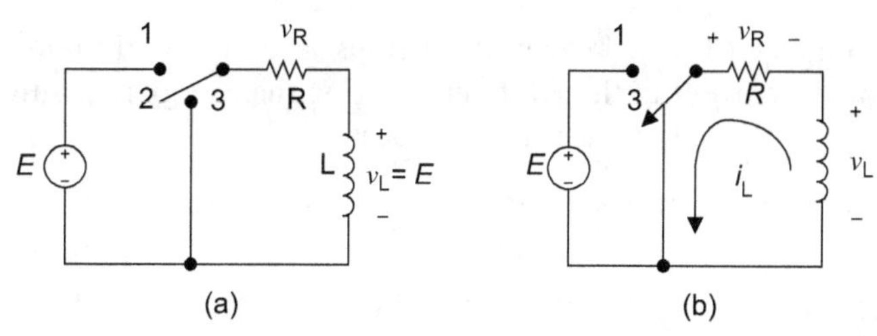

Figure 7.14 RL circuit

Without connecting the resistor R in this circuit, at the instant when the switch turns to position 3, the inductor will release the stored energy immediately. This might produce a spark on the switch and damage the circuit components. But if there is a resistor R in the circuit, the resistance in the circuit will increase the time required for releasing energy, the current in the circuit will take time to decay from the stored initial value to zero. This means the inductor releases the energy gradually, and the resistor absorbs the energy and converts it to heat energy. The current i_L curve of the energy release process in the RL circuit is illustrated in Figure 7.15.

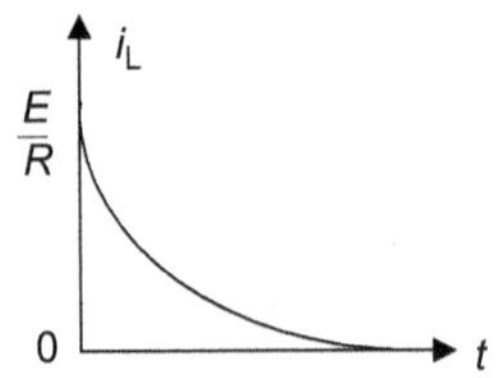

Figure 7.15 Energy release curve of the RL circuit

7.5.2 *Quantity analysis of the energy release process of an RL circuit*

The equations to calculate the inductor voltage v_L, resistor voltage v_R and circuit current i_L of the RL energy releasing circuit can be determined by the following mathematical analysis method.

Applying KVL to the circuit in Figure 7.14(b) will result in

$$v_L + v_R = 0 \quad \text{or} \quad v_L = -v_R \tag{7.15}$$

Substituting $v_L = L(\mathrm{d}i_L/\mathrm{d}t)$ and $v_R = Ri_L$ into (7.15) yields

$$L\frac{\mathrm{d}i_L}{\mathrm{d}t} = -Ri_L \tag{7.16}$$

- Determine the circuit current

Note: If you haven't learned calculus, then just keep in mind that (7.18) is the equation for the current in the RL circuit during the energy releasing, and skip the following mathematical derivation process.

Divide L on both sides in (7.16)

$$\frac{\mathrm{d}i_L}{\mathrm{d}t} = -\frac{Ri_L}{L}$$

Integrating the above equation on both sides yields

$$\int \frac{di_L}{i_L} = -\int \frac{R}{L}dt, \quad \ln|i_L| = -\frac{R}{L}t + \ln A$$

Rearrange:

$$\ln|i_L| - \ln A = -\frac{R}{L}t$$

Taking the natural exponent (e) on both sides results in

$$e^{\ln|i_L/A|} = e^{(-R/L)t} \quad \text{or} \quad \frac{i_L}{A} = e^{(-R/L)t}$$

Solve for i_L

$$i_L = Ae^{(-R/L)t} \tag{7.17}$$

Since energy has been stored in the inductor before it is been connected to the circuit in Figure 7.14(b), its initial condition or value should be

$$i_L(0-) = I_0$$

(I_0 can be any initial current, such as $I_0 = E/R$)

Since immediately before/after the switch is closed to position 3, i_L does not change, therefore,

$$i_L(0+) = i_L(0-) \quad \text{or} \quad i_L = I_0$$

When $t = 0$, substitute $i_L = I_0$ into (7.17) yields

$$I_0 = Ae^{(-R/L)\times 0} \quad \text{i.e.} \quad I_0 = A$$

Therefore,

$$i_L = I_0 e^{-t/\tau} \tag{7.18}$$

In the above equation, $\tau = L/R$ is the time constant for the RL circuit.

- Determine the resistor voltage
  Keep in mind that $i = i_L$ and apply Ohm's law to (7.18)

$$v_R = Ri = R(I_0 e^{-t/\tau}) = RI_0 e^{-t/\tau}$$

- Determine the inductor voltage
  Substituting (7.18) into $v_L + v_R = 0$ (as in (7.15)) results in

$$v_L = -v_R = -RI_0 e^{-t/\tau}$$

Energy releasing equations for an RL circuit

- Circuit current: $i_L = I_0 e^{-t/\tau}$
- Resistor voltage: $v_R = I_0 R e^{-t/\tau}$
- Inductor voltage: $v_L = -I_0 R e^{-t/\tau}$

In the above equations, t is the energy releasing time, $I_0 = E/R$ is the initial current for the inductor and $\tau = L/R$ is the time constant for the RL circuit.

These three equations mathematically indicate that inductor current, resistor voltage and inductor voltage decay exponentially from initial value I_0, $I_0 R$ and $-I_0 R$, respectively, to the final value zero. The curves of i_L, v_R and v_L versus time can be illustrated in Figure 7.16.

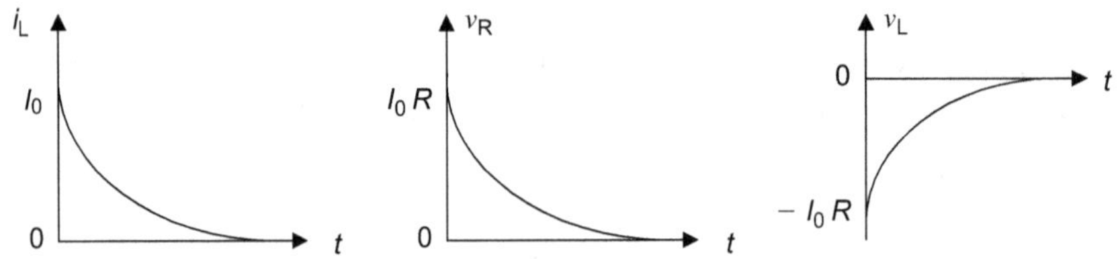

Figure 7.16 Curves of i_L, v_R and v_L versus time

7.5.3 RL time constant τ

In an RL circuit, the storing and releasing of energy is a gradual process that needs time. The time rate of this process depends on the values of the circuit inductance L and resistance R. The variation of R and L will affect the rate of the energy storing and releasing. The quotient of L and R is called the RL time constant $\tau = L/R$.

The RL time constant is the time interval required from the transient to the steady state or the energy storing/releasing time in an RL circuit.

RL time constant

$$\tau = \frac{L}{R}$$

Quantity	Quantity symbol	Unit	Unit symbol
Resistance	R	Ohm	Ω
Inductance	L	Henry	H
Time constant	τ	Second	s

The time constant τ represents the time the inductor current reaches (increases) to 63.2% of its final value (steady state); the time of the inductor current decays (decreases) below to 36.8% of the its initial value. The higher the value of L, the lower the R (or when the time constant τ increases), the longer the storing or releasing time, the lesser the i_L variation and the longer the time to reach the final or initial values. This can be shown in Figure 7.17.

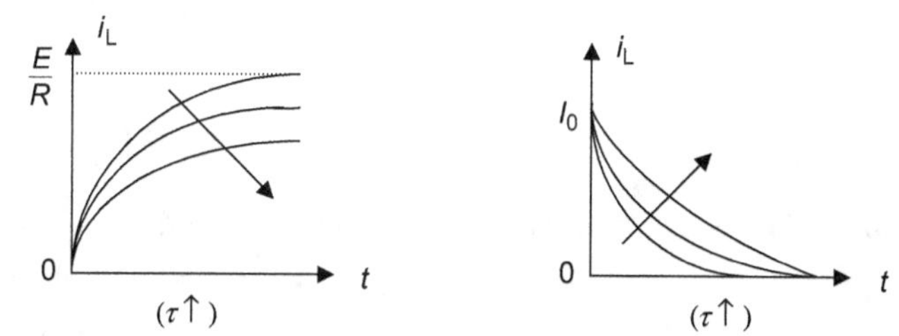

Figure 7.17 Effect of the time constant τ on i_L ($L\uparrow$ or $R\downarrow$)

7.5.4 The RL time constant and the energy storing and releasing

Similar to an RC circuit, the circuit current for an RL circuit can be determined when the time constant is 1 τ, 2 τ, ..., 5 τ, according to the equations of $i_L = (E/R)(1 - e^{-t/\tau})$ and $i_L = I_0 e^{-t/\tau}$, respectively. These results are summarized in Table 7.2 and Figure 7.18.

Table 7.2 Relationship between the time constant and the inductor current

RL energy storing/releasing time	**Increasing the inductor current (storing):** $i_L = \frac{E}{R}(1 - e^{-t/\tau})$	**Decreasing the inductor current (releasing):** $i_L = I_0 e^{-t/\tau}$
1 τ	63.2% of E/R	36.8% of I_0
2 τ	86.5% of E/R	13.5% of I_0
3 τ	95.0% of E/R	5% of I_0
4 τ	98.2% of E/R	1.8% of I_0
5 τ	99.3% of E/R	0.67% of I_0

Example 7.5: In the RL circuit of Figure 7.14, the resistance R is 100 Ω and the transient state has lasted 25 μs. Determine the inductance L.

Solution: The time of a transient state usually lasts 5τ, and this transient state is $5\tau = 25$ μs, $\therefore \tau = (25\,\mu s/5) = 5\,\mu s$.

The time constant $\tau = L/R$, $\therefore L = R\tau = (100\,\Omega)(5\,\mu s) = 500\,\mu H$.

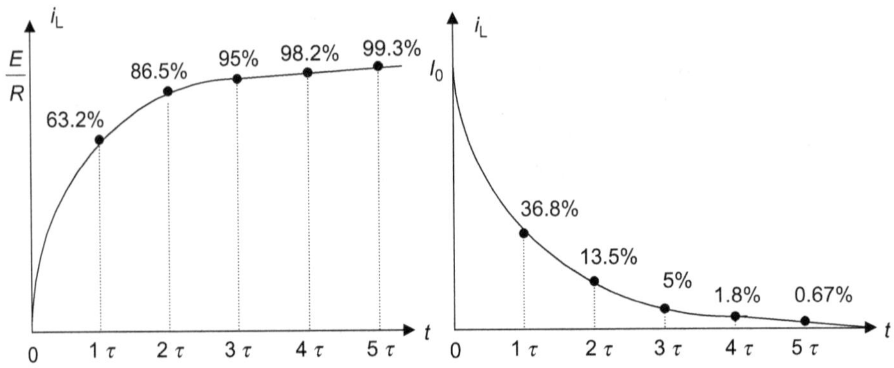

Figure 7.18 Relationship of inductor current and time constant

---

Example 7.6: In the circuit of Figure 7.14(b), $R = 2\ \text{k}\Omega$, $L = 40$ H, $E = 1$ V and $t = 0.2$ ms. Determine the circuit current i_L in this energy releasing circuit.

Solution:

$$\tau = \frac{L}{R} = \frac{40\,\text{H}}{2\,\text{k}\Omega} = 20\,\text{ms}$$

$$i_L = I_0 e^{-t/\tau} = \frac{E}{R} e^{-t/\tau} = \frac{1\,\text{V}}{2\,\text{k}\Omega} e^{(-0.2/20)\text{ms}} \approx 0.5\,\text{mA}$$

Summary

- First-order circuit:
  - The circuit that contains resistor(s), and a single energy storage element (L or C).
  - RL or RC circuits that are described by the first-order differential equations.
- Transient state: The dynamic state that occurs when the physical quantities have been changed suddenly.
- Steady state: An equilibrium condition that occurs when all physical quantities have stopped changing and all transients have finished.
- Step response: The circuit response when the initial condition of the L or C is zero, and the input is not zero in a very short time, i.e. the charging/storing process of the C or L.
- Source-free response: The circuit response when the input is zero, and the initial condition of L or C is not zero, i.e. the discharging/releasing process of the C or L.

- The initial condition:
  - $t = 0-$: the instant time before switching
  - $t = 0+$: the instant time after switching
  - $v_C(0+) = v_C(0-)$, $i_L(0+) = i_L(0-)$
  - Immediately before/after the switch is closed, v_C and i_L do not change instantly.
- The relationship between the time constants of *RC*/*RL* circuits and charging/storing or discharging/releasing:
- Summary of the first-order circuits (see p. 222).

Time	v_C and i_L increasing (charging/storing): $v_C = E(1 - e^{-t/RC})$, $i_L = \frac{E}{R}(1 - e^{-t/\tau})$	v_C and i_L decaying (discharging/releasing): $v_C = V_0 e^{-t/RC}$, $i_L = I_0 e^{-t/\tau}$
1 τ	63.2%	36.8%
2 τ	86.5%	13.5%
3 τ	95.0%	5%
4 τ	98.2%	1.8%
5 τ	99.3%	0.67%

Experiment 7: The first-order circuit (RC circuit)

Objectives

- Understand the capacitor charging/discharging characteristics in the RC circuit (the first-order circuit) by experiment.
- Construct an *RC* circuit, collect and evaluate experimental data to verify the capacitor charging/discharging characteristics in an RC circuit.
- Analyse and verify the capacitor's charging/discharging time by experiment.
- Analyse the experimental data, circuit behaviour and performance, and compare them to the theoretical equivalents.

Background information

- RC charging (the step response):

$$v_C = E(1 - e^{-t/\tau}), \quad v_R = Ee^{-t/\tau}$$

- RC discharging (the source-free response):

$$v_C = V_0 e^{-t/\tau}, \quad v_R = V_0 e^{-t/\tau}$$

- *RC* time constant τ.

$$\tau = RC$$

Summary of the first-order circuits

Circuits	Equations	Waveforms	Time constant
RC charging (step response)	$v_C = E(1 - e^{-t/\tau})$, $v_R = Ee^{-t/\tau}$, $i = \frac{E}{R}e^{-t/\tau}$	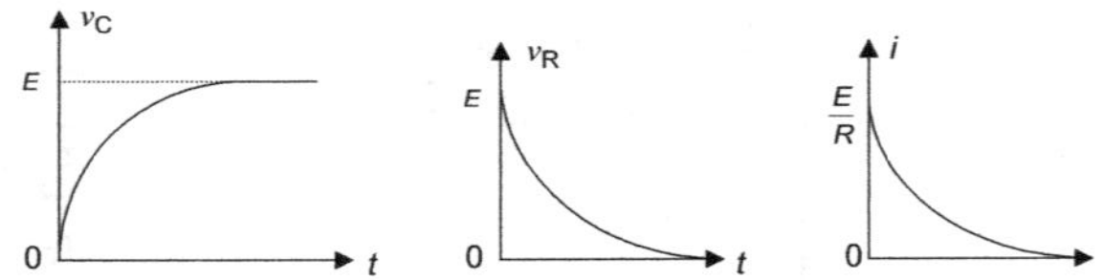	$\tau = RC$
RC discharging (source-free response)	$v_C = V_0e^{-t/\tau}$, $v_R = V_0e^{-t/\tau}$, $i = \frac{V_0}{R}e^{-t/\tau}$	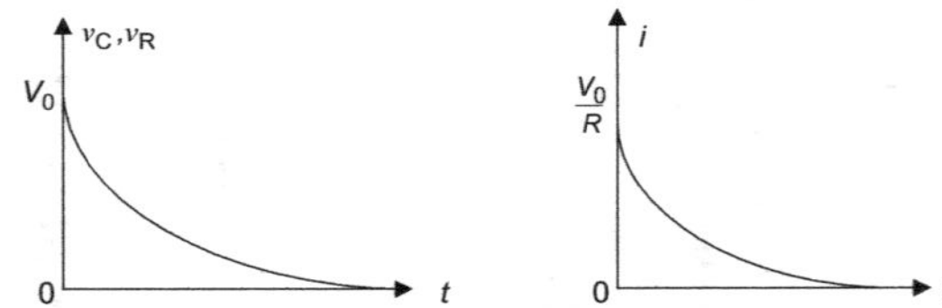	$\tau = RC$
RL storing (step response)	$i_L = \frac{E}{R}(1 - e^{-t/\tau})$ $v_R = E(1 - e^{-t/\tau})$ $v_L = Ee^{-t/\tau}$	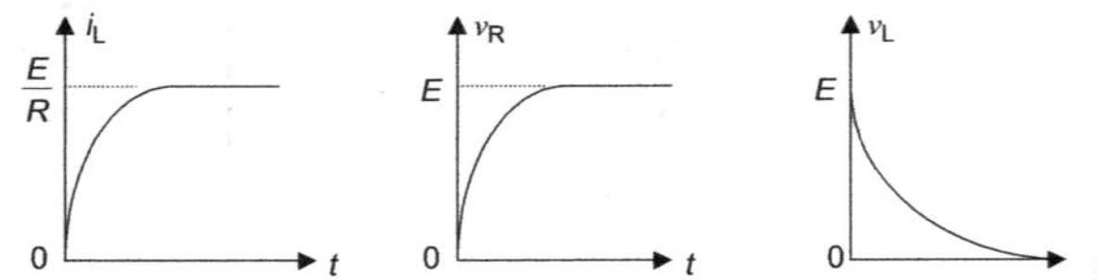	$\tau = L/R$
RL releasing (source-free response)	$i_L = I_0e^{-t/\tau}$ $v_R = I_0Re^{-t/\tau}$ $v_L = -I_0Re^{-t/\tau}$	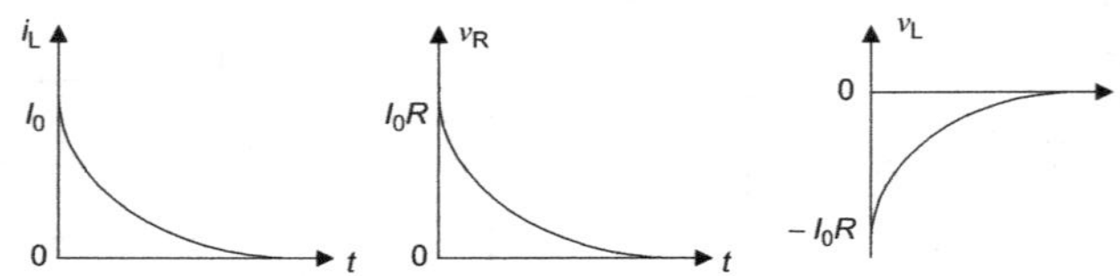	$\tau = L/R$

Equipment and components

- Multimeters (two)
- Breadboard
- DC power supply
- Stopwatch
- Z meter or LCZ meter
- Switch
- Resistors: 1 and 100 kΩ
- Capacitor: 100 μF electrolytic capacitor

Procedure

Part I: Charging/discharging process in an RC circuit

1. Use a jump wire to short circuit the 100 μF capacitor terminals to discharge it, then measure the value of capacitor using a Z meter or LCZ meter and record in Table L7.1.

Table L7.1

Capacitor	**C**
Nominal value	100 μF
Measured value	

2. Construct an *RC* circuit as shown in Figure L7.1 on the breadboard.

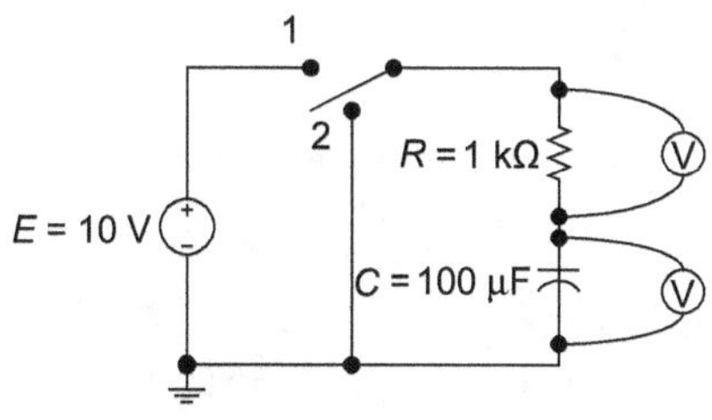

Figure L7.1 An RC circuit

3. Turn on the switch to position 1 in the circuit of Figure L7.1, and observe the needles of the two multimeters (voltmeter function). Wait until voltage across the capacitor reaches to steady state (does not change any more), and record the observation of capacitor voltage v_C and resistor voltage v_R in Table L7.2 (such as 0–10 V, etc.).

Table L7.2

Switching position	v_C	v_R
Turn on the switch to position 1		
Turn on the switch to position 2		

4. Turn on the switch to position 2 in the circuit of Figure L7.1, observe the needles of two multimeters (voltmeter function). Wait until voltage across the capacitor v_C decreased to 0 V, record the observations of capacitor voltage v_C and resistor voltage v_R in Table L7.1 (such as 10–0 V, etc.).

Part II: Capacitor's characteristics in DC circuit

1. Measure the resistors listed in Table L7.3 using the multimeter (ohmmeter function), and record the measured values in Table L7.3.

Table L7.3

Component	*R*	*C*
Nominal value	100 kΩ	100 μF
Measured value		

2. Use a jump wire to short circuit the 100 μF capacitor terminals to discharge it, then measure the value of capacitor using a Z meter or LCZ meter and record in Table L7.3.
3. Construct a circuit as shown in Figure L7.2 on the breadboard.

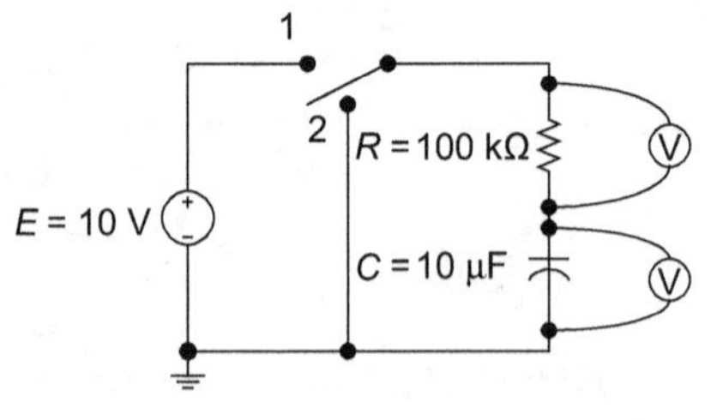

Figure L7.2 The RC circuit for Part II

4. Calculate the time constant τ for the circuit in Figure L7.2 (use measured *R* and *C* values). Record the value in Table L7.4.
5. Calculate the capacitor charging/discharging voltage v_C and v_R when $t = \tau$. Record the values in Table L7.4.

6. Turn on the switch to position 1, observe time required for capacitor voltage v_C charging to 6.3 V using both multimeter (voltmeter function) and stopwatch, this is the charging time constant τ. Also measure v_R at this time, and record τ, v_C and v_R in Table L7.4.

Table L7.4

	τ	v_C	v_R
Charging formula			
Discharging formula			
Calculated value for charging			
Calculated value for discharging			
Measured value for charging			
Measured value for discharging			

7. Keep the switch at position 1 and make sure it does not change, and observe capacitor voltage v_C using the multimeter (voltmeter function) until the capacitor voltage reaches and stays at 10 V ($v_C = 10$ V). Then turn on the switch to position 2, and observe the time required for the capacitor voltage to decrease to 3.6 V using both the multimeter (voltmeter function) and stopwatch (this is the discharging time constant τ). Also measure v_C and v_R at this time and record the values in Table L7.4.

Note: Since electrolyte capacitors may conduct leakage current, the measurement and calculation may be a little different, but it still can approximately verify the theory.

Conclusion

Write your conclusions below:

Chapter 8

Fundamentals of AC circuits

Objectives

After completing this chapter, you should be able to:

- understand the difference between DC and AC
- understand the definitions of AC phase shift, period, frequency, peak to peak, peak, RMS values, phasor, etc.
- understand the relationship of period and frequency
- understand and define three important components of sinusoidal waveform
- define the phase difference between sinusoidal voltage and current
- convert sinusoidal time-domain quantities to phasor-domain forms, and vice versa
- analyse the sinusoidal AC circuits using phasors
- study the effect of resistive, inductive, and capacitive elements in AC circuits

8.1 Introduction to alternating current (AC)

8.1.1 The difference between DC and AC

Previous chapters have studied DC (direct current) circuits. The DC power supply provides a constant voltage and current; hence, all resulting voltages and currents in DC circuit are constant and do not change with time. That is, the polarity of DC voltage and direction of DC current do not change, only their magnitude changes.

This chapter will discuss the alternating current (AC) circuits, in which the voltage alternates its polarity and the current alternates its direction periodically. Since the AC power supply provides an alternating voltage and current, the resulting currents and voltages in AC circuit also periodically switch their polarities and directions. Similar to DC circuits, an alternating voltage is called AC voltage and alternating current is called AC current.

Before the 19th century, the DC power supply was the main form of electrical energy to provide electricity. Since then, DC and AC have had constant competition; AC gradually showed its advantages and rapidly developed in the latter of the 19th century, and is still commonly used in current industries, businesses and homes throughout the world.

This is because the AC power can be more cost-effective for long-distance transmission from power plants to industrial, commercial or residential areas. This is why power transmission for electricity today is nearly all AC. It is also easy to convert from AC to DC, allowing for a wide range of applications.

8.1.2 DC and AC waveforms

The DC voltage and current do not change their polarity or direction over time, only their magnitude changes. A DC waveform (a graph of voltage and current versus time) is shown in Figure 8.1.

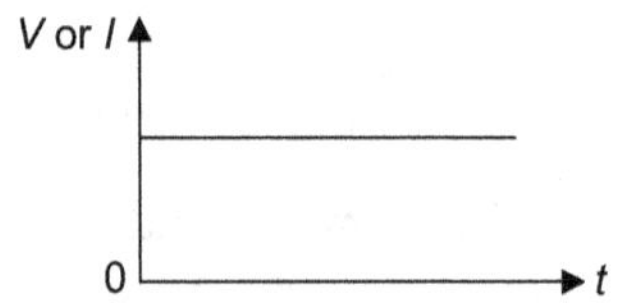

Figure 8.1 DC waveform

There is also a type of DC waveform known as the pulsing DC, in which the amplitude of DC pulse changes periodically from zero to the maximum with time, but its polarity or direction does not change with time (always above zero), so it still belongs to the DC category. Figure 8.2 shows some pulsating DC waveforms.

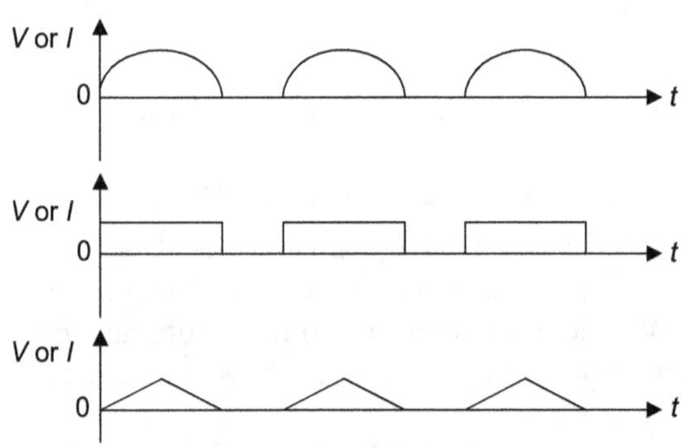

Figure 8.2 Pulsing DC waveforms

Direct current (DC)

- The polarity of DC voltage and direction of DC current do not change.
- The pulsing DC changes pulse amplitude periodically, but the polarity does not change.

AC voltage and current periodically change polarity or direction with time. A few examples of AC waveforms are shown in Figure 8.3.

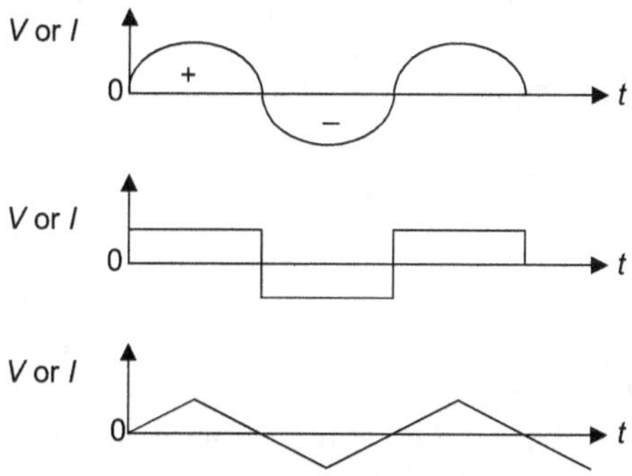

Figure 8.3 AC waveforms

The sinusoidal or sine AC wave is the most basic and widely used AC waveform, and is often referred to as AC, although other waveforms such as square wave, triangle wave, etc. also belong to AC. The sine AC wave energy is the type of power that is generated by the utility power industries around the world.

Sine AC voltage and current vary with sine (or we could use cosine by adding 90° to the sine wave) function, the symbol of AC source is —⊙—. AC quantities are represented by lowercase letters (*e*, v, i, etc.) and DC quantities use uppercase letters (*E*, *V*, *I*, etc.).

Alternating current (AC)

- The polarity of voltage and direction of AC current periodically change with time (such as sine wave, square wave, saw-tooth wave, etc.).
- Sine AC (or AC) varies over time according to sine (or cosine) function, and is the most widely used AC.

A sine function can be described as a mathematical expression of $f(t) = F_m \sin(\omega t + \psi)$. This is the expression of sine function in the time domain (the quantity versus time). Applying the expression of sine function to electrical quantities will obtain general expressions of AC voltage and current as follows:

Sinusoidal voltage: $v(t) = V_m \sin(\omega t + \psi)$
Sinusoidal current: $i(t) = I_m \sin(\omega t + \psi)$

8.1.3 Period and frequency

The waveform of a sinusoidal function is shown in Figure 8.4.

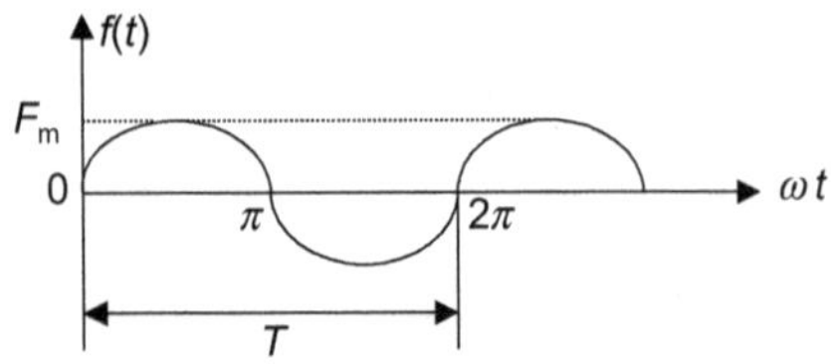

Figure 8.4 Sinusoidal waveform

- Period T: It is the time to complete one full cycle of the waveform, or the positive and negative alternations of one revolution. T is measured in seconds (s).
- Frequency f: It is the number of cycles of waveforms within 1 s. The frequency is measured in hertz (Hz).

For instance, in Figure 8.5, the number of complete cycles in 1 s is 2, so it has a frequency of 2 Hz.

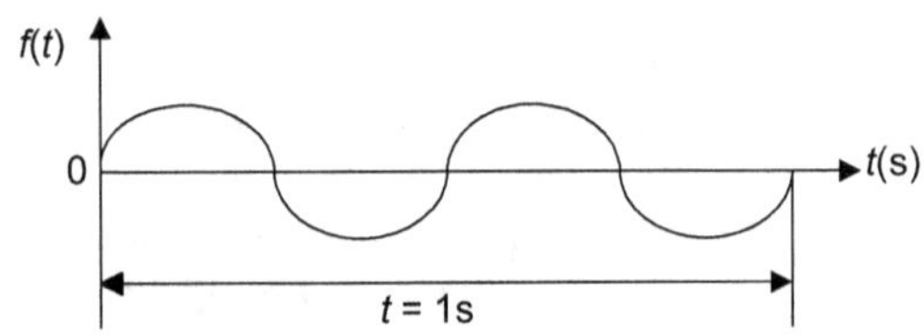

Figure 8.5 Frequency of sine waveform

- Relationship of T and f: The frequency f of the waveform is inversely proportional to period T of the waveform, i.e. $f = 1/T$.

Period and frequency
- Period T: Time to complete one full cycle.
- Frequency f: Number of cycles per second.
- $f = 1/T$

8.1.4 Three important components of a sine function

There are three important components in the expression of the sine function $f(t) = F_m \sin(\omega t + \psi)$: peak value F_m, angular velocity Ω and phase shift ψ.

- Peak value F_m: In the expression $f(t) = F_m \sin(\omega t + \psi)$, F_m is the peak value or amplitude of the sine wave (I_m for current or V_m for voltage). It is the distance from zero of the horizontal axis to the maximum point (positive or negative) that a waveform can reach during its entire cycle (Figure 8.6 (a)). It is measured in volts or amperes.

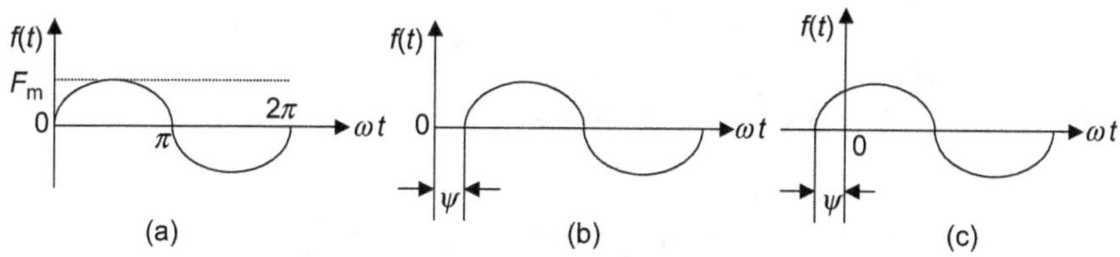

Figure 8.6 The peak value and phase of the sine wave

- Angular velocity ω (the Greek letter omega): Angular velocity or angular frequency of a sine wave reflects the rate of change of the rotation of the wave.

  Angular velocity = Rotating distance/Time

  (Same with the linear motion: Velocity = Distance/Time)

  Since the time required for a sine wave to complete one cycle is period T, the distance of one cycle is 2π as shown in Figure 8.4, so the angular velocity can be determined by

$$\omega = \frac{2\pi}{T}$$

The relationship between the angular velocity and frequency is

$$\omega = \frac{2\pi}{T} = 2\pi f \quad \left(f = \frac{1}{T}\right)$$

Since the angular velocity is directly proportional to the frequency, it is also called the angular frequency. It is measured in radian per second (rad/s).

- Phase ψ (the Greek letter phi): The phase or phase shift of a sine wave is an angle that represents the position of the wave shifted from a reference point at the vertical axis (0°). It is measured in degrees or radians. A sine wave may shift to the left or right of 0°. The range of phase shift is between $-\pi$ and $+\pi$.
  - If phase shift $\psi = 0$, the waveform of sine function $f(t) = F_m \sin\omega t$ starts from $t = 0$ as shown in Figure 8.6(a).
  - If phase shift ψ has a negative value ($\psi < 0$), the waveform of sine function $f(t) = F_m(\omega t - \psi)$ will shift to the right side of 0° as shown in Figure 8.6(b).
  - If phase shift ψ has a positive value ($\psi > 0$), the waveform of sine function $f(t) = F_m(\omega t + \psi)$ will shift to the left side of 0° as shown in Figure 8.6(c).

Three important components of sine function

$f(t) = F_m \sin(\omega t + \psi)$

- F_m: Peak value (amplitude)
- ω: Angular velocity or angular frequency
- $\omega = 2\pi/T = 2\pi f$ ($\pi = 180°$)
- ψ: Phase or phase shift
  - $\psi > 0$: Waveform shifted to the left side of 0°
  - $\psi < 0$: Waveform shifted to the right side of 0°

Example 8.1: Given a sinusoidal voltage $v(t) = 6\sin(25t - 30°)$V, determine its peak voltage, phase angle and frequency, and plot its waveform.

Solution:

Peak value: $V_m = 6$ V
Phase: $\psi = -30°$ ($\psi < 0$, waveform shifted to the right side of 0°)
Frequency: $f = 1/T$
Since $\omega = 2\pi/T$ and $\omega = 25$ rad/s

$$T = \frac{2\pi}{\omega} = \frac{2\pi \text{ rad}}{25 \text{ rad/s}} \approx 0.25\text{s} \quad f = \frac{1}{T} = \frac{1}{0.25 \text{ s}} = 4 \text{ Hz}$$

The waveform is shown in Figure 8.7.

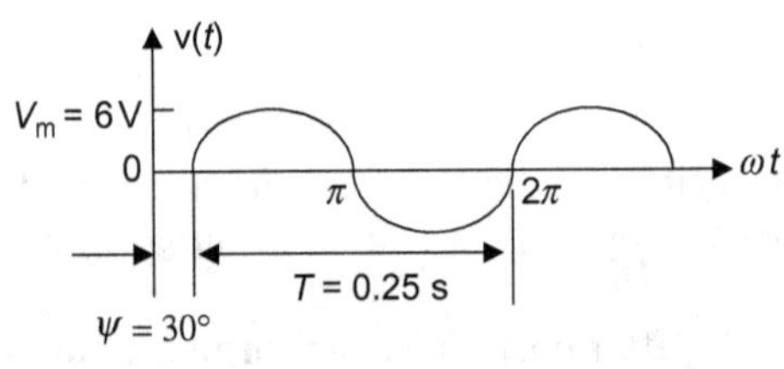

Figure 8.7 Waveform for Example 8.1

8.1.5 Phase difference of the sine function

For two different sine waves with the same frequency, the angular displacement of their phases is called phase difference and is denoted by ϕ (lowercase Greek letter phi). It is a phase angle by which one wave leads or lags another.

For instance, given the general expressions of sinusoidal voltage and current as

$$v(t) = V_m\sin(\omega t + \psi_v) \quad \text{and} \quad i(t) = I_m\sin(\omega t + \psi_i)$$

the phase difference between voltage and current is

$$\phi = (\omega t + \psi_v) - (\omega t + \psi_i) = \psi_v - \psi_i$$

- If $\phi = \psi_v - \psi_i = 0$, the two waveforms are in phase as shown in Figure 8.8(a).

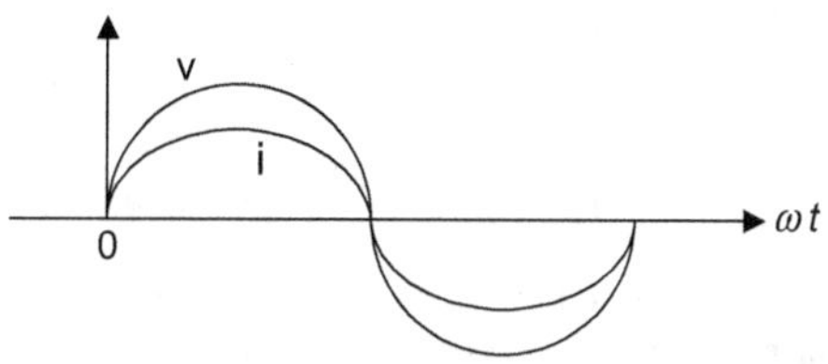

Figure 8.8(a) Two waveforms are in phase

- If $\phi = \psi_v - \psi_i > 0$, voltage leads current, or current lags voltage as shown in Figure 8.8(b).

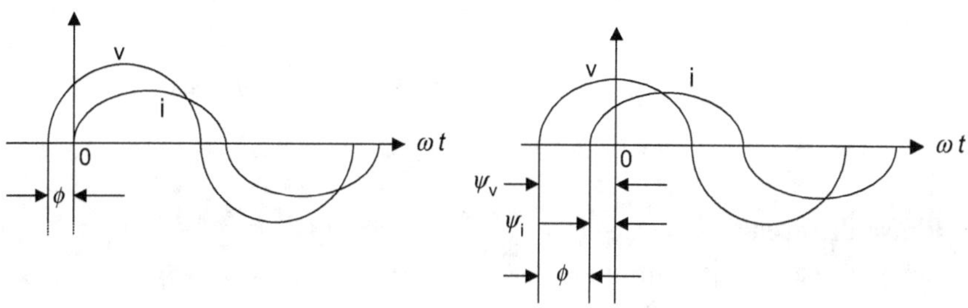

Figure 8.8(b) Current lags voltage

- If $\phi = \psi_v - \psi_i < 0$, current leads voltage, or voltage lags current, as shown in Figure 8.8(c).

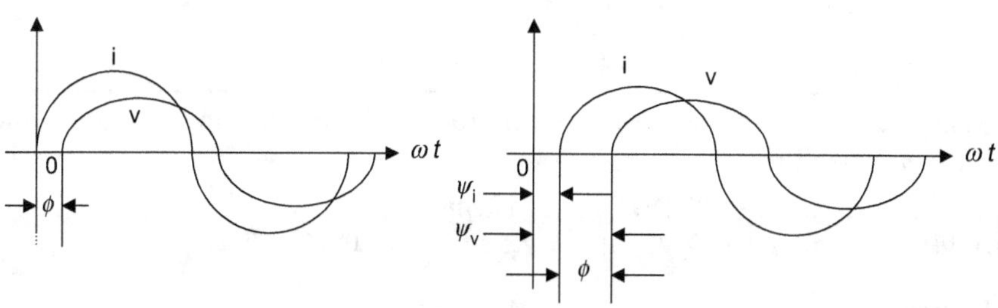

Figure 8.8(c) Current leads voltage

- If $\phi = \psi_v - \psi_i = \pm\pi/2$ (or $\pm 90°$), then voltage and current are orthogonal, or is a right angle (*orthos* means 'straight', and *gonia* means 'angle'). It is shown in Figure 8.8(d).

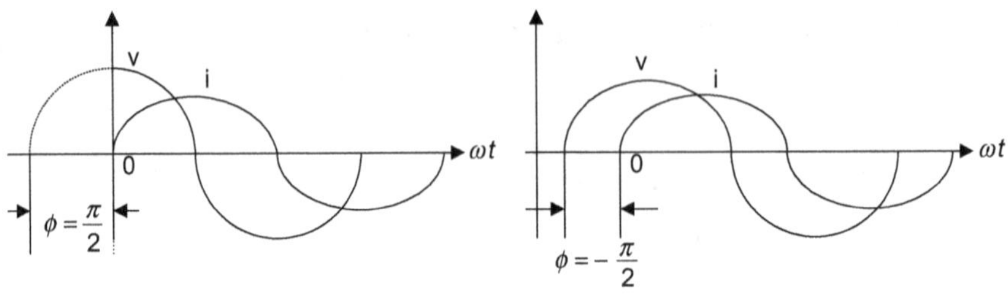

Figure 8.8(d) Voltage and current are orthogonal

- If $\phi = \psi_v - \psi_i = \pm\pi$ (or $\pm180°$), voltage and current are out of phase as shown in Figure 8.8(e).

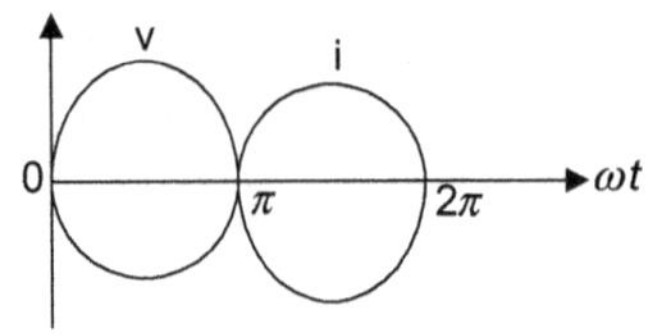

Figure 8.8(e) Voltage and current are out of phase

Phase difference $\phi = \psi_v - \psi_i$

For two waves with the same frequency such as $v(t) = V_m \sin(\omega t + \psi_v)$ and $i(t) = I_m \sin(\omega t + \psi_i)$:

- If $\phi = 0$: v and i are in phase
- If $\phi > 0$: v leads i
- If $\phi < 0$: v lags i
- If $\phi = \pm\pi/2$: v and i are orthogonal
- If $\phi = \pm\pi$: v and i are out of phase

Example 8.2: Determine the phase difference of the following functions and plot their waveforms.

(a) $v(t) = 20\sin(\omega t + 30°)V$, $i(t) = 12\sin(\omega t + 60°)A$
(b) $v(t) = 5\sin(\omega t + 60°)V$, $i(t) = 2.5\sin(\omega t + 20°)A$

Solution:

(a) $\phi = \psi_v - \psi_i = 30° - 60° = -30° \ < 0$
So voltage lags current 30° as shown in Figure 8.9(a).
(b) $\phi = \psi_v - \psi_i = 60° - 20° = 40° > 0$
So voltage leads current 40° as shown in Figure 8.9(b).

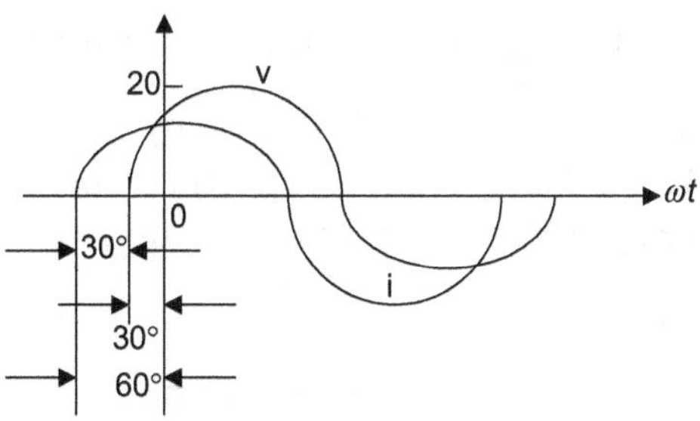

Figure 8.9(a) Figure for Example 8.2(a)

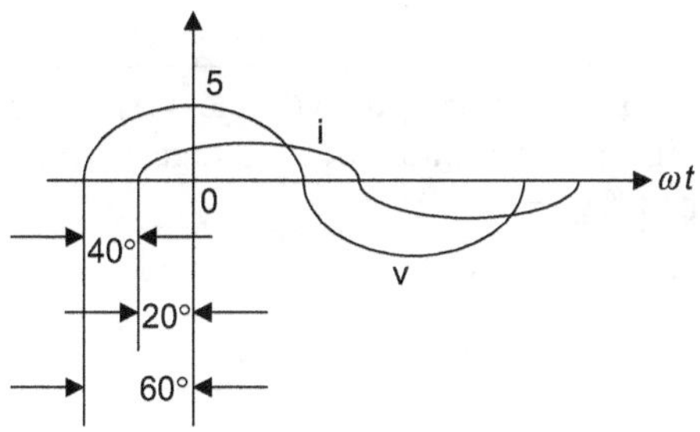

Figure 8.9(b) Figure for Example 8.2(b)

8.2 Sinusoidal AC quantity

A sinusoidal AC quantity such as AC voltage or current can be described in a number of ways. They can be described by their peak value, peak–peak value, instantaneous value, average value or root mean square (RMS) value. The different expressions will provide different ways to analyse the sinusoidal AC quantity, and it is also because a sinusoidal wave always varies periodically and there is no one single value that can truly describe it.

8.2.1 Peak and peak–peak value

As previously mentioned, the peak value of the sinusoidal waveform is one of three important components of the sine function, and is the amplitude or maximum value F_m in sine function $f(t) = F_m \sin(\omega t + \psi)$. The peak value is denoted by F_{pk} as shown in Figure 8.10.

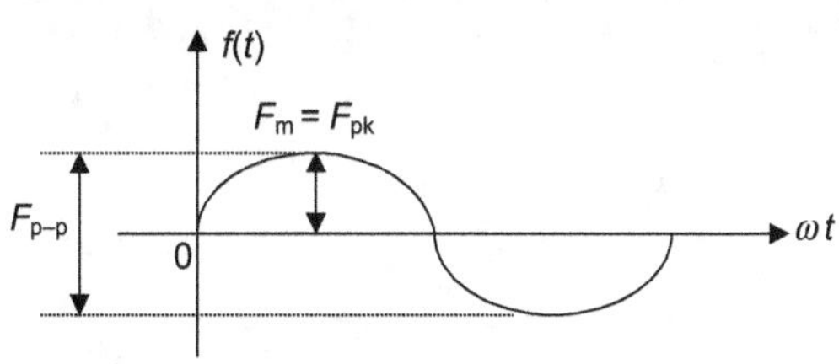

Figure 8.10 Peak and peak–peak value

The peak–peak value $F_{p\text{–}p}$ represents the distance from negative to positive peak, or minimum to maximum peak, or between peak and trough of the waveform, so $F_{p\text{–}p} = 2F_{pk}$ as shown in Figure 8.10.

To determine the maximum values that electrical equipment can withstand, the peak values or peak–peak values of the AC quantities should be considered.

8.2.2 *Instantaneous value*

The instantaneous value of the sinusoidal waveform $f(t)$ varies with time, and it is the value at any instant time t (or ωt) in any particular point of a waveform. Instantaneous values of the variables are denoted by lowercase letters, such as voltage v, current i, etc.

Example 8.3: Given a sinusoidal AC voltage $v(t) = V_m \sin \omega t$ as shown in Figure 8.11, determine the instantaneous voltage v_1 (voltage at 30°) and v_2 (voltage at 135°) when $V_m = 5$ V.

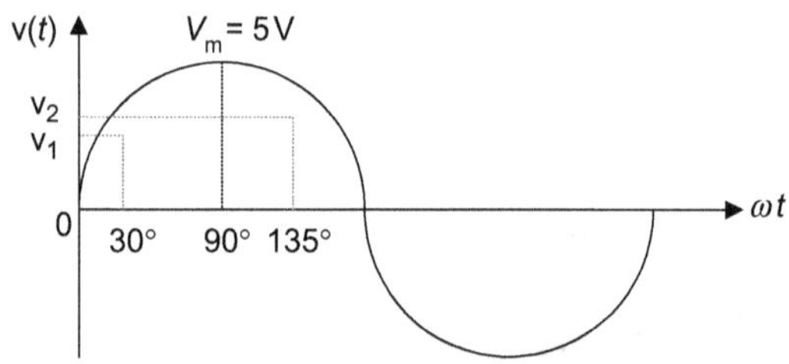

Figure 8.11 Figure for Example 8.3

Solution:

$$v_1 = V_m \sin \omega t = 5 \sin 30° = 2.5 \text{ V}$$
$$v_2 = V_m \sin \omega t = 5 \sin 135° \approx 3.54 \text{ V}$$

8.2.3 *Average value*

Because of the symmetry of the sinusoidal waveform, its average value in a complete full cycle is always zero. For a sinusoidal function $f(t) = F_m \sin(\omega t + \psi)$, its average value is defined as the average of its half-cycle (0 to π), as shown in Figures 8.12 and 8.13.

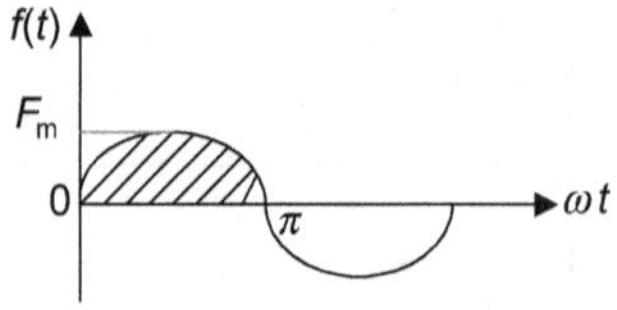

Figure 8.12 Average value

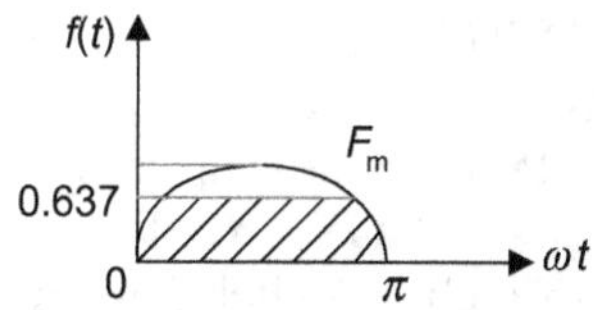

Figure 8.13 Average value

The average value of a half-cycle sinusoidal wave with a zero phase shift can be derived by using integration as follows:

Note: If you haven't learned calculus, then just keep in mind that $F_{avg} = 0.637F_m$ is the equation for the average value of a half-cycle sinusoidal wave, and skip the following mathematic derivation process.

$$F_{avg} = \frac{\text{Area}}{\pi} = \frac{1}{\pi}\int_0^{\pi} f(t)\mathrm{d}t = \frac{1}{\pi}\int_0^{\pi} F_m \sin\omega t \mathrm{d}\omega t$$

$$= \frac{F_m}{\pi}[-\cos\omega t]_0^{\pi} = -\frac{F_m}{\pi}[\cos\pi - \cos 0]$$

$$= -\frac{F_m}{\pi}(-1-1) = \frac{2F_m}{\pi} \approx 0.637F_m$$

$$F_{avg} \approx 0.637F_m$$

Therefore, the average value of a half-cycle sinusoidal wave is 0.637 times the peak value, as shown in Figure 8.13.

Peak value, peak–peak value, instantaneous value and average value

For a sinusoidal waveform:

- Peak value $F_{pk} = F_m$: The amplitude or maximum value
- Peak–peak value $F_{p\text{-}p}$: $F_{p\text{-}p} = 2F_{pk}$
- Instantaneous value $f(t)$: The value at any time at any particular point of the waveform
- Average value F_{avg}: $F_{avg} = 0.637F_m$

8.2.4 Root mean square (RMS) value

1. Applications of RMS value: RMS value (also referred to as the effective value) of the sinusoidal waveform is widely used in practice. For example, the values measured and displayed on instruments and the nominal ratings of the electrical equipment are RMS values. In North America, the single-phase AC voltage 110 V from the wall outlet is an RMS value.

2. The physical meaning of RMS value: For a sinusoidal waveform, the physical meaning of the AC RMS value is the heating effect of the sine wave. That is, an AC source RMS value will deliver the equivalent amount of average power to a load as a DC source. For instance, whether turning on the switch 1 (connect to DC) or switch 2 (connect to AC) in Figure 8.14, 20-V DC or 20-V AC RMS value will deliver the same amount of power (40 W) to the resistor (lamp). If the lamp is replaced by an electric heater, then the heating effect delivered by 20-V DC and 20-V AC RMS will be the same.

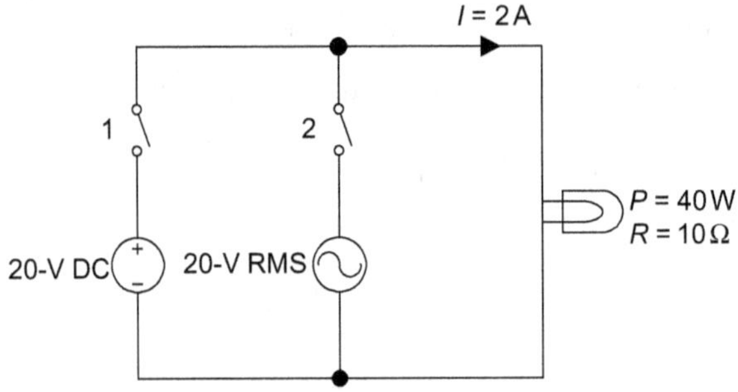

Figure 8.14 RMS value

3. Quantitative analysis of RMS value: The average power generated by an AC power supply is

$$p_{AC} = i_{AC}^2 R = (I_m \sin\omega t)^2 R = (I_m^2 \sin^2\omega t)R$$

$$p_{AC} = I_m^2 \left[\frac{1}{2}(1 - \cos 2\omega t)\right] R = \frac{I_m^2 R}{2} - \frac{I_m^2 R}{2}\cos 2\omega t$$

as $\sin^2\omega t = 1/2(1 - \cos 2\omega t)$.

Only the first part in the above power expression represents the average power of AC, since the average value of the second part in the power expression (a cosine function) is zero, i.e. $P_{AC} = (I_m^2 R)/2$.

- The average power generated by DC voltage is $P_{avg} = I^2 R$.
- RMS value of AC current: According to the physical meaning of RMS, the average AC power is equivalent to the average DC power when the AC source is an RMS value, so $(I_m^2 R)/2 = I^2 R$ or $I^2 = I_m^2/2$.

Taking the square root of both sides of the equation gives

$$I = \sqrt{\frac{I_m^2}{2}} = \frac{I_m}{\sqrt{2}} \approx 0.707 I_m \quad \text{or} \quad I_m = \sqrt{2} I \approx 1.414 I \qquad (8.1)$$

The current I in (8.1) is the RMS value of the AC current, and I_m is the peak value or amplitude of the AC current.

- RMS value of AC voltage: It can be obtained by the same approach by determining the RMS value of the AC current, i.e.

$$V = \frac{V_m}{\sqrt{2}} = 0.707\,V_m \qquad \text{or} \qquad V_m = \sqrt{2}\,\text{V} = 1.414\,\text{V} \tag{8.2}$$

The voltage V in (8.2) is the RMS value of AC voltage, and V_m is the peak value or amplitude of the AC voltage.

- RMS value of a periodical function $f(t)$: Equations (8.1) and (8.2) indicate the relationship between the RMS value and the peak value, which is related by $\sqrt{2}$. However, this relation only applies to the sine wave. For a non sine wave function $f(t)$, the following general equation can be used to determine its RMS value.

$$F = \sqrt{\frac{1}{T}\int_0^T f^2(t)\mathrm{d}t} \quad (T \text{ is the period of the function})$$

The name root mean square (RMS) is obtained from the above equation, in which the term $1/T$ denotes the average (mean), $f^2(t)$ denotes the square (square) and $\sqrt{\ }$ denotes the square root (root) value.

RMS value of AC function

- RMS value or effective value of AC: An AC source with RMS value will deliver the equivalent amount of power to a load as a DC source.
- $V = 0.707\,V_m$, $I = 0.707 I_m$ or $V_m = \sqrt{2}\,\text{V}$, $I_m = \sqrt{2}I$
- The general equation to calculate RMS value: $F = \sqrt{\frac{1}{T}\int_0^T f^2(t)\mathrm{d}t}$

8.3 Phasors

8.3.1 Introduction to phasor notation

Charles Proteus Steinmetz, a German-American mathematician and electrical engineer, developed the phasor notation in 1893. A *phasor* is a vector that contains both magnitude and direction or amplitude and phase information. It can be used to represent AC quantities. Since phasors have magnitudes and directions, they can be represented as complex numbers.

A phasor notation or phasor domain is a method that uses complex numbers to represent the sinusoidal quantities for analysing AC circuits. It can represent sine waves in terms of their peak value (magnitude) and phase angle (direction). The peak value can be easily converted to the RMS value.

The phasor notation can simplify the calculations for AC sinusoidal circuits; therefore, it is widely used in circuit analysis and calculations. Note that the phasor notation can be used for sinusoidal quantities only when all waveforms have the same frequency.

We have learned that a sinusoidal wave can be represented by its three important components: the peak value (or RMS value), the phase angle and the angular frequency. In an AC circuit, the AC source voltage and the current are the sinusoidal values with the same frequency, so the resulting voltages and currents in the circuit should also be sinusoidal values with the same frequency or angular frequency. Therefore, voltages and currents in an AC circuit can be analysed by using the phasor notation, i.e. they can be determined by the peak value or RMS value and the phase shift of the phasor notation.

Phasor

- A phasor is a vector that contains both amplitude and angle information, and it can be represented as complex number.
- Phasor notation is a method that uses complex numbers to represent the sinusoidal quantities for analysing AC circuits when all quantities have the same frequency.

The key for understanding the phasor notation is to know how to use complex numbers. Therefore, we will review some important formulas of complex numbers that you may have learned in previous mathematics courses.

8.3.2 Complex numbers review

The complex number has two main forms, the rectangular form and the polar form.

- Rectangular form: $\boldsymbol{A} = x + jy \quad (j = \sqrt{-1})$

  where x is the real part and y is the imagery part of the complex number $\boldsymbol{A}$; j is called the imagery unit.

Note: The symbol i is used to represent imagery unit in mathematics. Since i has been used to represent AC current in the circuit analysis, j is used to denote the imagery unit rather than i to avoid confusion.

- Polar form: $\boldsymbol{A} = a\angle\psi$

  This is the abbreviated form of the exponential form $A = ae^{j\psi}$, in which a is called modulus of the complex number, and the angle ψ is called argument of the complex number.

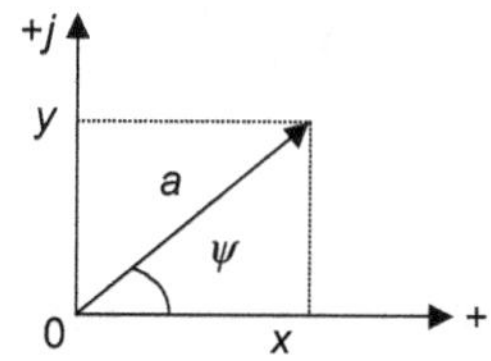

Figure 8.15 Complex number

- Convert rectangular form to polar form (refer to Figure 8.15).

  Let $A = x + jy = a\angle\psi$

  Applying $a = \sqrt{x^2 + y^2}$ (Pythagorus theory) gives

  $$A = x + jy = \sqrt{x^2 + y^2}\tan^{-1}y/x = a\angle\psi$$

- Convert polar form to rectangular or triangular form.
  $x = a\cos\psi$ and $y = a\sin\psi$ can be obtained from Figure 8.15.

  So $A = a\angle\psi = x + jy = a(\cos\psi + j\sin\psi)$.

Euler's formula can also be used for the conversion of triangular form to exponential form

$$e^{j\psi} = \cos\psi + j\sin\psi \quad \text{or} \quad ae^{j\psi} = a(\cos\psi + j\sin\psi)$$

- Operations on complex numbers: Given two complex numbers $A_1 = x_1 + jy_1 = a_1\angle\psi_1$ and $A_2 = x_2 + jy_2 = a_2\angle\psi_2$

  The basic algebraic operations of these two complex numbers are given as follows:
  Addition: $A_1 + A_2 = (x_1 + x_2) + j(y_1 + y_2)$
  Subtraction: $A_1 - A_2 = (x_1 - x_2) + j(y_1 - y_2)$
  Multiplication:

  - Polar form: $A_1 \cdot A_2 = a_1 \cdot a_2\angle(\psi_1 + \psi_2)$
  - Rectangular form: $A_1 \cdot A_2 = (x_1 + jy_1)(x_2 + jy_2) = (x_1x_2 - y_1y_2) + j(x_2y_1 + x_1y_2)$

    Here $j^2 = \sqrt{-1}\sqrt{-1} = (\sqrt{-1})^2 = -1$ is used.

  Division:

  - Polar form:

    $$\frac{A_1}{A_2} = \frac{a_1}{a_2}\angle(\psi_1 - \psi_2)$$

  - Rectangular form:

    $$\frac{A_1}{A_2} = \frac{x_1 + jy_1}{x_2 + jy_2} = \frac{(x_1 + jy_1)(x_2 - jy_2)}{(x_2 + jy_2)(x_2 - jy_2)} = \frac{x_1x_2 + y_1y_2}{x_2^2 + y_2^2} + j\frac{x_2y_1 - x_1y_2}{x_2^2 + y_2^2}$$

It will be much simpler to use the polar form on operations of multiplication and division.

Complex numbers

- Rectangular form: $\boldsymbol{A} = x + jy$
- Polar form: $\boldsymbol{A} = a\angle\psi$
- Conversion between rectangular and polar forms:

$$\boldsymbol{A} = x + jy = \sqrt{x^2 + y^2}\tan^{-1}\frac{y}{x} = a\angle\psi$$

$$\boldsymbol{A} = a\angle\psi = x + jy = a(\cos\psi + j\sin\psi)$$

- Addition and subtraction: $\boldsymbol{A}_1 \pm \boldsymbol{A}_2 = (x_1 \pm x_2) + j(y_1 \pm y_2)$
- Multiplication: $\boldsymbol{A}_1 \cdot \boldsymbol{A}_2 = a_1 \cdot a_2\angle(\psi_1 + \psi_2) = (x_1 + jy_1)(x_2 + jy_2)$
- Division: $\dfrac{\boldsymbol{A}_1}{\boldsymbol{A}_2} = \dfrac{a_1}{a_2}\angle(\psi_1 - \psi_2) = \dfrac{x_1 + jy_1}{x_2 + jy_2}$

8.3.3 Phasor

Using the phase notation to represent the sinusoidal function is based on Euler's formula $e^{j\phi} = \cos\phi + j\sin\phi$. For a sinusoidal function $f(t) = F_m\sin(\omega t + \psi)$, replacing ϕ with $(\omega t + \psi)$ in Euler's formula gives

$$e^{j(\omega t+\psi)} = \cos(\omega t + \psi) + j\sin(\omega t + \psi)$$

where $\cos(\omega t + \psi) = R_e[e^{j(\omega t+\psi)}]$ and $\sin(\omega t + \psi) = J_m[e^{j(\omega t+\psi)}]$.

'R_e[ ]' and 'J_m[ ]' stand for 'real part' and 'imaginary part' of complex numbers, respectively.

Therefore, sine function $f(t) = F_m\sin(\omega t + \psi) = J_m[F_m e^{j(\omega t+\psi)}] = J_m[F_m e^{j\psi} e^{j\omega t}]$.

That is, a sinusoidal function is actually taking the imaginary part of the complex number

$$f(t) = J_m[F_m e^{j\psi} e^{j\omega t}] \tag{8.3}$$

There are two terms in (8.3), $F_m e^{j\psi}$ and $e^{j\omega t}$. The second term $e^{j\omega t}$ is called the rotating factor that varies with time t, which will be discussed later. The first term is the phasor of the sinusoidal function

$$F_m e^{j\psi} = F_m\angle\psi = \boldsymbol{F}$$

So (8.3) of sine function can be written as

$$f(t) = F_m \sin(\omega t + \psi) = J_m[\boldsymbol{F} e^{j\omega t}]$$

Therefore, the first term in (8.3) is $\boldsymbol{F} = F_m \angle\psi$, where boldface letter $\boldsymbol{F}$ represents a phasor (vector) quantity, similar to the boldface that indicates a vector quantity in maths and physics. A phasor quantity can also be represented by a little dot on the top of the letter, such as $\dot{F} = F_m \angle\psi$. There is no difference between operations on phasors and complex numbers, since both of them are vectors.

If the sinusoidal currents and voltages in an AC circuit are represented by vectors with the complex numbers, this is known as phasors. The sinusoidal voltage $v(t) = V_m \sin(\omega t + \psi)$ and current $i(t) = I_m \sin(\omega t + \psi)$ in an AC circuit can be expressed in the phasor domain as:

- Peak value:

$$\dot{V} = V_m \angle\psi_v \text{ or } \boldsymbol{V} = V_m \angle\psi_v$$

$$\dot{I} = I_m \angle\psi_i \text{ or } \boldsymbol{I} = I_m \angle\psi_i$$

- RMS value:

$$\dot{V} = V \angle\psi_v \text{ or } \boldsymbol{V} = V \angle\psi_v$$

$$\dot{I} = I \angle\psi_i \text{ or } \boldsymbol{I} = I \angle\psi_i$$

8.3.4 Phasor diagram

Since a phasor is a vector that can be represented by a complex number, it can be presented with a rotating line in the complex plane as shown in Figure 8.16. The length of the phasor is the peak value F_m (or RMS value F). The angle between the rotating line and the positive horizontal axis is the phase angle ψ of the sinusoidal function. This diagram is called the phasor diagram.

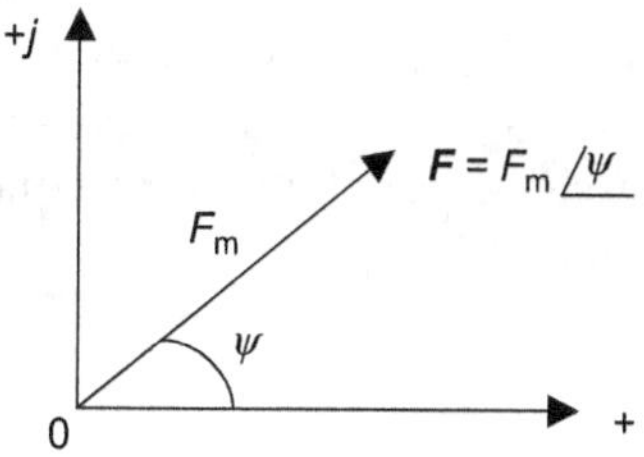

Figure 8.16 Phasor diagram

Phasor

Time domain	Phasor domain
$f(t) = F_m \sin(\omega t + \psi)$	$\boldsymbol{F}_m = F_m \angle\psi$ or $\dot{F}_m = F_m \angle\psi$ (Peak value) $\boldsymbol{F} = F\angle\psi$ or $\dot{F} = F\angle\psi$ (RMS value)
$v(t) = V_m \sin(\omega t + \psi)$	$\dot{V}_m = V_m \angle\psi_v$, $\dot{V} = V\angle\psi_v$
$i(t) = I_m \sin(\omega t + \psi)$	$\dot{I}_m = I_m \angle\psi_i$, $\dot{I} = I\angle\psi_i$

Example 8.4: Use the phasor notation to express the following voltage and current in which -10 and 12 are the peak values.

(a) $v = -10\sin(60t + 25°)$V

(b) $i = 12\sin(25t - 20°)$A

Solution:

(a) $\dot{V} = -10\angle 25°$V

(b) $\dot{I} = 12\angle -20°$A

Example 8.5: Use the instantaneous value to express the following voltage and current in which 120 and 12 are RMS values.

(a) $\dot{V} = 120\angle 30°$V

(b) $\dot{I} = 12\angle 0°$A

Solution:

(a) $v = 120\sqrt{2}\sin(\omega t + 30°)$ V

(b) $i = 12\sqrt{2}\sin\omega t$ A

8.3.5 *Rotating factor*

In the sinusoidal expression of $f(t) = F_m \sin(\omega t + \psi) = J_m[F_m e^{j\psi} e^{j\omega t}]$, the term $e^{j\omega t}$ varies with time t, known as the rotating factor or time factor. As time changes, it rotates counterclockwise at angular frequency ω in a radius F_m of the circle, as shown in Figure 8.17.

The rotating factor $e^{j\omega t}$ can be represented by Euler's formula

$$e^{j\omega t} = \cos\omega t + j\sin\omega t$$

when $\omega t = \pm 90°$: $e^{\pm j90°} = \cos(\pm 90°) + j\sin(\pm 90°) = \pm j$.

Therefore, $\pm 90°$ is also the rotating factor ($\pm j = \pm 90°$).

Rotating factor

$e^{j\omega t}$ or $\pm j = \pm 90°$

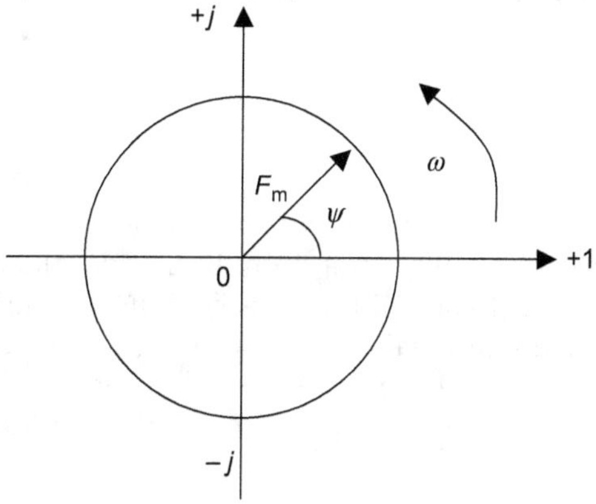

Figure 8.17 Rotating factor

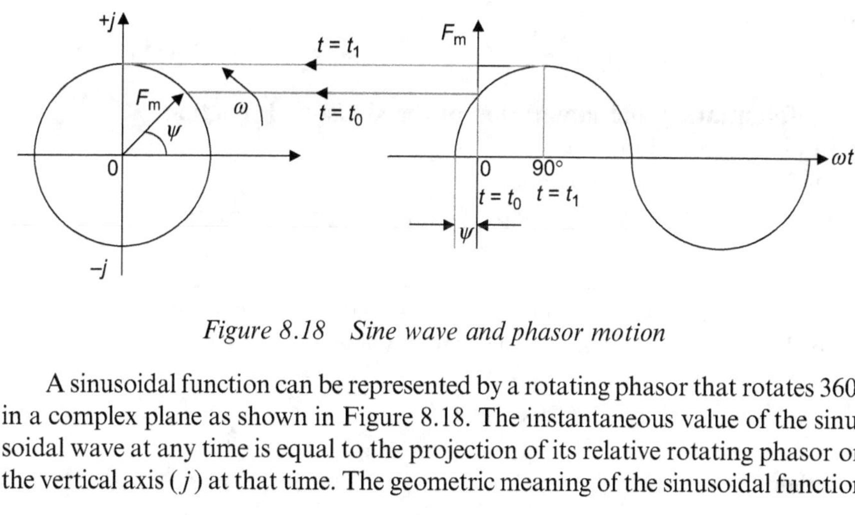

Figure 8.18 Sine wave and phasor motion

A sinusoidal function can be represented by a rotating phasor that rotates 360° in a complex plane as shown in Figure 8.18. The instantaneous value of the sinusoidal wave at any time is equal to the projection of its relative rotating phasor on the vertical axis (j) at that time. The geometric meaning of the sinusoidal function

$$f(t) = F_m \sin(\omega t + \psi) = I_m[F_m e^{j\psi} e^{j\omega t}]$$

represented by the rotational phasor motion can be seen from the following example.

Example 8.6: In Figure 8.18,

When $t = t_0 = 0$, the phasor is $\boldsymbol{F} = F_m \angle \psi$.
When $t = t_1$, the phasor is $\boldsymbol{F} = F_m \angle 90°$.
And it goes from ψ to 360°.

8.3.6 Differentiation and integration of the phasor

Note: Skip the following part and start from Example 8.8 if you haven't learned calculus.

For a sinusoidal function $f(t) = F_m \sin(\omega t + \psi)$, the derivative of the sinusoidal function with respect to time can be obtained by its phasor $\boldsymbol{F}$ multiplying with $j\omega$, i.e.

$$\frac{df(t)}{dt} \Leftrightarrow j\omega \boldsymbol{F}$$

This is equivalent to a phasor that rotates counterclockwise by 90° on the complex plane since $+j = +90°$. (Appendix B provides the details for how to derive the above differentiation of the sinusoidal function in phasor notation.)

The integral of the sinusoidal function with respect to time can be obtained from its phasor divided by $j\omega$, i.e.

$$\int f(t)dt = \frac{\dot{F}}{j\omega}$$

This is equivalent to a phasor that rotates clockwise on the complex plane by 90° (since $1/j = -j = -90°$).

Differentiation and integration of the sinusoidal function in phasor notation

Differentiation: $df(t)/dt \Leftrightarrow j\omega \boldsymbol{F}$ or $j\omega\dot{F}$ $\quad (+j = +90°)$

Integration: $\int f(t)dt \Leftrightarrow \boldsymbol{F}/j\omega$ or $(1/j\omega)\dot{F}$ $\quad (1/j = -j = -90°)$

Example 8.7: Convert the following sinusoidal time-domain expression to its equivalent phasor domain, and determine voltage $\dot{V}$ (or $\boldsymbol{V}$).

$$2v - 6\frac{dv}{dt} + 4\int v dt = 20\sin(4t + 30°)$$

Solution:

$$2\dot{V} - 6j\omega\dot{V} + 4\frac{\dot{V}}{j\omega} = 20\angle 30°$$

Since $\omega = 4$ in the original expression, so

$$2\dot{V} - 6j4\dot{V} + 4\frac{\dot{V}}{j4} = 20\angle 30°$$

$$\dot{V}(2 - 24j - j) = 20\angle 30°$$

$$\dot{V} = \frac{20\angle 30°}{2 - j25} \approx \frac{20\angle 30°}{25\angle -85.43°} = 0.8\angle 115.43°$$

Note: The complex number of the denominate is

$$Z = x + jy = 2 - j25 = \sqrt{x^2 + y^2}\tan^{-1}\frac{y}{x}$$

(Since x is positive and y is negative in $2 - j25$, the angle should be in the fourth quadrant, i.e. -85.43°.)

Example 8.8: Convert the phasor-domain voltage and current to their equivalent sinusoidal forms (time domain).

(a) $\dot{I} = j5e^{-j30^\circ}$ mA
(b) $\dot{V} = -6 + j8$ V

Solution:

(a)
$$\begin{aligned} \dot{I} = j5\angle -30^\circ \text{ mA} &= 5\angle 90^\circ \angle -30^\circ \text{ mA} \qquad (j = 90^\circ) \\ &= 5\angle(90^\circ - 30^\circ) \text{ mA} = 5\angle 60^\circ \text{ mA} \\ \mathrm{i}(t) &= 5\sin(\omega t + 60^\circ) \text{ mA} \end{aligned}$$

(b)
$$\dot{V} = -6 + j8 \text{ V} = \sqrt{(-6)^2 + 8^2}\tan^{-1}\angle\tfrac{8}{-6} \text{ V} \approx 10\angle 126.87^\circ \text{ V}$$

(Since y is positive and x is negative, it should be in the second quadrant.)

$$\mathrm{v}(t) = 10\sin(\omega t + 126.87^\circ) \text{ V}$$

If the phasors are used to express sinusoidal functions, the algebraic operations of sinusoidal functions of the same frequency can be replaced by algebraic operations of the equivalent phasors, which is shown in Example 8.9.

Example 8.9: Calculate the sum of the following two voltages

$$\mathrm{v}_1(t) = 2\sin(\omega t + 60^\circ) \text{ V} \quad \text{and} \quad \mathrm{v}_2(t) = 10\sin(\omega t - 40^\circ) \text{ V}$$

Solution: Convert the sinusoidal time-domain voltages to their equivalent phasor forms

$$\dot{V}_1 = 2\angle 60^\circ \text{ V and } \dot{V}_2 = 10\angle -40^\circ \text{ V}$$

So
$$\begin{aligned} \dot{V}_1 + \dot{V}_2 &= 2\angle 60^\circ + 10\angle -40^\circ \\ &= 2\cos 60^\circ + j2\sin 60^\circ + 10\cos(-40^\circ) + j10\sin(-40^\circ) \\ &\approx 1 + j1.732 + 7.66 - j6.43 \\ &= 8.66 - j4.698 \\ &= \sqrt{8.66^2 + (-4.698)^2}\tan^{-1}\left(\frac{-4.698}{8.66}\right) \\ &\approx 9.85\angle -28.48^\circ \text{ V} \end{aligned}$$

(Since y is negative and x is positive, it should be in the fourth quadrant.)

$$\therefore \quad \mathrm{v}(t) = 9.85\sin(\omega t - 28.48^\circ) \text{ V}$$

8.4 Resistors, inductors and capacitors in sinusoidal AC circuits

Any AC circuit may contain a combination of three basic circuit elements: resistor, inductors and capacitors. When these elements are connected to a sinusoidal AC voltage source, all resulting voltages and currents in the circuit are also sinusoidal and have the same frequency as AC voltage source. Therefore, they can all be converted from the sinusoidal time-domain form $f(t) = F_{\mathrm{m}}\sin(\omega t + \psi)$ to the phasor-domain form $\boldsymbol{F} = F_{\mathrm{m}}\angle\psi$.

8.4.1 Resistor's AC response

A resistor is connected to a sinusoidal voltage source as shown in Figure 8.19(a).

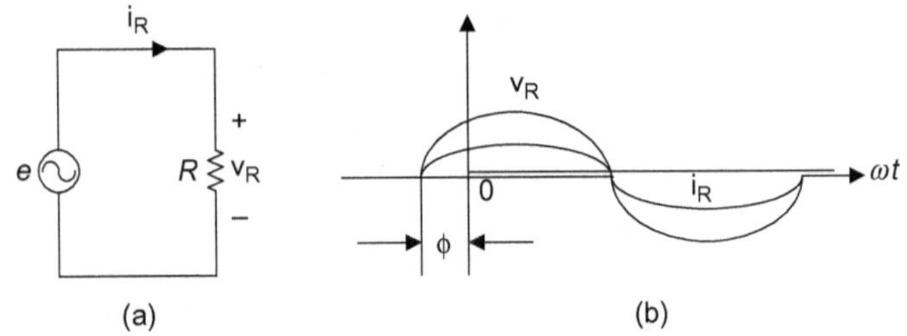

Figure 8.19 Resistor's AC response

Where the source voltage is

$$e = V_{\mathrm{m}}\sin(\omega t + \psi)$$

The sinusoidal current in the circuit can be obtained by applying Ohm's law for AC circuits (v = *R*i), i.e.

$$\mathrm{i_R} = \frac{e}{R} = \frac{V_{\mathrm{Rm}}}{R}\sin(\omega t + \psi) = I_{\mathrm{Rm}}\sin(\omega t + \psi)$$

where $I_{\mathrm{Rm}} = V_{\mathrm{Rm}}/R$ (peak values) or $I = V_{\mathrm{R}}/R$ (RMS value), and voltage across the resistor is the same as the source voltage, i.e. e $=$ $\mathrm{v_R}$ or $\mathrm{v_R} = V_{\mathrm{m}}\sin(\omega t + \psi)$.

The above sinusoidal expressions of resistor voltage $\mathrm{v_R}$ and current $\mathrm{i_R}$ indicate that voltage and current in the circuit have the same frequency f (since $\omega = 2\pi f$) and the same phase angle ψ (or $\mathrm{v_R}$ and $\mathrm{i_R}$ are in phase). This is also illustrated in Figure 8.19(b).

Assuming the initial phase angle is zero, i.e. $\psi = 0°$, then

$$\mathrm{i_R} = \frac{\mathrm{v_R}}{R} = I_{\mathrm{Rm}}\sin\omega t$$
$$\mathrm{v_R} = R\mathrm{i_R} = V_{\mathrm{Rm}}\sin\omega t$$

This is illustrated in Figure 8.19(c).

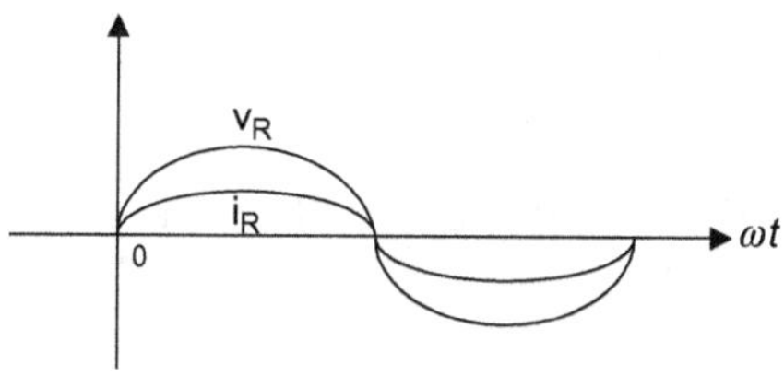

Figure 8.19(c) When $\psi = 0°$

Relationship of voltage and current of a resistor in an AC circuit

- Instantaneous values (time domain):

  $v_R = V_{Rm}\sin(\omega t + \psi)$

  $i_R = I_{Rm}\sin(\omega t + \psi)$

- Ohm's law:

  $V_{Rm} = I_{Rm}R$ (peak value)

  $V_R = I_R R$ (RMS value)

The sinusoidal expressions of resistor voltage (v_R) and current (i_R) are in the time domain. The peak and RMS values of the resistor voltage and the current in phasor domain also obey the Ohm's law as follows:

Peak value: $\dot{I}_{Rm} = \dot{V}_{Rm}/R$ or $\boldsymbol{V}_{Rm} = \boldsymbol{I}_{Rm}R$

RMS value: $\dot{I}_R = \dot{V}_R/R$ or $\boldsymbol{V}_R = \boldsymbol{I}_R R$

If it is expressed in terms of conductance, it will give

$$\dot{I}_R = G\dot{V}_R \quad (G = 1/R)$$

The relationship of the resistor voltage and current in an AC circuit can be presented by a phasor diagram illustrated in Figure 8.20(b).

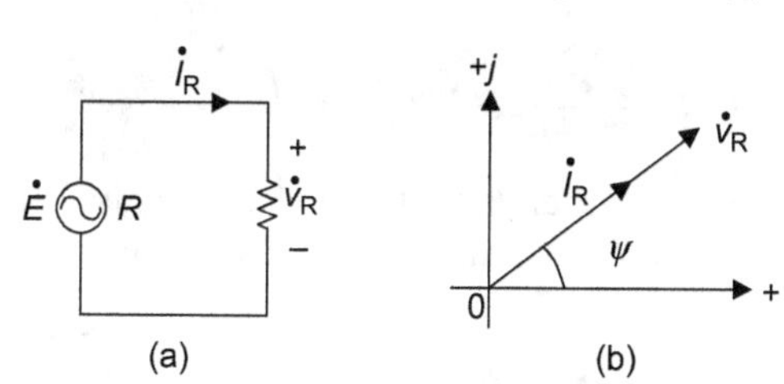

Figure 8.20 The phasor diagram of the AC resistive circuit

Example 8.10: If $R = 10\ \Omega$, $i_R = 6\sqrt{2}\sin(\omega t - 30^\circ)$A in Figure 8.20(a), determine the voltage across resistor in phasor domain.

Solution: $v_R = Ri_R = 10 \times 6\sqrt{2}\sin(\omega t - 30^\circ) = 60\sqrt{2}(\sin\omega t - 30^\circ)$
So $\dot{V}_{Rm} = 60\sqrt{2}\angle - 30^\circ$ V.

Resistor's AC response in phasor domain

- Ohm's law:
  Peak value: $\dot{V}_{Rm} = \dot{I}_{Rm}\ R$ or $V_{Rm} = I_{Rm}R$
  RMS value: $\dot{V}_R = \dot{I}_R\ R$ or $V_R = I_R R$
  Using conductance: $\dot{I}_R = G\dot{V}_R (G = 1/R)$
- Phasor diagram: $\xrightarrow{\dot{I}_R}\xrightarrow{\dot{V}_R}$
  (AC resistor voltage and current are in phase)

Note that we can use Ohm's law in AC circuits as long as the circuit quantities are consistently expressed, i.e. both the voltage and current are peak values, RMS values, instantaneous values, etc.

8.4.2 Inductor's AC response

If an AC voltage source is applied to an inductor as shown in Figure 8.21(a), the current flowing through the inductor will be

$$i_L = I_{Lm}\sin(\omega t + \psi) \tag{8.4}$$

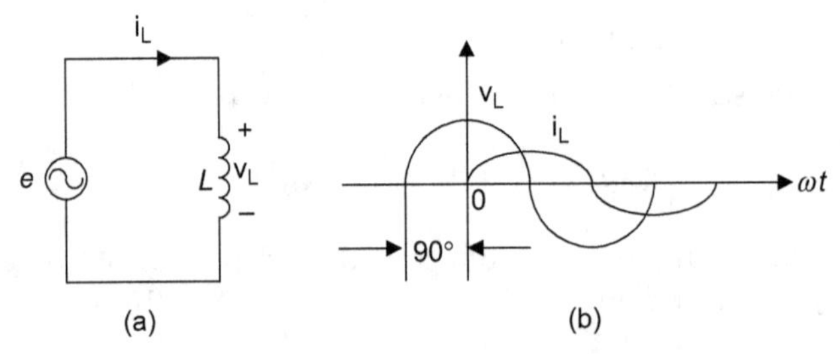

Figure 8.21 Inductor's AC response

We have learned from chapter 6 that the relationship between the voltage across the inductor and the current that flows through it is

$$v_L = L\frac{di}{dt} \tag{8.5}$$

Note: If you haven't learned calculus, then just keep in mind that $v_L = \omega L I_{Lm} \sin(\omega t + \psi + 90°)$ is the sinusoidal expression of the inductor voltage, and skip the following mathematic derivation process.

Substituting (8.4) into (8.5) and applying differentiation gives

$$v_L = L\frac{di_L}{dt} = L\frac{d[I_{Lm}\sin(\omega t + \psi)]}{dt} = \omega L I_{Lm}\cos(\omega t + \psi)$$

$$= \omega L I_{Lm}\sin(\omega t + \psi + 90°)$$

Therefore

$$v_L = \omega L I_{Lm}\sin(\omega t + \psi + 90°) \quad (8.6)$$

Note: $\cos\phi = \sin(\omega t + 90°)$

The sinusoidal expressions of the inductor voltage v_L and current i_L indicate that in an AC inductive circuit, the voltage and current have the same angular frequency (ω) and a phase difference. The inductor voltage v_L leads the current i_L by 90° as illustrated in Figure 8.21(b) if we assume that initial phase angle $\psi = 0°$.

The relationship between the voltage and current in an inductive sinusoidal AC circuit can be obtained from (8.6), which is given by

$$V_{Lm} = \omega L I_{Lm} \text{(peak value)} \quad \text{or} \quad V_L = \omega L I_L \text{(RMS value)}$$

This is also known as Ohm's law for an inductive circuit, where ωL is called inductive reactance and is denoted by X_L, i.e.

$$X_L = \omega L = 2\pi f L \quad (\omega = 2\pi f)$$

So $V_{Lm} = X_L I_{Lm}$ (peak value) and $V_L = X_L I_L$ (RMS value)

or $X_L = V_{Lm}/I_{Lm}$, $X_L = V_L/I_L$

where X_L is measured in ohms (Ω) and is the same as resistance R.

Recall that conductance G is the reciprocal of resistance R, and in an inductive circuit, the reciprocal of reactance is called inductive susceptance and is denoted by B_L, i.e. $B_L = 1/X_L$, and is measured in siemens (S) or mho (℧).

Relationship of voltage and current of inductor in an AC circuit

- Instantaneous values (time domain)

$$i_L = I_{Lm}\sin(\omega t + \psi)$$
$$v_L = X_L I_{Lm}\sin(\omega t + \psi + 90°)$$

- Ohm's law:

  $V_{Lm} = X_L I_{Lm}$ (peak value)
  $V_L = X_L I_L$ (RMS value)
- Inductive reactance: $X_L = \omega L = 2\pi f L$
- Inductive susceptance: $B_L = 1/X_L$

In an AC inductive circuit, the relationship between the voltage and current is not only determined by the value of inductance L in the circuit, but also related to the angular frequency ω. If an inductor has a fixed value in the circuit of Figure 8.21(a), inductance L in the circuit is a constant, and the higher the angular frequency ω, the greater the voltage across the inductor

$$V_L \uparrow = X_L I_L = (\omega \uparrow L) I_L$$

When $\omega \to \infty$, $V_L \to \infty$, i.e. when the angular frequency approaches to infinite, the inductor behaves as an open circuit in which the current is reduced to zero.

The lower the angular frequency ω, the lower the voltage across the inductor

$$V_L \downarrow = X_L I_L = (\omega \downarrow L) I_L$$

When $\omega = 0$, $V_L = 0$, i.e. the AC voltage across the inductor now is equivalent to a DC voltage since the frequency ($\omega = 2\pi f$) does not change any more. Recall that the inductor is equivalent to a short circuit at DC. In this case, the inductor is shortened because of zero voltage across the inductor.

This indicates that an inductor can pass the high-frequency signals (pass AC) and block the low-frequency signals (block DC).

Characteristics of an inductor

- An inductor can pass AC (open-circuit equivalent).
- An inductor can block DC (short-circuit equivalent).

The sinusoidal expressions of the inductor voltage v_L and current i_L are in the time domain. The peak and RMS values of the inductor voltage and the current in phasor domain also obey Ohm's law as follows:

Peak value: $\dot{V}_{Lm} = jX_L \dot{I}_{Lm}$ or $\boldsymbol{V}_{Lm} = jX_L \boldsymbol{I}_{Lm}$

RMS value: $\dot{V}_L = jX_L \dot{I}_L$ or $\boldsymbol{V}_L = jX_L \boldsymbol{I}_L$

This is because

$$\mathrm{v_L} = L\frac{\mathrm{d}i_\mathrm{L}}{\mathrm{d}t} \Leftrightarrow Lj\omega I_\mathrm{L} \text{ (differentiating: multiply by } j\omega)$$

So $\dot{V}_\mathrm{L} = (j\omega L)\dot{I}_\mathrm{L}$ or $\dot{V}_\mathrm{L} = jX_\mathrm{L}\dot{I}_\mathrm{L}$ $(X_\mathrm{L} = \omega L)$.

The relationship of the inductor voltage and current in an AC circuit can be presented by a phasor diagram illustrated in Figure 8.22(b and c). Figure 8.22(b) is when the initial phase angle is zero, i.e. $\psi = 0°$, and Figure 8.22(c) is when $\psi \neq 0°$ (the inductor current lags voltage by 90°).

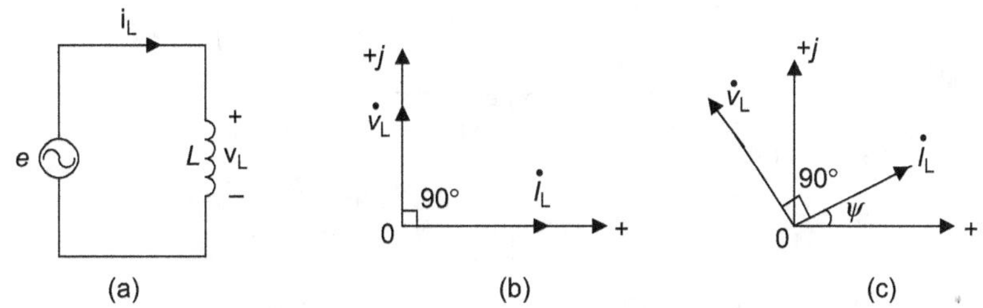

Figure 8.22 The phasor diagram of the AC inductive circuit

Inductor's AC response in phasor domain

- Ohm's law:
  Peak value: $\dot{V}_\mathrm{Lm} = jX_\mathrm{L}\dot{I}_\mathrm{Lm}$ or $\boldsymbol{V}_\mathrm{Lm} = jX_\mathrm{L}\boldsymbol{I}_\mathrm{Lm}$
  RMS value: $\dot{V}_\mathrm{L} = jX_\mathrm{L}\dot{I}_\mathrm{L}$ or $\boldsymbol{V}_\mathrm{L} = jX_\mathrm{L}\boldsymbol{I}_\mathrm{L}$
- Phasor diagram: 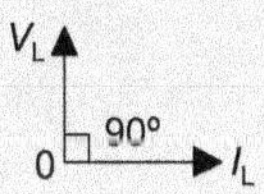

- Inductor voltage leads the current by 90°.

Example 8.11: In an AC inductive circuit, given $\mathrm{v_L} = 6\sqrt{2}\sin(60t + 35°)\mathrm{V}$ and L is 0.2 H, determine the current through the inductor in time domain.

Solution: Inductive reactance $X_\mathrm{L} = \omega L = (60\,\mathrm{rad/s})(0.2\,\mathrm{H}) = 12\,\Omega$

$$\dot{I}_\mathrm{Lm} = \frac{\dot{V}_\mathrm{Lm}}{jX_\mathrm{L}} = \frac{6\sqrt{2}\angle 35°\ \mathrm{V}}{j12\,\Omega} = \frac{6\sqrt{2}\angle 35°\ \mathrm{V}}{12\angle 90°\ \Omega} = 0.5\sqrt{2}\angle -55°\ \mathrm{A}$$

Convert the inductor current from the phasor domain to the time domain

$$\mathrm{i_L} = 0.5\sqrt{2}\sin(60t - 55°)\ \mathrm{A}$$

8.4.3 Capacitor's AC response

If an AC voltage source is applied to a capacitor as shown in Figure 8.23(a), the voltage across the capacitor will be

$$v_C = V_{Cm}\sin(\omega t + \psi)\text{V}$$

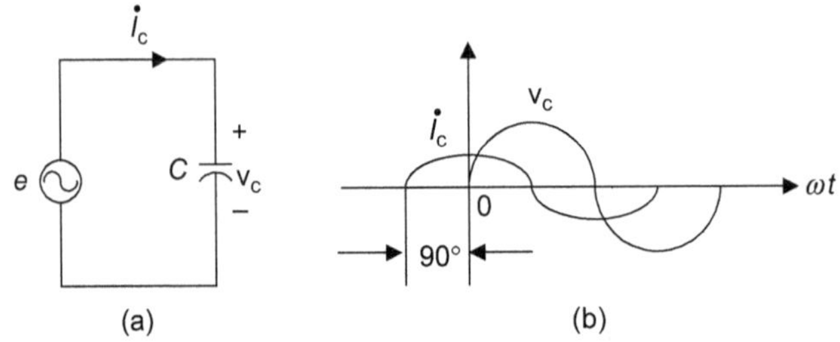

Figure 8.23 Capacitor's AC response

As we have learned from chapter 6, the relationship between the voltage across the capacitor and the current through it is

$$i_C = C\frac{dv_C}{dt}$$

Substituting v_C into the above expression and applying differentiation gives

$$i_C = C\frac{d[V_{Cm}\sin(\omega t + \psi)]}{dt} = \omega C V_{Cm}\sin(\omega t + \psi + 90^\circ)$$

That is

$$i_C = \omega C V_{Cm}\sin(\omega t + \psi + 90^\circ) \tag{8.7}$$

The sinusoidal expressions of the capacitor voltage v_C and current i_C indicated that in an AC capacitive circuit, the voltage and current have the same angular frequency (ω) and a phase difference. The capacitor current leads the voltage by 90° as illustrated in Figure 8.23(b), if we assume that the initial phase angle $\psi = 0^\circ$.

The relationship between voltage and current in an inductive sinusoidal AC circuit can be obtained from (8.7), which is given by

$$I_{Cm} = (\omega C)V_{Cm} \quad \text{(peak value)}$$

or

$$I_C = (\omega C)V_C \quad \text{(RMS value)}$$

This is also known as Ohm's law for a capacitive circuit, where ωC is called capacitive reactance that is denoted by the reciprocal of X_C, i.e.

$$X_C = \frac{1}{\omega C} = \frac{1}{2\pi f C} \quad (\omega = 2\pi f)$$

So

$$X_C = \frac{V_{Cm}}{I_{Cm}} \quad \text{(peak value)}$$

or

$$X_C = \frac{V_C}{I_C} \quad \text{(RMS value)}$$

X_C is measured in ohms (Ω) and that is the same as resistance R and inductive reactance X_L.

Recall that the inductive susceptance B_L is the reciprocal of the inductive reactance X_L. The reciprocal of capacitive reactance is called capacitive susceptance and is denoted by B_C, i.e. $B_C = 1/X_C$, and it is also measured in siemens or mho (same as B_L).

The relationship of voltage and current of capacitor in an AC circuit

- Instantaneous values (time domain):
  $v_C = V_{Cm}\sin(\omega t + \psi)$
  $i_C = \omega c V_{Cm}\sin(\omega t + \psi + 90°)$
- Ohm's law:
  $V_{Cm} = X_C I_{Cm}$ (peak value)
  $V_C = X_C I_C$ (RMS value)
- Capacitive reactance: $X_C = 1/\omega C = 1/2\pi f C$
- Capacitive susceptance: $B_C = 1/X_C$

Similar to an inductor, in an AC capacitive circuit not only is the relationship between voltage and current determined by the value of capacitive C in the circuit but it is also related to angular frequency ω. If there is a fixed capacitor in Figure 8.23(a), the conductance C in the circuit is a constant, and the higher the angular frequency ω, the lower the voltage across the capacitor.

$$V_C \downarrow = X_C I_C = \frac{I_C}{\omega \uparrow C}$$

When $\omega \to \infty$, $V_C \to 0$, i.e. when the angular frequency approaches infinite, the capacitor behaves as a short circuit in which the voltage across it will be reduced to zero.

The lower the angular frequency ω, the higher the voltage across the capacitor.

$$V_C \uparrow = \frac{I_C}{\omega \downarrow C}$$

When $\omega \to 0$, $V_C \to \infty$, i.e. the AC voltage across the capacitor now is equivalent to a DC voltage since the frequency ($\omega = 2\pi f$) does not change any more. Recall that a capacitor is equivalent to an open circuit at DC. In this case, the capacitor is open because there will be no current flowing through the capacitor.

This indicates that a capacitor can block the high-frequency signal (block AC) and pass the low-frequency signal (pass DC). The characteristics of a capacitor are opposite to those of an inductor.

Characteristics of a capacitor

- A capacitor can pass DC (short-circuit equivalent).
- A capacitor can block AC (open-circuit equivalent).

The sinusoidal expressions of the capacitor voltage v_C and current i_C are in the time domain. The peak and RMS values of the capacitor voltage and the current in phasor domain also obey Ohm's law as follows:

Peak value: $\dot{V}_{Cm} = -jX_C\dot{I}_{Cm}$ or $V_{Cm} = -jX_C\boldsymbol{I}_{Cm}$
RMS value: $\dot{V}_C = -jX_C\dot{I}_C$ or $V_C = -jX_C\boldsymbol{I}_C$

This is because $i_C = C\dfrac{dv_C}{dt} \Leftrightarrow Cj\omega V_C$ (differentiating: multiply by $j\omega$).

So $\dot{I}_C = j\omega C\dot{V}_C = j(1/X_C)\dot{V}_C \quad (X_C = 1/\omega C)$

or $\dot{V}_C = -jX_C\dot{I}_C \quad (1/j = -j)$.

The relationship of the capacitor voltage and current in an AC circuit can be presented by a phasor diagram and is illustrated in Figure 8.24(b and c). Figure 8.24(b) is when the initial phase angle is zero, i.e. $\psi = 0°$ (capacitor voltage lags current by 90°), and Figure 8.24(c) is when $\psi \neq 0°$

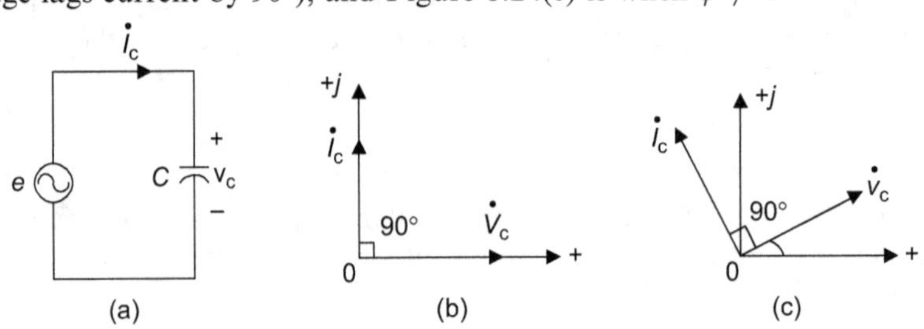

Figure 8.24 The phasor diagram of an AC capacitive circuit

Capacitor's AC response in phasor domain

- Ohm's law:
  Peak value: $\dot{V}_{Cm} = -jX_C\dot{I}_{Cm}$ or $\boldsymbol{V}_{Cm} = -jX_C\boldsymbol{I}_{Cm}$
  RMS value: $\dot{V}_C = -jX_C\dot{I}_C$ or $\boldsymbol{V}_C = -jX_C\boldsymbol{I}_C$
- Phasor diagram: 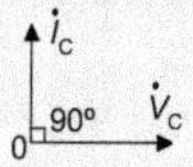

  Capacitor current leads voltage by 90°.

Example 8.12: Given a capacitive circuit in which $v_C = 50\sqrt{2}\sin(\omega t - 20°)$V, capacitance is 5 μF and frequency is 500 Hz, determine the capacitor current in the time domain.

Solution:

$$\omega = 2\pi f = 2\pi(500\,\text{Hz}) \approx 3142\,\text{rad/s}$$

$$X_C = \frac{1}{\omega C} = \frac{1}{(3\,142\,\text{rad/s})(5 \times 10^{-6}\,\text{F})} \approx 63.65\,\Omega$$

$$\therefore\ I_{Cm} = \frac{V_{Cm}}{X_C} = \frac{50\sqrt{2}\,\text{V}}{63.65\,\Omega} \approx 786\sqrt{2}\,\text{mA}$$

$$i_C = 786\sqrt{2}\sin(\omega t - 20° + 90°) = 786\sqrt{2}\sin(\omega t + 70°)\text{mA}$$

Summary

- Direct current (DC)
  - The polarity of DC voltage and direction of DC current do not change.
  - The pulsing DC changes the amplitude of the pulse, but does not change the polarity.
- Alternating current (AC)
  - The voltage and current periodically change polarity with time (such as sine wave, square wave, saw-tooth wave, etc.).
  - Sine AC varies over time according to the sine function, and is the most widely used AC.
- Period and frequency
  - Period T is the time to complete one full cycle of the waveform.
  - Frequency f is the number of cycles of waveforms within 1 s: $f = 1/T$.
- Three important components of the sinusoidal function $f(t) = F_m\sin(\omega t + \psi)$
  - F_m: Peak value (amplitude)
  - ω: Angular velocity (or angular frequency) $\omega = 2\pi/T = 2\pi f$
  - ψ: Phase or phase shift

- $\psi > 0$: Waveform shifted to the left side of 0°
- $\psi < 0$: Waveform shifted to the right side of 0°

- Phase difference ϕ: For two waves with the same frequency such as

$$\mathrm{v}(t) = V_{\mathrm{m}}\sin(\omega t + \psi_{\mathrm{v}}), \quad \mathrm{i}(t) = I_{\mathrm{m}}\sin(\omega t + \psi_{\mathrm{i}})$$

$$\phi = \psi_{\mathrm{v}} - \psi_{\mathrm{i}}$$

  - If $\phi = 0$: v and i in phase
  - If $\phi > 0$: v leads i
  - If $\phi < 0$: v lags i
  - If $\phi = \pm\pi/2$: v and i are orthogonal
  - If $\phi = \pm\pi$: v and i are out of phase
- Peak value, peak–peak value, instantaneous value and average value of sine waveform.
  - Peak value $F_{\mathrm{pk}} = F_{\mathrm{m}}$: the amplitude
  - Peak–peak value $F_{\mathrm{p-p}}$: $F_{\mathrm{p-p}} = 2F_{\mathrm{pk}}$
  - Instantaneous value $f(t)$: value at any time at any particular point of the waveform
  - Average value: average value of a half-cycle of the sine waveform $F_{\mathrm{avg}} = 0.637F_{\mathrm{m}}$
- RMS value (or effective value) of AC sinusoidal function
  - If an AC source delivers the equivalent amount of power to a resistor as a DC source, which is the effective or RMS value of AC.

$$V = \frac{V_{\mathrm{m}}}{\sqrt{2}} = 0.707\,V_{\mathrm{m}}, \quad I = \frac{I_{\mathrm{m}}}{\sqrt{2}} = 0.707\,I_{\mathrm{m}}$$

  - The general formula to calculate RMS value is

$$F = \sqrt{\frac{1}{T}\int_0^{\mathrm{T}} f^2(t)\mathrm{d}t}$$

- Complex numbers
  - Rectangular form: $\boldsymbol{A} = x + jy$
  - Polar form: $\boldsymbol{A} = a\angle\psi$
  - Conversion between rectangular and polar forms:

$$\boldsymbol{A} = x + jy = \sqrt{x^2 + y^2}\tan^{-1}\frac{y}{x} = a\angle\psi$$

$$\boldsymbol{A} = a\angle\psi = x + jy = a(\cos\psi + j\sin\psi)$$

  - Addition and subtraction: $\boldsymbol{A}_1 \pm \boldsymbol{A}_2 = (x_1 \pm x_2) + j(y_1 \pm y_2)$
  - Multiplication: $\boldsymbol{A}_1 \cdot \boldsymbol{A}_2 = a_1 \cdot a_2\angle(\psi_1 + \psi_2) = (x_1 + jy_1)(x_2 + jy_2)$

- Division:

$$\frac{A_1}{A_2} = \frac{a_1}{a_2} \angle(\psi_1 - \psi_2) = \frac{x_1 + jy_1}{x_2 + jy_2}$$

- Phasor
  - A phasor is a vector that contains both amplitude and angle information, and can be represented as a complex number.
  - The phasor notation is a method that uses complex numbers to represent the sinusoidal quantities for analysing AC circuits when all quantities have the same frequency.

Time domain	**Phasor domain**
$f(t) = F_{\mathrm{m}} \sin(\omega t + \psi)$	$\boldsymbol{F}_{\mathrm{m}} = F_{\mathrm{m}} \angle\psi$ or $\dot{F}_{\mathrm{m}} = F_{\mathrm{m}} \angle\psi$ (peak value) $\boldsymbol{F} = F\angle\psi$ or $\dot{F} = F\angle\psi$ (RMS value)
$\mathrm{v}(t) = V_{\mathrm{m}} \sin(\omega t + \psi)$	$\dot{V}_{\mathrm{m}} = V_{\mathrm{m}} \angle\psi_{\mathrm{v}}$, $\dot{V} = V\angle\psi_{\mathrm{v}}$
$\mathrm{i}(t) = I_{\mathrm{m}} \sin(\omega t + \psi)$	$\dot{I}_{\mathrm{m}} = I_{\mathrm{m}} \angle\psi_{\mathrm{i}}$, $\dot{I} = I\angle\psi_{\mathrm{i}}$

- Rotation factor: $e^{j\omega t}$ or $\pm j = \pm 90°$
- Differentiation and integration of the sinusoidal function in phasor notation:
  - Differentiation: $\mathrm{d}f(t)/\mathrm{d}t = j\omega \boldsymbol{F}$ or $j\omega\dot{F}$ $(+j = +90°)$
  - Integration: $\int f(t)\mathrm{d}t = \boldsymbol{F}/j\omega$ or $(1/j\omega)\dot{F}$ $(1/j = -j = -90°)$
- Characteristics of the inductor and capacitor:

Element	**DC ($\omega = 0$)**	**AC ($\omega \to \infty$)**	**Characteristics**
Inductor	Short circuit	Open circuit	Pass DC and block AC
Capacitor	Open circuit	Short circuit	Pass AC and block DC

- Three basic elements in an AC circuit

Element	**Time domain**	**Phasor domain**	**Resistance and reactance**	**Conductance and susceptance**	**Phasor diagram**
Resistor	$\mathrm{v_R} = R\mathrm{i_R}$	$\dot{V}_{\mathrm{R}} = \dot{I}_{\mathrm{R}} R$	R	$G = 1/R$	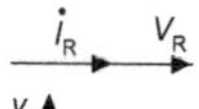
Inductor	$\mathrm{v_L} = L(\mathrm{di}/\mathrm{d}t)$	$\dot{V}_{\mathrm{L}} = jX_{\mathrm{L}}\dot{I}_{\mathrm{L}}$	$X_{\mathrm{L}} = \omega L$	$B_{\mathrm{L}} = 1/X_{\mathrm{L}}$	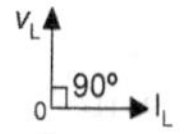
Capacitor	$\mathrm{i_C} = C(\mathrm{dv_C}/\mathrm{d}t)$	$\dot{V}_{\mathrm{C}} = -jX_{\mathrm{C}}\dot{I}_{\mathrm{C}}$	$X_{\mathrm{C}} = 1/\omega C$	$B_{\mathrm{C}} = 1/X_{\mathrm{C}}$	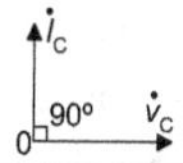

Experiment 8: Measuring DC and AC voltages using the oscilloscope

Objectives

- Become familiar with the operations of a function generator.
- Become familiar with the settings and correction of an oscilloscope.
- Become familiar with the operations of an oscilloscope.
- Become familiar with the method to measure DC and AC voltages with an oscilloscope.

Background information

1. Function generator: The function generator is an electronic equipment that can generate various types of waveforms that can have different frequencies and amplitudes. A function generator can be used as an AC voltage source to provide time-varying signals such as sine waves, square waves, triangle waves, etc.
2. Oscilloscope: The oscilloscope is one of the most important experimental and measurement instruments available for testing electric and electronic circuits. Its main function is to display waveforms to observe and analyse voltage, frequency, period and phase difference of DC or AC signals.
   The oscilloscope is a complex testing equipment and it is important to be familiar with its operations. There are various types of oscilloscopes that may look different, but most of their controls (knobs and buttons) in Table L8.1 have similar functions. Figure L8.1 shows the front panel of an oscilloscope. We will use this oscilloscope as an example for a brief description of the operations of the oscilloscope.

Table L8.1 The main controls of an oscilloscope

Display	**Horizontal control**	**Vertical control**	**Selecting switch**	**Probe**
INTENSITY	Time base setting (TIME/DIV)	Volts per division (VOLTS/DIV)	Channel coupling (CH I–DUAL–CH II)	× 1
FOCUS	Horizontal position control (X-POS ↔)	Vertical position control (Y-POS)	Input coupling (DC–GND–AC)	× 10

- Intensity control (INTENSITY): It can adjust the brightness of the display.
- Focus control (FOCUS). It can adjust the sharpness and clarity of the display.

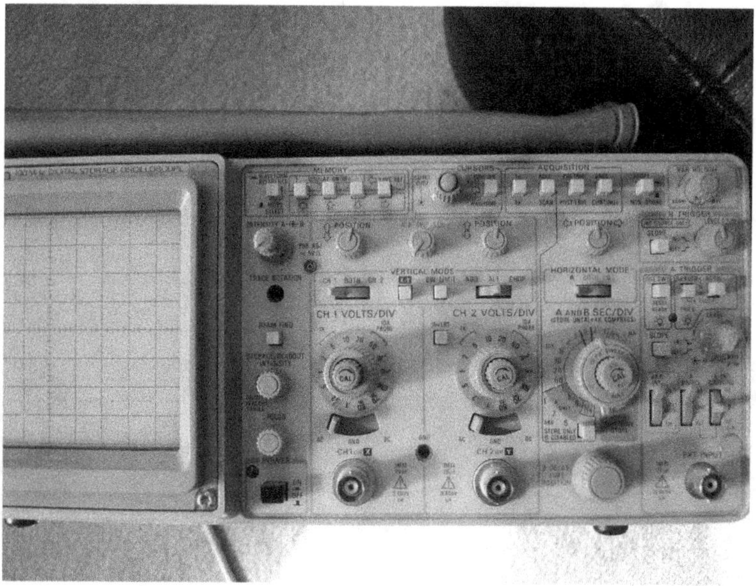

Figure L8.1 An oscilloscope

- Time base control (TIME/DIV – seconds per division): It can set up the length of time displayed per horizontal square (division) on the screen.
- Volts per division selector (VOLTS/DIV – volts per division): It can set up the waveform amplitude value per vertical square (division) on the screen.

$$\text{Measured amplitude} = (\text{Number of vertical divisions}) \times (\text{VOLTS/DIV})$$

Note: There is a small calibration (CAL) knob in the centre of both the VOLTS/DIV and TIME/DIV knobs. It should be in the fully clockwise position for the accuracy of the measurement.

- Horizontal position control (X-POS↔): It can adjust the horizontal position of the waveform.
- Vertical position control (Y-POS↕): It can adjust the vertical position of the waveform.
- Channel coupling (CH I–DUAL–CH II):
  CH I: Displays the input signal from channel I.
  CH II: Displays the input signal from channel II.
  DUAL: Displays the input signals from both channels I and II.
- Input coupling (DC–GND–AC): The connection from the test circuit to the oscilloscope.

DC: The DC position can display both DC and AC waveforms (the AC signal is superimposed on the DC waveform).
AC: The AC position blocks the DC waveform and only displays AC waveform.
GND: The GND position has a horizontal line on the screen that represents zero reference.

- × 1 Probe: Can measure and read the signal directly but may load the circuit under test and distort the waveform.
- × 10 Probe: Needs to multiply by 10 for each measured reading (more accurate).

Equipment and components

- Digital multimeter
- Breadboard
- DC power supply
- Oscilloscope
- Function generator

Procedure

Part I: Measure DC voltage using an oscilloscope

1. Set up the oscilloscope controls to the following positions:
   - Channel coupling: CH I or CH II
   - Input coupling: Set up to GND and adjust the trace to the central reference line (0 V) first, then switch to DC
   - TIME/DIV: 1 ms/DIV
   - Trigger: Auto
     (The trigger can stabilize repeating waveforms and capture single-shot waveforms.)
2. Connect a circuit as shown in Figure L8.2. The negative terminal of DC power, ground of the oscilloscope probe, and negative terminal of the multimeter (voltmeter function) should be connected together.

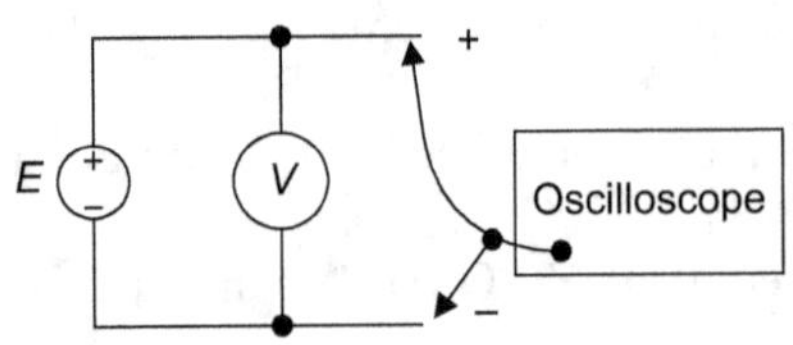

Figure L8.2 Measuring DC voltage using an oscilloscope

3. Set up the oscilloscope probe to × 1, adjust VOLTS/DIV of the oscilloscope to 1 V/DIV, and adjust DC power supply to 3 V. The voltmeter reading should be 3 V now. The DC wave on the oscilloscope screen

occupies three vertical grids (squares) at this time, so the voltage measured by the oscilloscope is also 3 V.

$$(3\ \text{vertical divisions}) \times (1\ \text{V/DIV}) = 3\ \text{V}$$

4. Keep the oscilloscope probe at × 1, adjust VOLTS/DIV of the oscilloscope to 0.5 V/DIV, and adjust DC power supply to 4 V. The DC wave on the oscilloscope screen occupies eight vertical divisions at this time.

   $$(8\ \text{vertical divisions}) \times (0.5\ \text{V/DIV}) = 4\ \text{V}$$

   Read the value on the voltmeter, and record it in Table L8.2.

Table L8.2

Probe	DC power supply (V)	Vertical division (DIV)	VOLTS/DIV (V/DIV)	Voltmeter (V)	Oscilloscope (V)
× 1	Example: 3	3	1	3	3
	4	8	0.5		4
	5		2		
× 10	8				
	12				
	16				

5. Keep the oscilloscope probe at × 1, adjust VOLTS/ DIV of the oscilloscope to 2 V/DIV, and adjust DC power supply to 5 V. Read the voltage value on the voltmeter and oscilloscope, and record them in Table L8.2.
6. Set up the oscilloscope probe to × 10, adjust DC power supply to 8, 12 and 16 V, respectively, and adjust VOLTS/DIV to suitable positions. Read the voltage values on the voltmeter and oscilloscope, and record them in Table L8.2.

Part II: AC measurements using an oscilloscope

1. Replace the DC power supply by a function generator in Figure L8.2. The ground of the function generator, ground of the oscilloscope probe and negative terminal of multimeter (voltmeter function) should be connected together.
   - Set up the function generator:

     Waveform: sine

     Frequency: 1.5 kHz

     DC offset: 0 V

     Amplitude knob: minimum (Fully counter clockwise)
   - Set up the oscilloscope:

     VOLTS/DIV: 0.5 V/DIV

     Channel coupling: CH I

     TIME/DIV: 0.2 ms/DIV

Input coupling: Set up to GND and adjust the trace to the central reference line (0 V) first, then switch to AC

2. Adjust the amplitude knob of the function generator until that sine wave on the vertical division of the oscilloscope screen occupies six divisions (squares). The voltage amplitude at this time is

$$V_{P-P} = (6\,\text{DIV}) \times (0.5\,\text{V/DIV}) = 3\,\text{V}$$

Note that the reading of the multimeter is RMS value, and it can be converted to the peak value comparing with the waveform obtained from the oscilloscope.

3. Adjust the horizontal position control of the oscilloscope (X-POS) until the sine wave on the oscilloscope screen occupies four horizontal divisions.

   - Determine the period of the sine wave T:

     $$\text{Period } (T) = (\text{Number of horizontal divisions}) \times (\text{TIME/DIV})$$

     $$T = (4\,\text{divisions}) \times (0.2\,\text{ms/DIV}) = 0.8\,\text{ms}$$

   - Determine the frequency f:

     $$f = \frac{1}{T} = \frac{1}{0.8\,\text{ms}} = 1.25\,\text{kHz}$$

4. Adjust the horizontal position control of the oscilloscope (X-POS) until the sine wave on the oscilloscope screen occupies six horizontal divisions (adjust the frequency knob on the function generator if necessary). Determine the period T and frequency f of the sine wave, and record the values in Table L8.3 (keep TIME/DIV = 0.2 ms/DIV).

Table L8.3

	Period T	**Frequency f**
Step 4		
Step 5		

5. Adjust TIME/DIV of the oscilloscope to 0.5 ms/DIV, and adjust the horizontal position control of the oscilloscope (X-POS) until the sine wave on the oscilloscope screen occupies five horizontal divisions. Determine the period T and frequency f of the sine wave, and record the values in Table L8.3.

Conclusion

Write your conclusions below:

Chapter 9
Methods of AC circuit analysis

Objectives

After completing this chapter, you will be able to:

- understand concepts and characteristics of the impedance and admittance of AC circuits
- define the impedance and admittance of resistor R, inductor L and capacitor C
- determine the impedance and admittance of series and parallel AC circuits
- apply the voltage divider and current divider rules to AC circuits
- apply KCL and KVL to AC circuits
- understand the concepts of instantaneous power, active power, reactive power, apparent power, power triangle and power factor
- apply the mesh analysis, node voltage analysis, superposition theorem and Thevinin's and Norton's theorems, etc. to analyse AC circuits

9.1 Impedance and admittance

9.1.1 Impedance

In the previous chapter, we had learned that the phasor forms of relationship between voltage and current for resistor, inductor and capacitor in an AC circuit are as follows:

$$\dot{V}_{\mathrm{R}} = \dot{I}_{\mathrm{R}}R, \quad \dot{V}_{\mathrm{L}} = \mathrm{j}\dot{I}_{\mathrm{L}}X_{\mathrm{L}}, \quad \dot{V}_{\mathrm{C}} = -\mathrm{j}\dot{I}_{\mathrm{C}}X_{\mathrm{C}}$$

The above equations can be changed to a ratio of voltage and current

$$\frac{\dot{V}_{\mathrm{R}}}{\dot{I}_{\mathrm{R}}} = R, \quad \frac{\dot{V}_{\mathrm{L}}}{\dot{I}_{\mathrm{L}}} = \mathrm{j}X_{\mathrm{L}}, \quad \frac{\dot{V}_{\mathrm{C}}}{\dot{I}_{\mathrm{C}}} = -\mathrm{j}X_{\mathrm{C}}$$

The ratio of voltage and current is the impedance of an AC circuit, and it can be generally expressed as $Z = \dot{V}/\dot{I}$. This equation is also known as Ohm's law of AC circuits.

The physical meaning of the impedance is that it is a measure of the opposition to AC current in an AC circuit. It is similar to the concept of resistance in DC circuits, so the impedance is also measured in ohms. The impedance can be extended to the inductor and capacitor in an AC circuit. It is a complex number that describes both the amplitude and phase characteristics.

The impedances of resistor, inductor and capacitor are as follows:

$$Z_R = R = \frac{\dot{V}_R}{\dot{I}_R}, \quad Z_L = jX_L = \frac{\dot{V}_L}{\dot{I}_L}, \quad Z_C = -jX_C = \frac{\dot{V}_C}{\dot{I}_C}$$

Impedance *Z*

- Z is a measure of the opposition to AC current in an AC circuit.
- Ohm's law in AC circuits: $Z = \dot{V}/\dot{I}$.

Quantity	Quantity Symbol	Unit	Unit symbol
Impedance	Z	Ohm	Ω

9.1.2 Admittance

Recall that the conductance G is the inverse of resistance R, and it is a measure of how easily current flows in a DC circuit. It is more convenient to use the conductance in a parallel DC circuit. Similarly, the admittance is the inverse of impedance Z, it is denoted by Y, $Y = 1/Z$, and is measured in siemens (S). The admittance is a measure of how easily a current can flow in an AC circuit. It can be expressed as the ratio of current and voltage of an AC circuit, i.e. $Y = \dot{I}/\dot{V}$. It is more convenient to use the admittance in a parallel AC circuit.

Admittance *Y*

- Y is the measure of how easily current can flow in an AC circuit.
- Y is the inverse of impedance: $Y = 1/Z$.
- Ohm's law in AC circuits: $\dot{I} = \dot{V}Y$.

Quantity	Quantity Symbol	Unit	Unit symbol
Admittance	Y	Siemens	S

The admittance of resistor, inductor and capacitor are as follows:

$$Y_R = \frac{1}{R}, \quad Y_L = \frac{1}{jX_L} = -j\frac{1}{X_L}, \quad Y_C = \frac{1}{-jX_C} = j\frac{1}{X_C} \quad \left(j = -\frac{1}{j}\right)$$

9.1.3 Characteristics of the impedance

Since the impedance is a vector quantity, it can be expressed in both polar form and rectangular form (complex number) as follows:

$$Z = z\angle\phi = R + \mathrm{j}X = z(\cos\phi + \mathrm{j}\sin\phi) \tag{9.1}$$

The rectangular form is the sum of the real part and the imaginary part, where the real part of the complex is the resistance R, and the imaginary part is the reactance X. The reactance is the difference of inductive reactive and capacitive reactance, i.e.

$$X = X_{\mathrm{L}} - X_{\mathrm{C}}$$

The lower case letter z in (9.1) is the magnitude of the impedance, which is

$$z = \sqrt{R^2 + X^2}$$

The corresponding angle ϕ between the resistance R and reactance X is called the impedance angle and can be expressed as follows:

$$\phi = \tan^{-1}\frac{X}{R}$$

The relationship between R, X and Z in the expression of the impedance is a right triangle, and can be described using the Pythagoras' theorem. This can be illustrated in Figure 9.1(a).

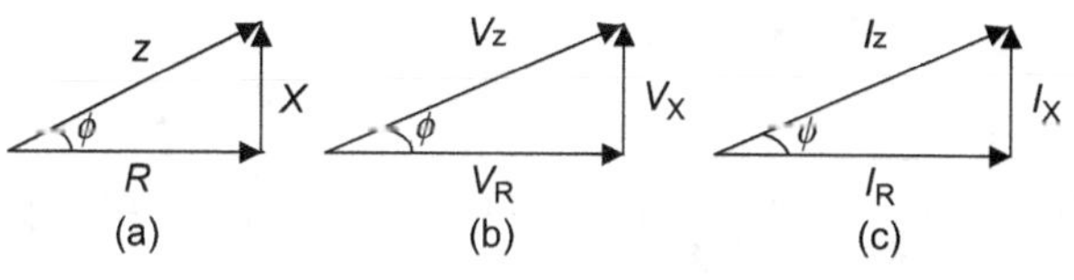

Figure 9.1 Impedance, voltage and current triangles

Figure 9.1(a) is an impedance triangle. If we multiply each side of the quantity in the impedance triangle by current $\dot{I}$ the following expressions will be obtained:

$$\dot{V}_{\mathrm{z}} = Z\dot{I}_{\mathrm{z}},\ \dot{V}_{\mathrm{X}} = \dot{I}_{\mathrm{X}}X,\ \dot{V}_{\mathrm{R}} = \dot{I}_{\mathrm{X}}R$$

These can form another triangle that is called the voltage triangle, which is illustrated in Figure 9.1(b). If we divide each side of the value by voltage $\dot{V}$ in the impedance triangle, the following expressions will be obtained:

$$\dot{I}_{\mathrm{z}} = \frac{\dot{V}_{\mathrm{z}}}{Z},\quad \dot{I}_{\mathrm{X}} = \frac{\dot{V}_{\mathrm{X}}}{X},\quad \dot{I}_{\mathrm{R}} = \frac{\dot{V}_{\mathrm{R}}}{R}$$

The above expression can form another triangle that is called the current triangle, and it is illustrated in Figure 9.1(c). The characteristics of the impedance triangle in Figure 9.1(a) can be summarized as follows:

- If $X > 0$ or $X = X_L - X_C > 0$, $X_L > X_C$: The reactance X is above the horizontal axis, and the impedance angle $\phi > 0$. The circuit is more inductive as shown in Figure 9.2(a).
- If $X < 0$ or $X = X_L - X_C < 0$, $X_C > X_L$: The reactance X is below the horizontal axis, and the impedance angle $\phi < 0$. The circuit is more capacitive as shown in Figure 9.2(b).
- If $X = 0$ or $X = X_L - X_C = 0$, $X_C = X_L$: The impedance angle $\phi = 0$, the circuit will look like a purely resistive circuit ($z = R$) as shown in Figure 9.2(c).

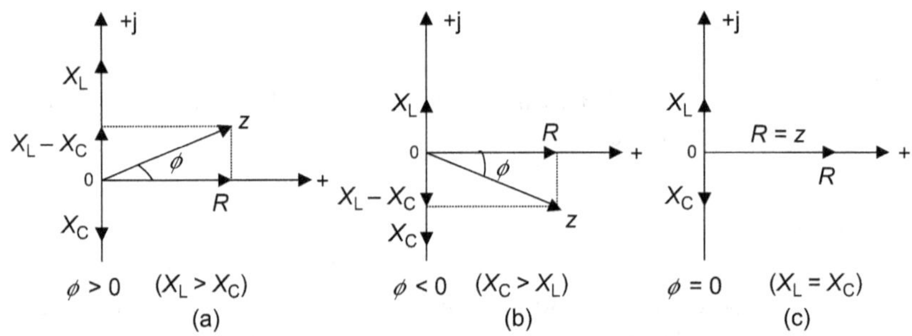

Figure 9.2 The phasor diagrams of the impedance

Example 9.1: Determine the impedance Z in the circuit of Figure 9.3 and plot the phasor diagram of the impedance.

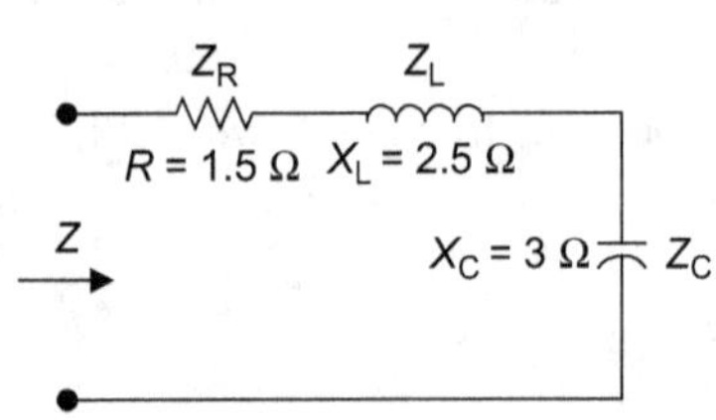

Figure 9.3 Circuit for Example 9.1

Solution: The impedances in series in an AC circuit behave like resistors in series.

So,

$$\begin{aligned} Z &= Z_R + Z_L + Z_C = R + jX = R + j(X_L - X_C) \\ &= 1.5\,\Omega + j(2.5 - 3)\Omega = 1.5\,\Omega - j0.5\,\Omega \\ &= \sqrt{1.5^2 + (-0.5)^2}\tan^{-1}\frac{-0.5}{1.5} \approx 1.58\angle - 18.44°\,\Omega \end{aligned}$$

Note: Since the imaginary term is -0.5 on y-coordinate, and the real term is $+1.5$ on the x-coordinate, the impedance angle for this circuit is located in the 4th quadrant.

The circuit for Example 9.1 is more capacitive since $X_C > X_L$, and $\phi < 0$ as shown in Figure 9.4.

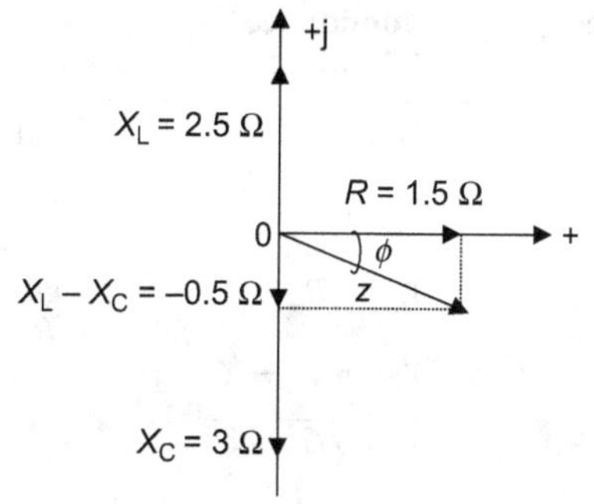

Figure 9.4 Impedance angle for Example 9.1

9.1.4 *Characteristics of the admittance*

The admittance is also a complex number; it can be expressed in both polar and rectangular forms as follows:

$$Y = y\angle\phi_y = G + jB = y(\cos\phi + j\sin\phi) \tag{9.2}$$

The real part of the complex is the conductance G, and the imaginary part is called the susceptance B. The susceptance is measured in the same way as the admittance, i.e. siemens (S). The susceptance is the difference of the capacitive susceptance and inductive susceptance, i.e. $B = B_C - B_L$.

The lower case letter y in (9.2) is the magnitude of the admittance, i.e. $y = \sqrt{G^2 + B^2}$

The corresponding angle ϕ between the conductance G and susceptance B is called the admittance angle and can be expressed as

$$\phi_y = \tan^{-1}\frac{B}{G}$$

The admittance of resistor, inductor and capacitor are as follows:

$$Y_R = \frac{1}{R} = G, \quad Y_L = \frac{1}{jX_L} = -j\frac{1}{X_L}, \quad Y_C = \frac{1}{-jX_C} = j\frac{1}{X_C}$$

The impedance, admittance, susceptance and their relationship can be summarized as given in Table 9.1.

Table 9.1 Impedance and admittance

Component	**Impedance** $Z = \dot{V}/\dot{I}$	**Admittance** $Y = 1/Z$	**Conductance and susceptance**
Resistor (R)	$Z_R = R$	$Y_R = G$	Conductance: $G = 1/R$
Inductor (L)	$Z_L = jX_L$	$Y_L = -jB_L$	Inductive susceptance: $B_L = 1/X_L$
Capacitor (C)	$Z_C = -jX_C$	$Y_C = jB_C$	Capacitive susceptance: $B_C = 1/X_C$
	$Z = z\angle\phi = R + jX$	$Y = y\angle\phi_y = G + jB$	Reactance: $X = X_L - X_C$
	$z = \sqrt{R^2 + X^2}$	$y = \sqrt{G^2 + B^2}$	Susceptance: $B = B_C - B_L$
	$\phi = \tan^{-1}\frac{X}{R}$	$\phi_y = \tan^{-1}\frac{B}{G}$	

The characteristics of the admittance triangle in Figure 9.5 can be summarized as follows:

$$\left(X_L = \omega L, \quad X_C = \frac{1}{\omega C}, \quad j = \frac{1}{-j}\right)$$

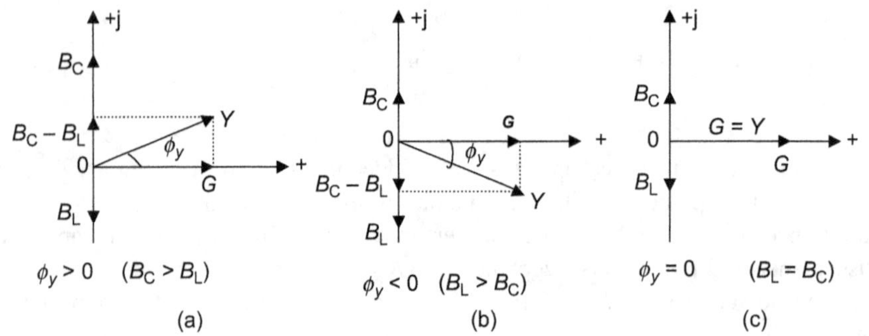

Figure 9.5 The phasor diagrams of the admittance

- If $B > 0$ or $B = (B_C - B_L) > 0$, $B_C > B_L$: The susceptance B is above the horizontal axis, the admittance angle $\phi_y > 0$, and the circuit is more capacitive as shown in Figure 9.5(a).
- If $B < 0$ or $B = (B_C - B_L) < 0$, $B_L > B_C$: The susceptance B is below the horizontal axis, the admittance angle $\phi_y < 0$, and the circuit is more inductive as shown in Figure 9.5(b).
- If $B = 0$ or $B = (B_C - B_L) = 0$, $B_L = B_C$: The admittance angle $\phi_y = 0$, the circuit will look like a purely resistive circuit ($Y = G$) as shown in Figure 9.5(c).

Characteristics of impedance and admittance

- If $X > 0$, $\phi > 0$, $B < 0$, $\phi_y < 0$: The circuit is more inductive.
- If $X < 0$, $\phi < 0$, $B > 0$, $\phi_y > 0$: The circuit is more capacitive.
- If $X = 0$, $\phi = 0$, $B = 0$, $\phi_y = 0$: The circuit is purely resistive.

Example 9.2: Determine the admittance in the circuit of Figure 9.6 and plot the phasor diagrams of the admittance.

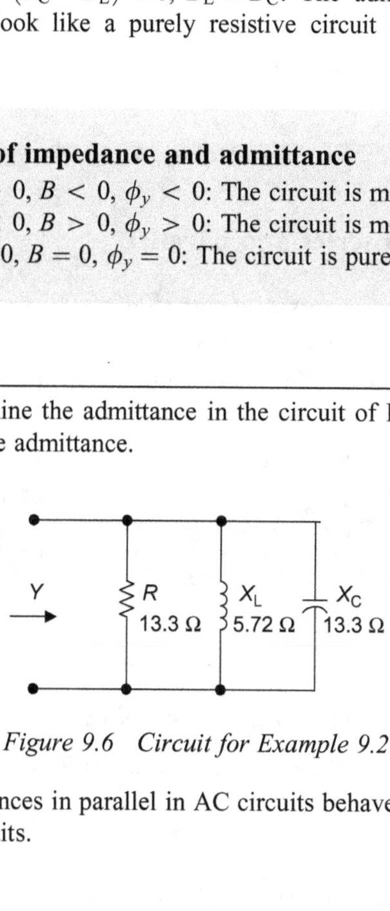

Figure 9.6 Circuit for Example 9.2

Solution: The admittances in parallel in AC circuits behave like the conductances in parallel in DC circuits.
So,

$$Y = Y_R + Y_L + Y_C = G + jB = G + j(B_C - B_L)$$
$$= \frac{1}{13.3\,\Omega} + j\left(\frac{1}{13.3\,\Omega} - \frac{1}{5.72\,\Omega}\right) \approx 0.075\text{ S} + j(0.075 - 0.175)\text{S}$$
$$= 0.075\text{ S} - j0.1\text{ S} = \sqrt{0.075^2 + (-0.1)^2}\tan^{-1}\frac{-0.1}{0.075} = 0.125\angle - 53.13°\text{ S}$$

Note: The admittance angle for this circuit is located in the fourth quadrant since the imaginary term is -0.1 and the real term is $+0.075$.

Since $B_L > B_C$ ($B_L = 0.175$, $B_C = 0.075$) and $\phi_y < 0$, the circuit is more inductive as shown in Figure 9.7.

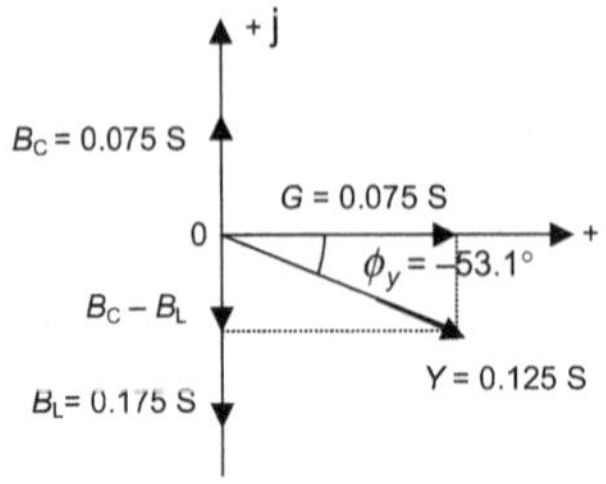

Figure 9.7 Admittance angle for Example 9.2

9.2 Impedance in series and parallel

9.2.1 Impedance of series and parallel circuits

The impedances in series and parallel AC circuits behave like resistors in series and parallel DC circuits, except the phasor form (complex number) is used. The equivalent impedance (or total impedance) for a series circuit in Figure 9.8 is given as

$$Z_{eq} = Z_1 + Z_2 + \cdots + Z_n$$

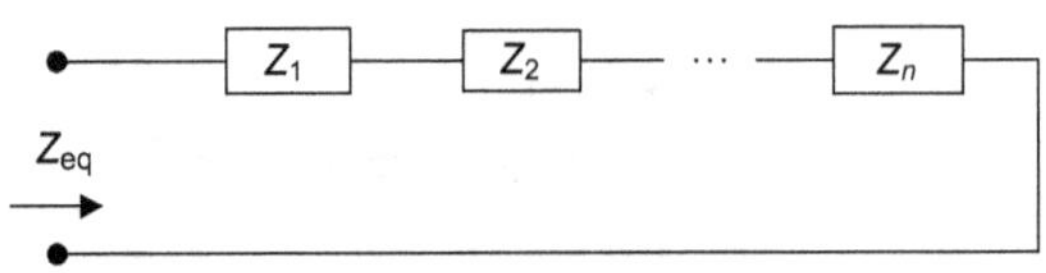

Figure 9.8 Impedance of a series circuit

The equivalent impedance (or total impedance) for a parallel circuit in Figure 9.9 is given as

$$Z_{eq} = \frac{1}{(1/Z_1) + (1/Z_2) + \cdots + (1/Z_n)} = Z_1//Z_2//\ldots//Z_n$$

$$Y_{eq} = Y_1 + Y_2 + \cdots + Y_n$$

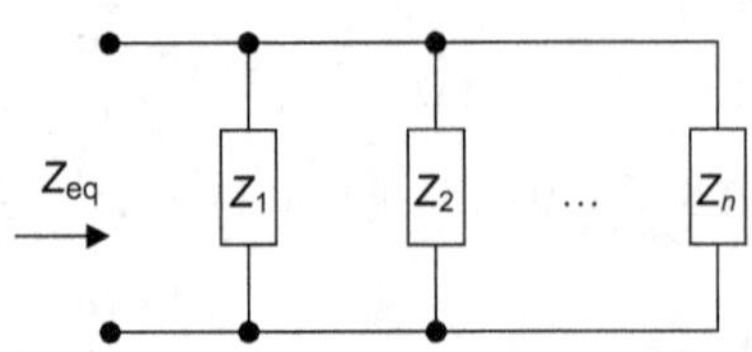

Figure 9.9 Impedance of a parallel circuit

The equivalent impedance is the reciprocal of equivalent admittance, $Z_{eq} = 1/Y_{eq}$. If only have two impedances in parallel, the equivalent impedance is given as

$$Z_{eq} = \frac{Z_1 Z_2}{Z_1 + Z_2} = Z_1//Z_2$$

Impedances in series and parallel

- Impedances in series: $Z_{eq} = Z_1 + Z_2 + \cdots + Z_n$
- Impedances in parallel:
  $Z_{eq} = Z_1//Z_2//\cdots//Z_n, \quad Y_{eq} = Y_1 + Y_2 + \cdots + Y_n$
- Two impedances in parallel: $Z_{eq} = \dfrac{Z_1 Z_2}{Z_1 + Z_2}$

To determine the equivalent impedance in series and parallel AC circuits, use the same method that determines the equivalent resistance in series and parallel DC circuits.

9.2.2 *Voltage divider and current divider rules*

The voltage divider and current divider rules in phasor form in AC circuits are very similar to the DC circuits as follows:

$$\dot{V}_1 = \frac{Z_1}{Z_1 + Z_2}\dot{E}, \quad \dot{V}_2 = \frac{Z_2}{Z_1 + Z_2}\dot{E}$$

$$\dot{I}_1 = \frac{Z_2}{Z_1 + Z_2}\dot{I}_T, \quad \dot{I}_2 = \frac{Z_1}{Z_1 + Z_2}\dot{I}_T$$

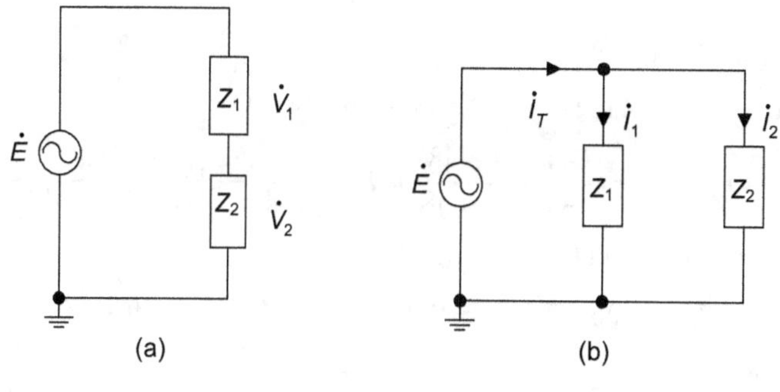

Figure 9.10 Voltage and current dividers

9.2.3 *The phasor forms of KVL and KCL*

The phasor forms of Kirchhoff's voltage law (KVL) and Kirchhoff's current law (KCL) also hold true in AC circuits.

Phasor forms of KVL and KCL

- KVL: $\Sigma\dot{V} = 0$ or $\dot{V}_1 + \dot{V}_2 + \cdots + \dot{V}_n = \dot{E}$
- KCL: $\Sigma\dot{I} = 0$ or $\dot{I}_{in} = \dot{I}_{out}$

The following examples show how to use the above equations in series-parallel AC circuits.

Example 9.3: Determine the following values for the circuit in Figure 9.11.

- the input equivalent impedance Z_{eq} and
- the current $\dot{I}_3$ in the branch of R_L and X_L.

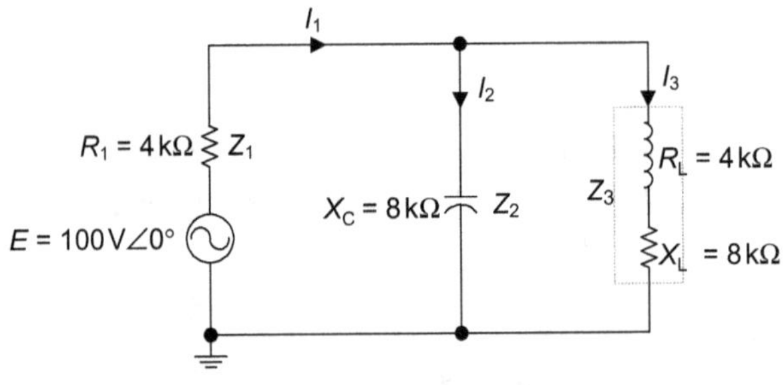

Figure 9.11 Circuit for Example 9.3

Solution:

(a) $Z_{eq} = Z_1 + Z_2//Z_3$

$$Z_1 = R_1 = 4\,k\Omega$$

$$Z_2 = -jX_C = -j8\,k\Omega$$

$$Z_3 = R_L + jX_L = 4\,k\Omega + j8\,k\Omega \approx 8.94\angle 63.44°\,k\Omega$$

$$Z_2//Z_3 = \frac{Z_2 Z_3}{Z_2 + Z_3} = \frac{(-j8)(4 + j8)}{-j8 + 4 + j8}\,k\Omega = \frac{64 - j32}{4}\,k\Omega \approx \frac{71.55\angle -26.57}{4\angle 0°}\,k\Omega$$

$$\approx (16 - j8)k\Omega$$

$$= 17.9\angle -26.57°\,k\Omega = 17.9[\cos(-26.57°) + j\sin(-26.57°)]k\Omega$$

$$Z_{eq} = Z_1 + Z_2//Z_3$$

$$= [4 + (16 - j8)]k\Omega = (20 - j8)k\Omega \approx 21.54\angle -21.8°\,k\Omega$$

(b) $\dot{I}_3 = \dfrac{Z_2}{Z_2 + Z_3}\dot{I}_1$

Here $\dot{I}_1 = \dfrac{\dot{E}}{Z_{eq}} = \dfrac{100\angle 0°\ \mathrm{V}}{21.54\angle -21.8°\ \Omega} \approx 4.64\angle 21.8°\ \mathrm{mA}$

$$\begin{aligned}\therefore\ \dot{I}_3 &= \frac{Z_2}{Z_2 + Z_3}\dot{I}_1\\ &= (4.64\angle 21.8°)\mathrm{mA}\frac{8\angle -90°\mathrm{k}\Omega}{(-\mathrm{j}8 + 4 + \mathrm{j}8)\mathrm{k}\Omega} = \frac{37.12\angle -68.2}{4\angle 0°}\mathrm{mA}\\ &= 9.28\angle -68.2°\mathrm{mA}\end{aligned}$$

Example 9.4: Determine the voltage across the inductor L for the circuit in Figure 9.12.

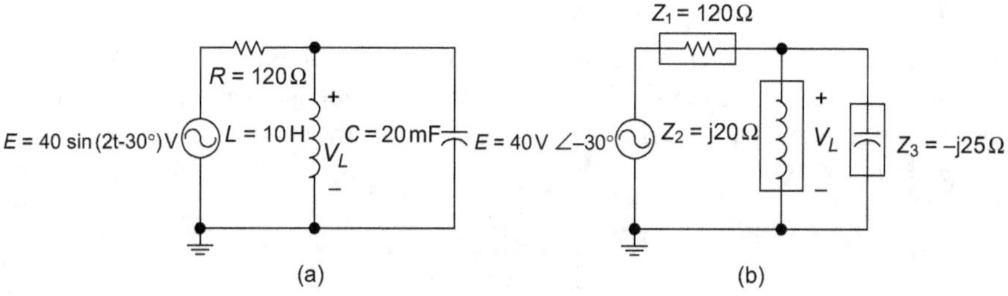

Figure 9.12 Circuits for Example 9.4

Solution:

- Convert the time domain to the phasor domain as shown in Figure 9.12(b) first.

$$Z_1 = R = 120\,\Omega$$

$$Z_2 = \mathrm{j}X_L = \mathrm{j}(\omega L) = \mathrm{j}(2 \times 10\,\mathrm{H}) = \mathrm{j}20\,\Omega$$

$$Z_3 = -\mathrm{j}X_C = -\mathrm{j}\frac{1}{\omega C} = -\mathrm{j}\frac{1}{2 \times 20\,\mathrm{mF}} = -\mathrm{j}25\,\Omega$$

$$e = 40\sin(2t - 30°)\mathrm{V} \Rightarrow \dot{E} = 40\angle -30°\mathrm{V}$$

- $\dot{V}_L = \dot{E}\dfrac{Z_2//Z_3}{Z_1 + Z_2//Z_3}$ $\quad (Z_2//Z_3 = ?)$

$$Z_2//Z_3 = \frac{Z_2 Z_3}{Z_2 + Z_3} = \frac{\mathrm{j}20(-\mathrm{j}25)}{\mathrm{j}20 - \mathrm{j}25}\,\Omega = \frac{500}{-\mathrm{j}5}\,\Omega = \mathrm{j}100\,\Omega$$

$$\therefore \dot{V}_L = \dot{E} = \frac{Z_2//Z_3}{Z_1 + Z_2//Z_3} = (40\angle -30°)V\frac{j100\,\Omega}{(120 + j100)\Omega}$$
$$\approx \frac{4000\angle 60°}{156.2\angle 39.8°} V$$
$$\approx 25.61\angle 20.2° \text{ V}$$

After converting the phasor form to the time form gives

$$v_L = 25.61\sin(2t + 20.2°)V$$

9.3. Power in AC circuits

There are different types of power in AC circuits, such as instantaneous power, active power, reactive power and apparent power.

9.3.1 *Instantaneous power p*

The instantaneous power p is the power dissipated in a component of an AC circuit at any instant time. It is the product of instantaneous voltage v and current i at that particular moment (Figure 9.13), i.e. instantaneous power can be expressed as

$$p = vi$$

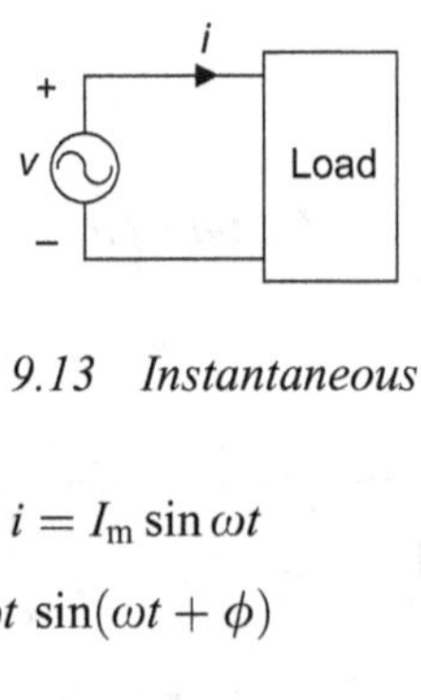

Figure 9.13 Instantaneous power

If $v = V_m\sin(\omega t + \phi)$ and $i = I_m\sin\omega t$

Then, $p = vi = V_m I_m \sin\omega t\,\sin(\omega t + \phi)$

$$\because\ -\sin x\sin y = \frac{1}{2}[\cos(x + y) - \cos(x - y)]$$

$$\therefore p = -\frac{1}{2}V_m I_m[\cos(2\omega t + \phi) - \cos\phi]$$
$$= VI\cos\phi - VI\cos(2\omega t + \phi) \quad (V_m = \sqrt{2}V, I_m = \sqrt{2}I)$$
$$= VI\cos\phi - VI(\cos 2\omega t\cos\phi - \sin 2\omega t\,\sin\phi)$$

$[\because \cos(x + y) - \cos x\cos y \quad \sin x\sin y]$

Therefore, instantaneous power is given as

$$p = VI\cos\phi(1 - \cos 2\omega t) + VI\sin\phi\sin 2\omega t$$

The waveform of the instantaneous power can be obtained from the product of instantaneous voltage and current at each point on their waveforms as shown in Figure 9.14.

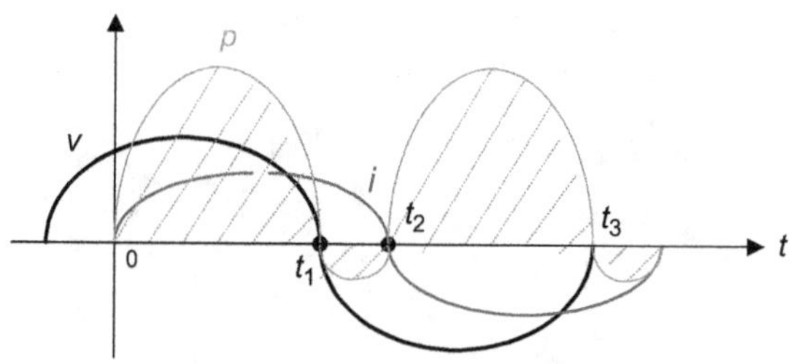

Figure 9.14 The waveform of instantaneous power

Such as:

- at time $t = 0 : i = 0, p = vi = 0$
- at time $t = t_1 : v = 0, p = vi = 0$
- between time $0 \sim t_1 : v > 0$ and $i > 0, \therefore p = vi > 0$
- between time $t_1 \sim t_2 : v < 0$ and $i > 0, \therefore p = vi < 0$
- between time $t_2 \sim t_3 : v < 0$ and $i < 0, \therefore p = vi > 0$

When instantaneous power p is >0 (p is positive), the component stores energy provided by the source. When instantaneous power p is <0 (p is negative), the component returns the stored energy to the source.

Instantaneous power *p*

p is the product of instantaneous voltage and current at any instant time:

$$p = vi = VI\cos\phi\,(1 - \cos 2\omega t) + VI(\sin\phi\sin 2\omega t)$$

$p > 0$: The component absorbs (stores) energy.
$p < 0$: The component returns (releases) energy.

- Instantaneous power for a resistive component p_R: Since voltage and current in a purely resistive circuit is in phase, i.e. $\phi = 0$, substituting this into the equation of the instantaneous power gives

$$\begin{aligned} p_R = vi &= VI\cos 0°(1 - \cos 2\omega t) + VI\sin 0° \sin 2\omega t \\ &= VI - VI\cos 2\omega t = VI(1 - \cos 2\omega t) \end{aligned} \tag{9.3}$$

The first part VI in (9.3) is average power dissipated in the resistive load ($p > 0$, the load absorbs power). The second part in (9.3) is a sinusoidal quantity with a double frequency 2ω, this indicates that when voltage and current waveforms oscillate one full cycle in one period of time, power waveform will oscillate two cycles as illustrated in Figure 9.15. The mathematical expression and the waveform all show that instantaneous power of a resistive load is always positive, or a resistor always dissipates power, indicating that the resistor is an energy consuming element.

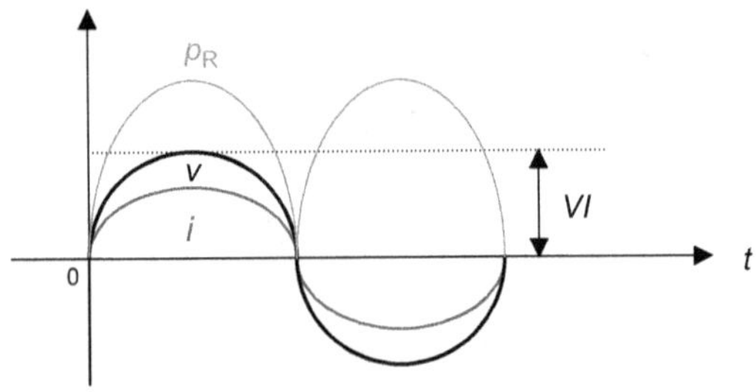

Figure 9.15 The waveform of instantaneous power for a R load

- Instantaneous power for capacitive and inductive components: In a purely inductive load circuit, voltage leads current by 90°. In a purely capacitive circuit, voltage lags current by 90°. Substituting $\phi = \pm 90°$ into the equation of instantaneous power gives

$$p = VI\cos(\pm 90°)(1 - \cos 2\omega t) + VI\sin(\pm 90°)\sin 2\omega t = \pm VI\sin 2\omega t \tag{9.3}$$

The instantaneous power for inductive and capacitive loads can be obtained from (9.3) as follows:

- Instantaneous power for an inductive load: $p_L = VI\sin 2\omega t$
- Instantaneous power for a capacitive load: $p_C = -VI\sin 2\omega t$

The diagrams of instantaneous power for inductive and capacitive loads are illustrated in Figure 9.16.

As seen from (9.3) and waveforms in Figure 9.16, both the instantaneous powers of inductive and capacitive loads are sinusoidal quantities with a double frequency 2ω. They have an average value of zero over a complete cycle since the positive and negative waveforms will cancel each other out. When instantaneous power is positive, the component stores energy; when instantaneous power is

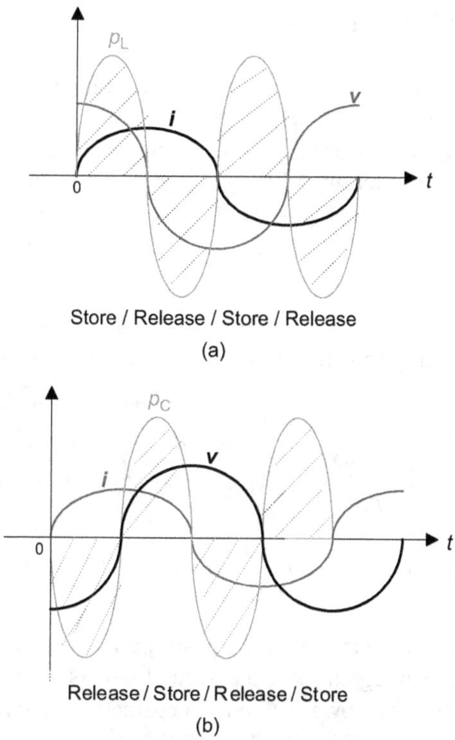

Figure 9.16 The waveforms of instantaneous power for L and C loads

negative, the component releases energy. Therefore, the inductor and capacitor do not absorb power, they convert or transfer energy between the source and elements. This also indicates that the inductor and capacitor are energy storage elements.

Instantaneous power for R, L and C components

- $p_R = VI - VI\cos 2\omega t$
- $p_L = VI\sin 2\omega t$
- $p_C = -VI\sin 2\omega t$

9.3.2 Active power P (or average power)

The active power is also known as average power, which is the product of the RMS voltage and RMS current in an AC circuit. It is actually the average power dissipation on the resistive load, i.e. the average power within one period of time (one full cycle) for a sinusoidal power waveform in an AC circuit.

The active power is also called true or real power since the power is *really* dissipated by the load resistor, and it can be converted to useful energy such as heat or light energy, etc. Electric stoves and lamps are examples of this kind of resistive load.

The instantaneous power always varies with time and is difficult to measure, so it is not very practical to use. Since it is the actual power dissipated in the load, average or active power P is used more often in AC sinusoidal circuits. Average power is easy to measure by an AC power metre (an instrument to measure AC power) in an AC circuit. Average power is the average value of instantaneous power in one period of time. It can be obtained from integrating for instantaneous power in one period of time.

Note: If you haven't learned calculus, then just keep in mind that $P = VI\cos\phi$ is the equation for average power, and skip the following mathematical derivation process.

$$P = \frac{1}{T}\int_0^T p(t)dt = \frac{1}{T}\int_0^T [VI\cos\phi(1-\cos 2\omega t) + VI\sin\phi\sin 2\omega t]d\omega t$$

$$P = \frac{1}{T}VI\cos\phi(\omega t)\Big|_0^T + VI\sin\phi\frac{1}{T}\int_0^T \sin 2\,\omega t\, d\omega t = VI\cos\phi \tag{9.4}$$

where ϕ is a constant, and ωt is a variable, so the first part of the integration is a constant $VI\cos\phi$. The integration of the second part is zero (integrating for sine function), since the average power value for a cosine function in one period of time is zero.

Therefore, active or average power P is a constant. It consists of the product of RMS values of voltage and current VI and $\cos\phi$ where $\cos\phi$ is called power factor and it will be discussed at the end of this section.

When active power $P > 0$, the element absorbs power; when active power $P < 0$, the element releases power.

- When $\phi = 0°$, the voltage and current are in phase, the circuit is a purely resistive circuit, and $P_R = VI\cos 0° = VI \quad (\because \cos 0° = 1)$
  Therefore,

$$P_R = VI = I^2R = \frac{V^2}{R}$$

or

$$P_R = VI = \frac{V_m}{\sqrt{2}}\frac{I_m}{\sqrt{2}} = \frac{1}{2}V_mI_m$$

- When $\phi = 90°$, the voltage leads the current by 90°, the circuit is a purely inductive circuit, and $P_L = VI\cos 90° = 0 \quad (\because \cos 90° = 0)$, i.e. $P_L = 0$

- When $\phi = -90°$, the current leads the voltage by 90°, the circuit is a purely capacitive circuit, and $P_C = VI\cos(-90°) = 0 \quad [\because \cos(-90°) = 0]$, i.e. $P_C = 0$.

Active power *P* (or average power, real power and true power)
The active power is the average value of the instantaneous power that is actually dissipated by the load.

$$P = VI\cos\phi$$

$$\text{When } \phi = 0° \qquad P_R = VI = \frac{1}{2}V_m I_m = I^2 R = \frac{V^2}{R}$$

$$\text{When } \phi = 90° \qquad P_L = 0$$

$$\text{When } \phi = -90° \quad P_C = 0$$

Quantity	**Quantity symbol**	**Unit**	**Unit symbol**
Active power	*P*	Watt	W

9.3.3 Reactive power Q

Since the effect of charging/discharging in a capacitor C and storing/releasing energy from an inductor L is that energy is only exchanged or transferred back and forth between the source and the component and will not do any real work for the load. So the average power dissipated on the load is zero. The reactive power Q can describe the maximum velocity of energy transferring between the source and the storage element L or C.

The first part in (9.4) is active or average power. The integration of the second part of (9.4) is zero, and that is the reactive power. While energy is converting between the source and energy store elements, the load will do not do any actual work, and average power dissipated on the load will be zero. Also, because the physical meaning of the reactive power is the *maximum* velocity of energy conversion between the energy storing element and the source, the peak value of the second part is reactive power, denoted as Q. It can be expressed mathematically as $Q = VI\sin\phi$, and measured in volt-amperes reactive (Var).

- When $\phi = 0°$, the circuit is a purely resistive circuit:

$$Q_R = VI\sin 0° = 0 \quad (\because \sin 0° = 0)$$

- When $\phi = 90°$, the circuit is a purely inductive circuit:

$$Q_L = VI \sin 90° = VI \quad (\because \sin 90° = 1)$$

Substituting $V = IX_L$ or $I = \dfrac{V}{X_L}$ into Q_L gives

$$Q_L = VI = I^2 X_L = \frac{V^2}{X_L}$$

- When $\phi = -90°$, the circuit is a purely capacitive circuit:

$$Q_C = VI \sin(-90°) = -VI \quad [\sin(-90°) = -1]$$

Substituting $V = IX_C$ or $I = \dfrac{V}{X_C}$ into Q_C gives

$$Q_C = -VI = -I^2 X_C = -\frac{V^2}{X_C}$$

Since Q_L is positive ($Q_L > 0$) and Q_C is negative ($Q_C < 0$), the inductor absorbs (consumes) reactive power, and the capacitor produces (releases) reactive power.

Reactive power *Q*

Q is the maximum velocity of energy conversion between the source and energy storing element.

$$Q = VI \sin \phi$$

When $\phi = 0°$ $\quad Q_R = 0$

When $\phi = 90°$ $\quad Q_L = VI = I^2 X_L = \dfrac{V^2}{X_L}$

When $\phi = -90°$ $\quad Q_C = -VI = -I^2 X_C = -\dfrac{V^2}{X_C}$

Quantity	Quantity symbol	Unit	Unit symbol
Reactive power	*Q*	Volt-amperes reactive	Var

9.3.4 *Apparent power S*

When the voltage *V* across a load produces a current *I* in the circuit of Figure 9.17, the power produced in the load is the product of voltage and current *VI*. If the load *Z* includes both the resistor and storage element inductor or capacitor, then *VI* will be neither a purely active power nor a purely reactive power. Since *VI* is the expression of the power equation, it is called apparent power. Apparent power is

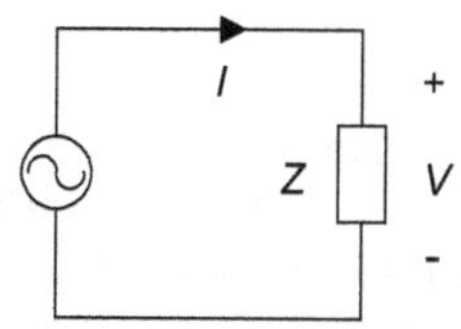

Figure 9.17 Apparent power

the maximum average power rating that a source can provide to the load or maximum capacity of an AC source and is denoted as S.

The mathematical expression of apparent power is the product of the source current and voltage, i.e. $S = IV$ and is measured in VA (volt-amperes).

Substituting $I = V/Z$ or $V = IZ$ into apparent power S equation gives:

$$S = I^2 Z = \frac{V^2}{Z}$$

Usually the power listed on the nameplates of electrical equipment is the apparent power.

Apparent power S

S is the maximum average power rating that a source can provide to an AC circuit.

$$S = IV = I^2 Z = V^2/Z$$

where S represents apparent power, measured in VA.

Different types of power in AC circuits are summarized in Table 9.2.

Table 9.2 Powers in AC circuits

Power	General expression	R	L	C
Instantaneous power	$p = VI\cos\phi(1 - \cos 2\omega t) + VI\sin\phi \sin 2\omega t$	$p_R = VI - VI\cos 2\omega t$	$p_L = VI\sin 2\omega t$	$p_L = -VI\sin 2\omega t$
Active power	$P = VI\cos\phi$	$P_R = VI = 1/2(V_m I_m) = I^2R = V^2/R$	$P_L = 0$	$P_C = 0$
Reactive power	$Q = VI\sin\phi$	$Q_R = 0$	$Q_L = VI = I_2 X_L = V^2/X_L$	$Q_C = -VI = -I_2 X_C = V^2/X_C$
Apparent power		$S = VI = I^2Z = V^2/Z$		

9.3.5 Power triangle

We have discussed three different powers in AC circuits, the active power, reactive power and apparent power. Now the question is what are the relationships between these three powers. These three powers are actually related to one another in a right triangle, is called the power triangle, and can be derived as follows.

For a series resistor, inductor and capacitor circuit, if the circuit is more inductive ($X = X_L - X_C > 0$, $\phi > 0$), then the impedance triangle, voltage triangle and current triangle can be illustrated as shown in Figure 9.18(a–c) (refer to section 9.1).

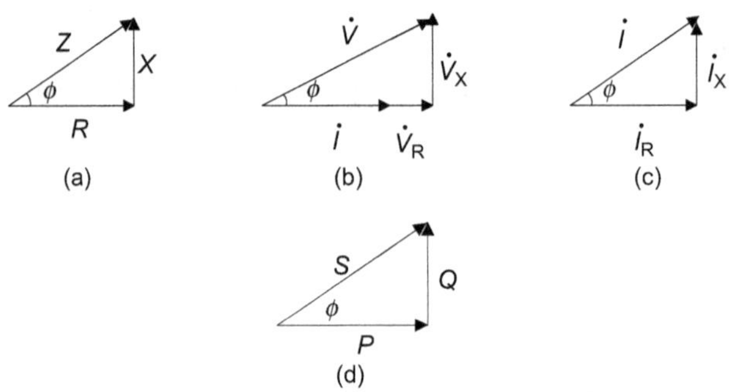

Figure 9.18 Circuit triangles for a more inductive circuit

If we multiply all quantities on each side of the voltage triangle by the current I, it will yield $VI = S$, $IV_X = Q$ and $V_R I = P$ and this can be illustrated as a power triangle as shown in Figure 9.18(d).

If the circuit load is more capacitive ($X_C > X_L$, $\phi < 0$), the circuit triangles will be opposite to the inductive circuit triangles as shown in Figure 9.19.

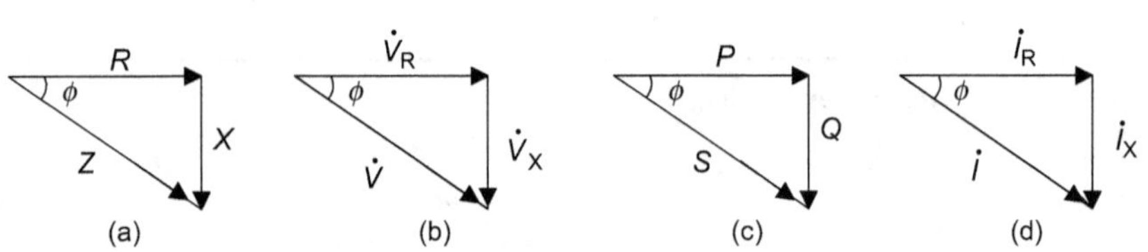

Figure 9.19 Circuit triangles for a more capacitive circuit

The impedance triangle indicates that it has an angle ϕ between resistance R and impedance Z of the circuit. It is called the impedance angle; ϕ is also in the power triangle. Later on, we'll introduce the power factor $\cos\phi$, and ϕ is also called the power factor angle.

The relationship between different powers in the power triangle can be obtained from the Pythagoras' theorem, i.e. $S = \sqrt{P^2 + Q^2}$

If expressed by complex numbers, it will give: $\dot{S} = P + \mathrm{j}Q$

This is known as the phasor power. The phasor apparent power can also be expressed as

$$\dot{S} = \dot{V}\dot{I} = \dot{I}^2 Z = \frac{\dot{V}^2}{Z}$$

The impedance angle ϕ can be obtained from circuit triangles (either inductive or capacitive circuit) and can be expressed as

$$\phi = \tan^{-1}\frac{Q}{P} = \tan^{-1}\frac{X}{R} = \tan^{-1}\frac{\dot{V}_{\mathrm{X}}}{\dot{V}_{\mathrm{R}}} = \tan^{-1}\frac{\dot{I}_{\mathrm{X}}}{\dot{I}_{\mathrm{R}}}$$

Active power P and reactive power Q can be expressed with the impedance angle ϕ and obtained from the power triangle in Figures 9.18 or 9.19 as:

$$P = S\cos\phi \quad \text{and} \quad Q = S\sin\phi$$

Power triangle

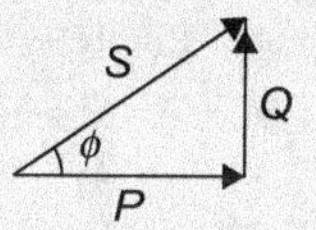

$$P = S\cos\phi, \quad Q = S\sin\phi, \quad S = \sqrt{P^2 + Q^2},$$

Impedance angle:

$$\phi = \tan^{-1}\frac{Q}{P} = \tan^{-1}\frac{X}{R} = \tan^{-1}(\dot{V}_{\mathrm{X}}/\dot{V}_{\mathrm{R}}) = \tan^{-1}(\dot{I}_{\mathrm{X}}/\dot{I}_{\mathrm{R}})$$

Phasor power:

$$\dot{S} = \dot{V}\dot{I} = \dot{I}^2 Z = \dot{V}^2/Z, \quad \dot{S} = P + jQ$$

9.3.6 Power factor (PF)

The ratio of active power P and apparent power S is called the power factor PF, and represented by $\cos\phi$. It also can be obtained from the power triangle as

$$\text{PF} = \frac{P}{S} \cos\phi$$

For a purely resistive circuit ($\phi = 0°$), the reactive power Q is zero, so the apparent power S is equal to the active power P, i.e.

$$S = \sqrt{P^2 + Q^2} = \sqrt{P^2 + 0^2} = P$$

and the power factor is 1, i.e. $\cos\phi = P/S = P/P = 1$

This is the maximum value for the power factor $\cos\phi$.

For a purely reactive load ($\phi = \pm 90°$), active power P in the circuit is zero, so the power factor is also zero, i.e. $\cos\phi = P/S = 0/S = 0$.

Therefore, the range of the power factor $\cos\phi$ is between 0 and 1, and the impedance angle ϕ is between $0°$ and $\pm 90°$.

The power factor is an important factor in circuit analysis. The circuit source will produce active power P to the load, and the amount of the active power P can be determined by the power factor $\cos\phi$. This is indicated in the equation of $P = S\cos\phi$.

If the power factor $\cos\phi$ of the load is the maximum value of 1, the active power produced by the source is the maximum capacity of the source, and all the energy supplied by the source will be consumed by the load ($P = S$, $\because \cos\phi = 1$).

If the power factor $\cos\phi$ decreases, the active power P produced by the source will also decrease accordingly ($P \downarrow = S\cos\phi \downarrow$).

So increasing the power factor can increase the real power in a circuit. But how to increase the power factor of a circuit? A method called power-factor correction can be used. This method can increase the power factor and does not affect the load voltage and current. Since most of the loads of the electrical systems are inductive loads (such as the loads that are driven by a motor), an inductive load in parallel with a capacitor (Figure 9.20(b)) can increase the power factor of the load.

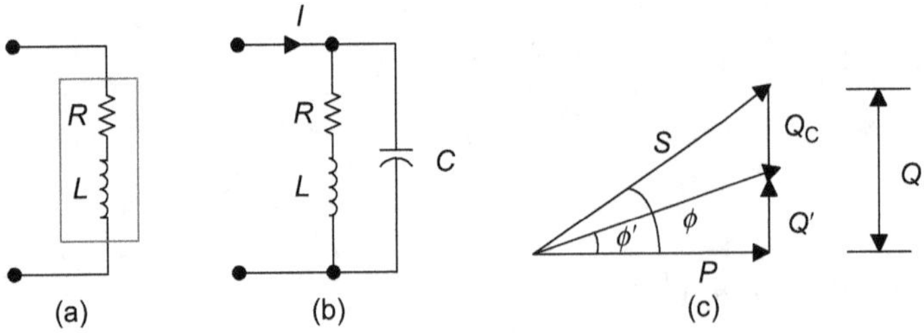

Figure 9.20 Increasing the power factor

The power triangle in Figure 9.20(c) indicates that when a capacitor C is in parallel with the inductive load, the reactive power Q in the circuit will be reduced to Q' ($Q' = Q - Q_C$). Therefore, the impedance angle will reduce from ϕ to ϕ', and the power factor $\cos\phi$ will increase to $\cos\phi'$.

Since $\phi\downarrow \rightarrow \cos\phi\uparrow$, for instance $\cos 30° = 0.866$ is $> \cos 60° = 0.5$, the total current I will also decrease, since $I\downarrow = P/(V\cos\phi\uparrow)$ $(P = S\cos\phi = VI\cos\phi)$. This can reduce the source current and line power loss (I^2R). This is why increasing the power factor has a significant meaning.

Power factor (cos ϕ)

- $\cos\phi = P/S$ $(0 \le \cos\phi \le 1, \cos\phi$ – dimensionless).
- When $\cos\phi = 1$: All energy supplied by the source is consumed by the load.
- Power-factor correction: An inductive load in parallel with a capacitor can increase $\cos\phi$.

9.3.7 *Total power*

When calculating the total power in a complicated series–parallel circuit, determine the active power P and reactive power Q in each branch first, and the sum of all the active powers is the total active power P_T. The difference between Q_{LT} and Q_{CT} is the total reactive power Q_T. Q_{LT} is the sum of all reactive powers for the inductors and Q_{CT} is the sum of all reactive powers for the capacitors. The total apparent power S can be determined by using Q_T and P_T using the Pythagoras' theorem, i.e.

$$S_T = \sqrt{P_T^2 + Q_T^2}$$

The total power factor can be determined by using the total active and reactive power, i.e.

$$PF_T = \cos\phi_T = \frac{P_T}{S_T}$$

Total power

- Total active power: $P_T = P_1 + P_2 + \cdots + P_n$
- Total reactive power:
  $Q_T = Q_{LT} - Q_{CT} = (Q_{L_1} + Q_{L_2} + \cdots) - (Q_{C_1} + Q_{C_2} + \cdots)$
  where Q_{LT} is the total reactive power for inductors and Q_{CT} the total reactive power for capacitors.

- Total apparent power: $S_T = \sqrt{P_T^2 + Q_T^2}$
- Total power factor: $PF_T = \cos\phi_T = P_T/S_T$

Example 9.5: Determine the total power factor $\cos\phi$ in the circuit of Figure 9.21 and plot the power triangle for this circuit.

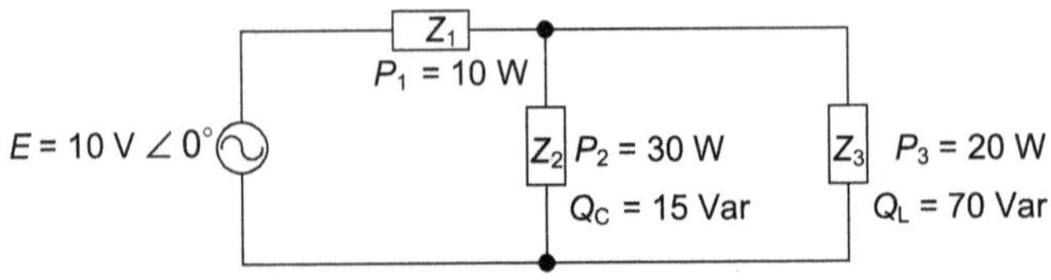

Figure 9.21 Circuit for Example 9.5

Solution:

- Total power factor $PF_T = \cos\phi_T = \dfrac{P_T?}{S_T?}$, $S_T = \sqrt{P_T?^2 + Q_T?^2}$ (the symbol '?' indicates an unknown).

  Total active power: $P_T = P_1 + P_2 + P_3 = (10 + 30 + 20)\text{W} = 60\text{ W}$

  Total reactive power: $Q_T = Q_{LT} - Q_{CT} = (70 - 15)\text{Var} = 55\text{ Var}$

  Total apparent power: $S_T = \sqrt{P_T^2 + Q_T^2} = \sqrt{60^2 + 55^2} = 81.39\text{ VA}$

  Total power factor: $PF_T = \cos\phi_T = \dfrac{P_T}{S_T} = \dfrac{60\text{ W}}{81.39\text{ VA}} \approx 0.74$
- Impedance angle: $\phi = \cos^{-1}\phi_T = \cos^{-1} 0.74 \approx 42.3°$
- The power triangle is shown in Figure 9.22.

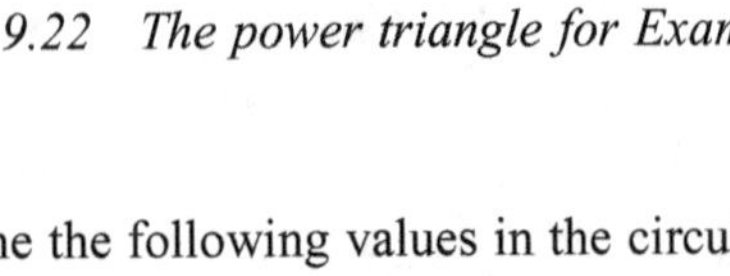

Figure 9.22 The power triangle for Example 9.5

Example 9.6: Determine the following values in the circuit shown in Figure 9.23:

- the total power P_T, Q_T and S_T for the circuit
- power factor $\cos\phi$
- power triangle
- source current I
- the capacitance C needed to increase the power factor $\cos\phi$ to 0.87
- the source current I' after increasing the power factor

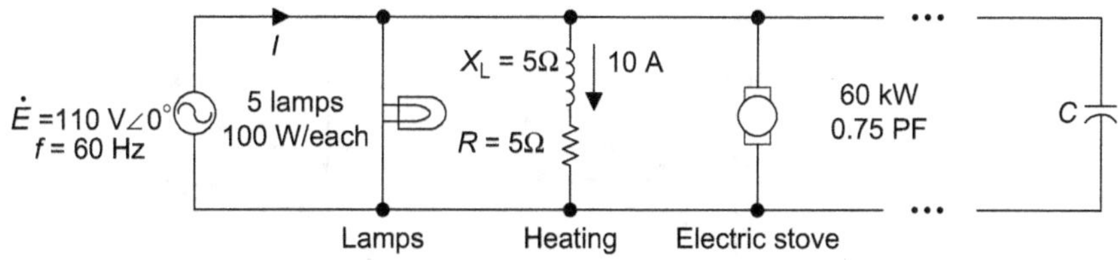

Figure 9.23 Circuit for Example 9.6

Solution:

(a) Lamp: $P_1 = 5 \times 100\,\text{W} = 500\,\text{W}$

Heating: $P_2 = I^2R = (10\,\text{A})^2(5\,\Omega) = 500\,\text{W}$

$Q_2 = I^2X_L = (10\,\text{A})^2(5\,\Omega) = 500\,\text{Var}$

Electric stove: $P_3 = 6\,\text{kW} = 6\,000\,\text{W},\ \phi = \cos^{-1}0.75 \approx 41.4°$

$Q_3 = P_3 \tan\phi = (6\,000\,\text{W})(\tan 41.4°) \approx 5\,290\,\text{Var}, \quad (\phi = \tan^{-1}(Q/P))$

Total power: $P_T = P_1 + P_2 + P_3 = (500 + 500 + 6\,000)\text{W} = 7\,000\,\text{W}$

$Q_T = Q_1 + Q_2 + Q_3 = 0 + 500\,\text{Var} + 5\,290\,\text{Var} = 5\,790\,\text{Var}$

$S_T = \sqrt{P_T + Q_T} = \sqrt{7\,000^2 + 5\,790^2} \approx 9\,084.3\,\text{VA}$

(b) Power factor $\text{PF}_T = \cos\phi_T = \dfrac{P_T}{S_T} = \dfrac{7\,000}{9\,084.3} \approx 0.77$

(c) Power triangle (as shown in Figure 9.24): $\phi = \cos^{-1} 0.77 \approx 39.7°$

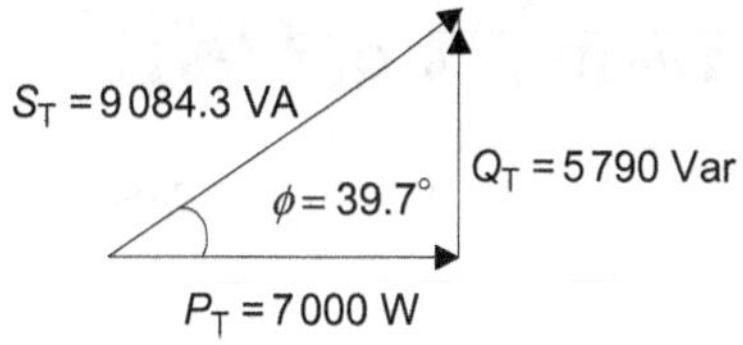

Figure 9.24 Power triangle for Example 9.6

(d) Source current I: $\because S_T = EI$

$$\therefore I = \frac{S_T}{E} = \frac{9\,084.3\,\text{VA}}{110\,\text{V}} \approx 82.6\,\text{A}$$

Therefore, $\dot{I} = 82.6\angle - 39.7°\text{A}$ (Voltage leads current or current lags voltage in the inductive load, so $\phi = -39.7°$.)

(e) The capacitance C that needs to increase the power factor to 0.87 can be determined by the following way:

$$C = \frac{1}{2\pi f X_C?} \Rightarrow Q_C = -\frac{V^2}{X_C} \Rightarrow X_C = -\frac{V^2}{Q_C?} \Rightarrow \left(X_C = \frac{1}{2\pi fC}\right)$$

$$Q_T = Q_C + Q_T'? \Rightarrow Q_T' = P_T \tan\phi'? \Rightarrow \phi' = \cos^{-1} 0.87$$

(as shown in Figure 9.25)

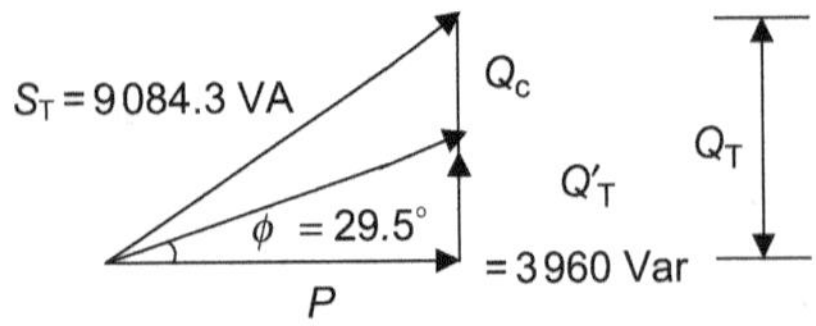

Figure 9.25 New power factor angle

Therefore, to increase the power factor to 0.87, the power factor angle should be reduced to $\phi' = \cos^{-1} 0.87 \approx 29.5°$.
The reactive power can be determined from the above expression as

$$Q'_T = P_T \tan \phi' = 7\,000 \tan 29.5° \approx 3\,960 \text{ Var}$$

The new power factor angle ϕ' is shown in Figure 9.25. Q_C can be obtained from Figure 9.25:

$$Q_C = Q_T - Q'_T = 5\,790 - 3\,960 = 1\,830 \text{ Var}$$

$$X_C = \frac{-V^2}{Q_C} = -\frac{E^2}{Q_C} = \frac{-110^2 \text{ V}}{1\,830 \text{ Var}} \approx 6.61\,\Omega$$

(The voltage across X_L and R is equal to E.)

Therefore

$$C = \frac{1}{2\pi f X_C} = \frac{1}{2\pi(60\text{ Hz})(6.61\,\Omega)} \approx 0.0004 \text{ F} = 400\,\mu\text{F}$$

That is the capacitance C needed to increase the power factor to 0.87 should be 400 μF.

(f) The source current I' after increasing the power factor can be determined by the following expression:

$$P = S\cos\phi = IE\cos\phi$$

So,

$$I' = \frac{P_T}{E\cos\phi'} = \frac{7000\text{ W}}{110\text{ V}\cos 29.5°} \approx 73.1\text{ A}$$

Comparing with the original source current I = 82.6 A from step (d), after a capacitor is in parallel and the power factor is increased, the source current is I' = 73.1 A. So the source current can decrease 9.5 A ($I - I'$ = 82.6 A − 73.1 A = 9.5 A). This can reduce the line power loss (I^2R) and utilize the capacity of the source more efficiently.

9.4 Methods of analysing AC circuits

All analysis methods that we have learned for analysing DC circuits with one or two more sources can also be used for analysing AC circuits, such as the branch current analysis, mesh analysis, node voltage analysis, superposition theorem, Thevenin's and Norton's theorems, etc. But the phasor form will be used to represent the circuit quantities. Since these analysis methods have been discussed in detail for DC circuits (chapters 4 and 5), some examples will be presented to use these methods in AC circuits or networks. Reviewing chapters 4 and 5 before reading the following contents is highly recommended.

9.4.1 Mesh current analysis

The procedure for applying the mesh current analysis method in an AC circuit:

1. Identify each mesh and label the reference directions for each mesh current clockwise.
2. Apply KVL around each mesh of the circuit, and the numbers of KVL equations should be equal to the numbers of mesh (windowpanes). Sign each self-impedance voltage as positive and each mutual-impedance voltage as negative in KVL equations.
   - Self-impedance: An impedance that only has one mesh current flowing through it.
   - Mutual-impedance: An impedance that is located on the boundary of two meshes and has two mesh currents flowing through it.
3. Solve the simultaneous equations resulting from step 2, and determine each mesh current.

Note:

- Convert the current source to the voltage source first in the circuit, if there is any.
- If the circuit has a current source, the source current will be the same with the mesh current, so the number of KVL equations can be reduced.

The procedure for applying the mesh current analysis method in an AC circuit is demonstrated in the following example.

Example 9.7: Use the mesh current analysis method to determine the mesh current I_1 in the circuit of Figure 9.26.

Solution: Convert the current source to the voltage source (connect R and X to Z) as shown in Figure 9.26(b). There, $\dot{E}_2 = \dot{I}R_3 = (2.5\,\text{A}\angle 0°)(4\,\Omega) = 10\,\text{V}\angle 0°$.

1. Label all the reference directions for each mesh current $\dot{I}_1$ and $\dot{I}_2$ (clockwise), as shown in Figure 9.26(b).
2. Write KVL around each mesh (windowpane), and the number of KVL is equal to the number of meshes (there are two meshes in Figure 9.26(b)).

Sign each self-impedance voltage as positive, and each mutual-impedance voltage as negative in KVL ($\Sigma\dot{V} = \Sigma\dot{E}$).

Mesh 1: $(Z_1 + Z_2)\dot{I}_1 - Z_2\dot{I}_2 = -\dot{E}_1$

Mesh 2: $-Z_2\dot{I}_1 + (Z_2 + Z_3)\dot{I}_2 = -\dot{E}_2$

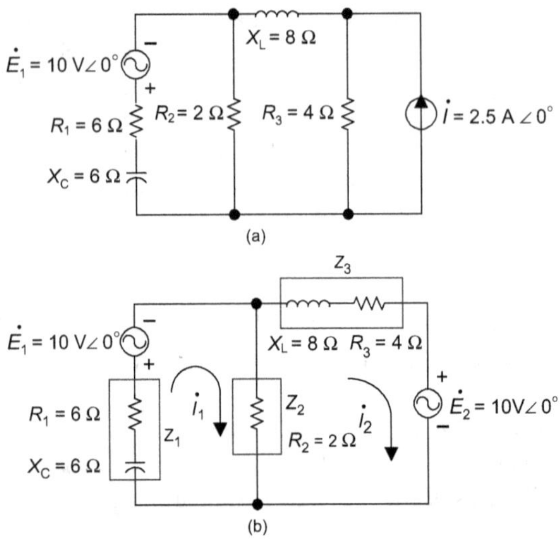

Figure 9.26 Circuits for Example 9.7

Substitute the following values of Z_1, Z_2 and Z_3 into the above equations:

$Z_1 = (6 - j6)\Omega$, $Z_2 = 2\,\Omega$ and $Z_3 = (4 + j8)\Omega$

so, $(8 - j6)\dot{I}_1 - 2\dot{I}_2 = -10\,V$

$$-2\dot{I}_1 + (6 + j8)\dot{I}_2 = -10\,V$$

3. Solve the simultaneous equations resulting from step 2 using the determinant method, and determine the mesh current $\dot{I}_1$:

$$\dot{I}_1 = \frac{\begin{vmatrix} -10\angle 0° & -2 \\ -10\angle 0° & 6+j8 \end{vmatrix}}{\begin{vmatrix} 8-j6 & -2 \\ -2 & 6+j8 \end{vmatrix}} \approx 1.18\angle -151.9°\,A$$

9.4.2 *Node voltage analysis*

The following is the procedure for applying the node analysis method in an AC circuit.

1. Label the circuit:
   - Label all the nodes and choose one of them to be the reference node.
   - Assign an arbitrary reference direction for each branch current (this step can be skipped if using the inspection method).

2. Apply KCL to all $n - 1$ nodes except for the reference node (n is the number of nodes).
   - Method 1: Write KCL equations and apply Ohm's law to the equations (assign a positive sign (+) to the self-impedance voltage and entering node current, and negative sign (−) for the mutual-impedance voltage and exiting node current).
   - Method 2: Convert voltage sources to current sources and write KCL equations using the inspection method.
3. Solve the simultaneous equations and determine each nodal voltage.

The procedure for applying the node voltage analysis method in an AC circuit is demonstrated in the following example.

Example 9.8: Write node equations for the circuit in Figure 9.27(a).

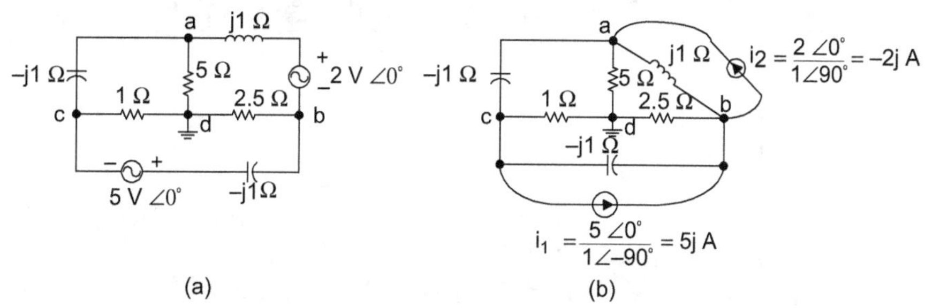

Figure 9.27 Circuits for Example 9.8

1. Label nodes a, b, c and d, and choose ground d to be the reference node as shown in Figure 9.27(a).
2. Convert two voltage sources to current sources from Figure 9.27(a) to Figure 9.27(b), and write KCL equations to $n - 1 = 4 - 1 = 3$ nodes by inspection (method 2). KCL equations are shown in Table 9.3.
3. Three equations can solve three unknowns that are node voltages.

9.4.3 Superposition theorem

The following is the procedure for applying the superposition theorem in an AC circuit:

1. Turn off all power sources except one, i.e. replace the voltage source with the short circuit (placing a jump wire), and replace the current source with an open circuit. Redraw the original circuit with a single source.
2. Analyse and calculate this circuit by using the single source method, and repeat steps 1 and 2 for the other power sources in the circuit.
3. Determine the total contribution by calculating the algebraic sum of all contributions due to single sources.

Table 9.3 KVL Equations for example 9.8

	$\dot{V}_a$		$\dot{V}_b$		$\dot{V}_c$		$\dot{I}$
Node a	$\left(\frac{1}{5}+\frac{1}{-j1}+\frac{1}{j1}\right)\dot{V}_a$	$-$	$(1/j1)\dot{V}_b$	$-$	$(1/-j1)\dot{V}_c$	$=$	$-2j$
Node b	$(1/-j1)\dot{V}_a$	$+$	$\left(\frac{1}{2.5}+\frac{1}{j1}+\frac{1}{-j1}\right)\dot{V}_b$	$-$	$(1/-j1)\dot{V}_c$	$=$	$5j-(-2j)$
Node c	$(-1/-j1)\dot{V}_a$	$-$	$(1/-j1)\dot{V}_b$	$+$	$\left(\frac{1}{1}+\frac{1}{-j1}+\frac{1}{-j1}\right)\dot{V}_c$	$=$	$-5j$
After simplifying							
	$0.2\,\dot{V}_a$	$+$	$j\dot{V}_b$	$-$	$j\dot{V}_c$	$=$	$-j2$
	$j\dot{V}_a$	$+$	$0.4\dot{V}_b$	$-$	$j\dot{V}_c$	$=$	$j7$
	$-j\dot{V}_a$	$-$	$j\dot{V}_b$	$+$	$(1+2j)\dot{V}_c$	$=$	$-j5$

(The result should be positive when the reference polarity of the unknown in the single source circuit is the same with the original circuit, otherwise it should be negative.)

The procedure for applying the superposition theorem in an AC circuit is demonstrated in the following example.

Example 9.9: Determine $\dot{V}_c$ in circuit as shown in Figure 9.28(a) by using the superposition theorem.

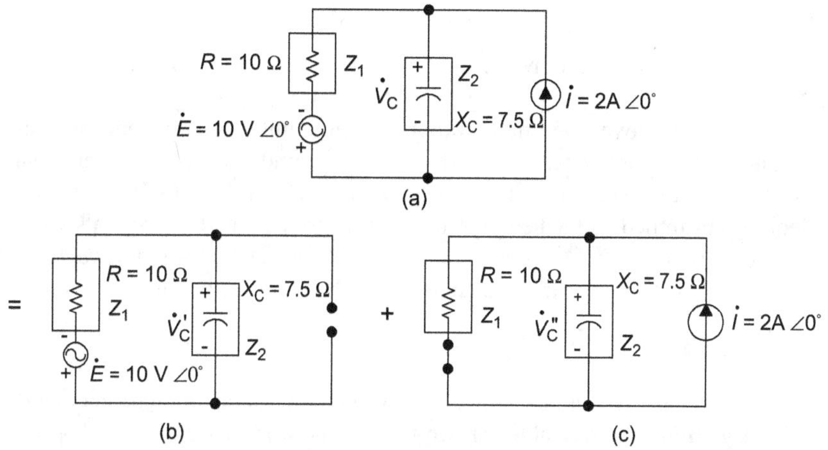

Figure 9.28 Circuits for Example 9.9

Solution:

1. Choose $\dot{E}$ to apply to the circuit first, and use an open circuit to replace the current source $\dot{I}$ as shown in Figure 9.28(b), and calculate $\dot{V}'_c$:

$$\dot{V}'_{\mathrm{C}} = \dot{E}\frac{Z_2}{Z_1 + Z_2} = -10\mathrm{V}\angle 0^\circ \frac{7.5\,\Omega\angle -90^\circ}{10\,\Omega - \mathrm{j}7.5\,\Omega}$$
$$= \frac{-75\angle -90^\circ}{12.5\angle -36.87^\circ}\mathrm{V} = -6\angle -53.13^\circ\,\mathrm{V}$$

2. When the current source $\dot{I}$ is applied to the circuit only and the voltage source $\dot{E}$ is replaced by a jump wire, the circuit is as shown in Figure 9.28(c). Calculate $\dot{V}''_c$ in Figure 9.28(c):

$$\dot{V}''_{\mathrm{C}} = \dot{I}(Z_1//Z_2)$$

$$Z_1//Z_2 = \frac{Z_1 Z_2}{Z_1 + Z_2} = \frac{10(-\mathrm{j}7.5)}{10 - \mathrm{j}7.5}\,\Omega$$
$$\approx \frac{75\angle -90^\circ}{12.5\angle -36.87^\circ}\,\Omega = 6\angle -53.13^\circ\,\Omega$$

$$\dot{V}''_{\mathrm{C}} = \dot{I}(Z_1//Z_2) = (2\angle 0^\circ\,\mathrm{A})(6\angle -53.13^\circ\,\Omega) = 12\angle -53.13^\circ\,\mathrm{V}$$

3. Calculate the sum of voltages $\dot{V}'_C$ and $\dot{V}''_C$:

$$\begin{aligned}\dot{V}_C &= \dot{V}'_C + \dot{V}''_C = -6\angle-53.13^\circ\text{ V} + 12\angle-53.13^\circ\text{ V}\\ &= [-6\cos(-53.13^\circ) - 6\text{j}\sin(-53.13^\circ)\\ &\quad + 12\cos(-53.13^\circ) + 12\text{j}\sin(-53.13^\circ)]\text{V}\\ &\approx [-3.6 + \text{j}4.8 + 7.2 - \text{j}9.6]\text{V} = (3.6 - \text{j}4.8)\text{V} = 6\angle-53.13^\circ\text{V}\end{aligned}$$

9.4.4 Thevenin's and Norton's theorems

The following is the procedure for applying Thevenin's and Norton's theorems in an AC circuit.

1. Open and remove the load branch (or any unknown current or voltage branch) in the network, and mark the letter a and b on the two terminals.
2. Determine the equivalent impedance Z_{TH} or Z_N: It should be equal to the equivalent impedance when you look at it from the a and b terminals when all sources are turned off or equal to zero. (A voltage source should be replaced by a short circuit, and a current source should be replaced by an open circuit.)

   i.e. $Z_{TH} = Z_N = Z_{ab}$

3.
   - Determine Thevenin's equivalent voltage V_{TH}: It equals the open circuit voltage from the original linear two-terminal network of a and b, i.e. $V_{TH} = V_{ab}$.
   - Determine Norton's equivalent current I_N: It equals the short circuit current for the original linear two-terminal network of a and b, i.e. $I_N = I_{sc}$.

4. Plot Thevenin's or Norton's equivalent circuits, and connect the load (or unknown current or voltage branch) to a and b terminals of the equivalent circuit. Then the load voltage or current can be calculated.

The procedure for applying Thevenin's and Norton's theorems method in an AC circuit is demonstrated in the following example.

Example 9.10: Determine the current $\dot{I}_L$ in the load branch of Figure 9.29(a) by using Thevenin's theorem, and use Norton's theorem to check the answer.

Solution:

1. Open the load branch and remove Z_L, and label a and b on the terminals of the load branch as shown in Figure 9.29(b).
2. Determine Thevenin's equivalent impedance Z_{TH} (the voltage source $\dot{E}$ is replaced by a short circuit) in Figure 9.29(b):

$$Z_{TH} = Z_{ab} = Z_3 + Z_4 + Z_1//Z_2$$

$$\begin{aligned}Z_{TH} &= \left[1 - \text{j}1 + \frac{-\text{j}2.5(2.5 + \text{j}2.5)}{\text{j}2.5 + (2.5 + \text{j}2.5)}\right]\Omega = \left[1 - \text{j}1 + \frac{6.25 - \text{j}6.25}{2.5}\right]\Omega\\ &= (1 - \text{j}1 + 2.5 - \text{j}2.5)\Omega = (3.5 - \text{j}3.5)\Omega \approx 4.95\angle-45^\circ\Omega\end{aligned}$$

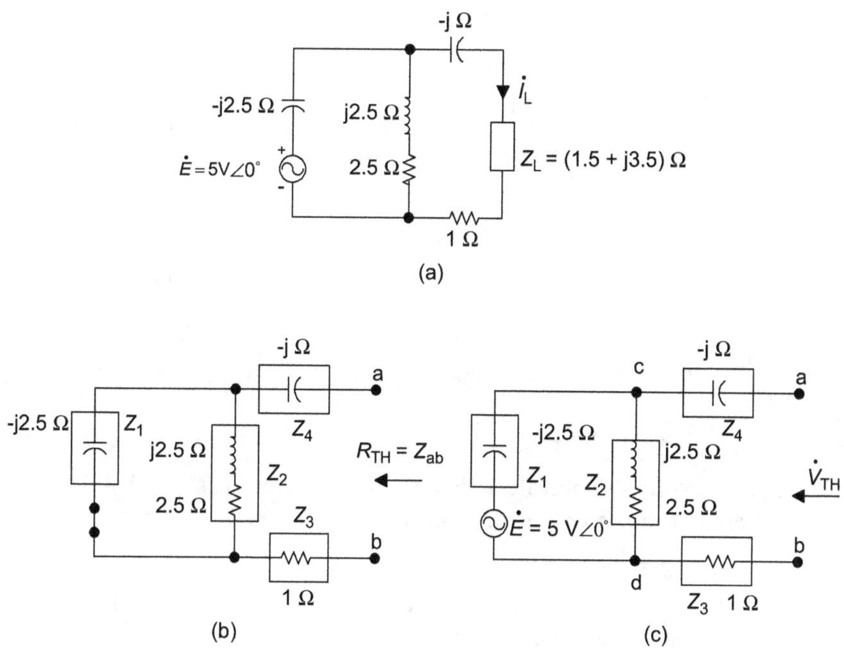

Figure 9.29 Circuits for Example 9.10(a)–(c)

3. Determine Thevenin's equivalent voltage $\dot{V}_{TH}$ by using Figure 9.29(c) to calculate the open circuit voltage across terminals a and b:

$$\dot{V}_{TH} = \dot{V}_{ab} = \dot{V}_{cd}$$

Since $\dot{I} = 0$ for Z_3 and Z_4 in Figure 9.29(c), voltages across Z_3 and Z_4 are also zero

$$\therefore\ \dot{V}_{TH} = \dot{V}_{cd} = \dot{E}\frac{Z_2}{Z_1 + Z_2} = 5\angle 0^\circ\ \text{V}\frac{2.5 + \text{j}2.5}{-\text{j}2.5 + (2.5 + \text{j}2.5)}\ \Omega$$
$$\approx 5\angle 0^\circ(1.414\angle 45^\circ) \approx 7.07\angle 45^\circ\ \text{V}$$

4. Plot Thevenin's equivalent circuit as shown in Figure 9.29(d). Connect the load Z_L to a and b terminals of the equivalent circuit and calculate the load current $\dot{I}_L$.

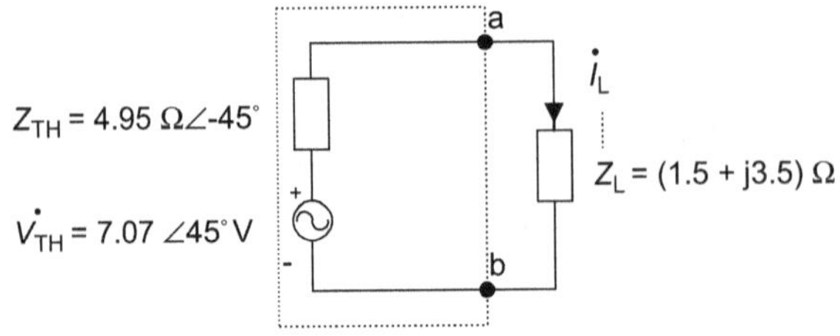

Figure 9.29(d) Thevenin's equivalent circuit for Example 9.10

$$\dot{I}_L = \frac{\dot{V}_{TH}}{Z_{TH} + Z_L} = \frac{7.07\angle 45^\circ \text{ V}}{(3.5 - j3.5)\Omega + (1.5 + j3.5)\Omega} = \frac{7.07\angle 45^\circ \text{ V}}{5\,\Omega}$$

$$\approx 1.4\angle 45^\circ \text{ A}$$

5. Determine Norton's equivalent circuit in Figure 9.29(a) as seen by Z_L.
   - Norton's equivalent impedance Z_N:

     $Z_N = Z_{TH} = 3.5 - j3.5 = 4.95\ \angle -45^\circ$
   - Norton's equivalent current $\dot{I}_N$: It is equal to the short circuit current for the original two-terminal circuit of a and b (as shown in Figure 9.29(e, I)).

     $\dot{I}_N = \dot{I}_{SC} = \dot{I}\dfrac{Z_2}{Z_2 + (Z_3 + Z_4)}$ (the current-divider rule).

     There,

$$\dot{I} = \frac{\dot{E}}{Z_1 + Z_2//(Z_3 + Z_4)} = \frac{5\angle 0^\circ \text{ V}}{\left[-j2.5 + \dfrac{(2.5 + j2.5)(1 - j1)}{(2.5 + j2.5) + (1 - j1)}\right]\Omega}$$

$$= \frac{5\angle 0^\circ \text{V}}{(-j2.5 + (j5)/(3.5 + j1.5))\Omega} \approx 1.54\angle 68.2^\circ \text{A}$$

Therefore,

$$\dot{I}_N = \dot{I}\frac{Z_2}{Z_2 + (Z_3 + Z_4)}$$

$$= 1.54\angle 68.2^\circ \text{A}\frac{(2.5 + j2.5)\Omega}{[2.5 + j2.5 + (1 - j1)]\Omega} \approx 1.43\angle 90^\circ \text{A}$$

6. Use Norton's theorem to check the load current $\dot{I}_L$: Determine the load current $\dot{I}_L$ on the terminals of a and b in Figure 9.29(e, II) by using Norton's equivalent circuit.

$$\dot{I}_L = \dot{I}_N\frac{Z_N}{Z_N + Z_L} = 1.43\angle 90^\circ \text{ A}\frac{4.95\angle -45^\circ\ \Omega}{[(3.5 - j3.5) + (1.5 + j3.5)]\Omega}$$

$$= 1.43\angle 90^\circ \text{ A}\frac{4.95\angle -45^\circ}{5} \approx 1.4\angle 45^\circ \text{ A}$$

Therefore, $\dot{I}_L$ is the same by Norton's theorem as the method by using Thevenin's theorem (checked).

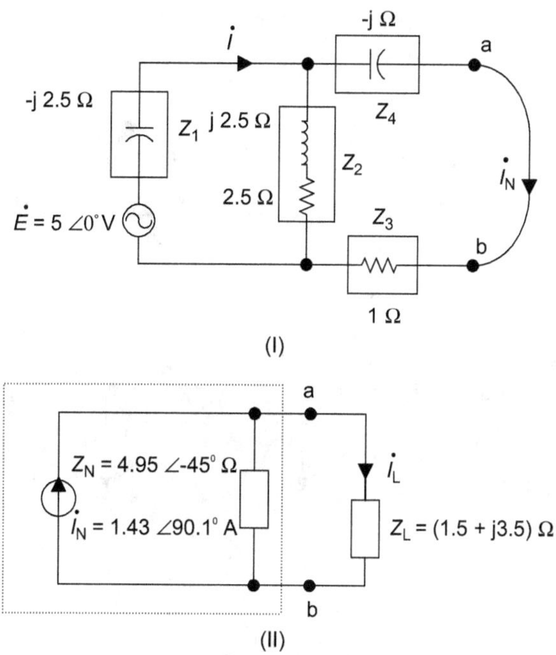

Figure 9.29(e) Norton's equivalent circuit for Example 9.10

Summary

Impedance and admittance

Component	Impedance $Z = \dot{V}/\dot{I}$	Admittance $Y = 1/Z$	Conductance and susceptance
R	$Z_R = R$	$Y_R = G$	Conductance: $G = 1/R$
L	$Z_L = jX_L$	$Y_L = -jB_L$	Inductive susceptance: $B_L = 1/X_L$
C	$Z_C = -jX_C$	$Y_C = jB_C$	Capacitive susceptance: $B_C = 1/X_C$
	$Z = z \angle\phi = R + jX$ $z = \sqrt{R^2 + X^2}$ $\phi = \tan^{-1}\dfrac{X}{R}$	$Y = y \angle \phi_y = G + jB$ $y = \sqrt{G^2 + B^2}$ $\phi_y = \tan^{-1}\dfrac{B}{G}$	Reactance: $X = X_L - X_C$ and susceptance: $B = B_C - B_L$

$$\left(X_L = \omega L, \quad X_C = \frac{1}{\omega C}, \quad j = \frac{1}{-j}\right)$$

- Impedance, voltage, current and power triangles
  - For a more inductive circuit:

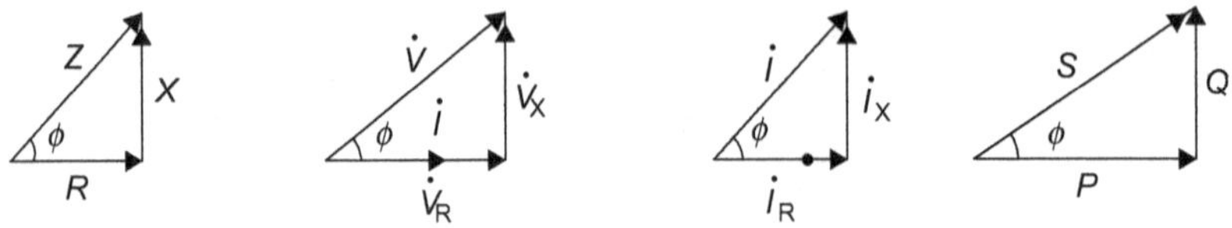

  - For a more capacitive circuit:

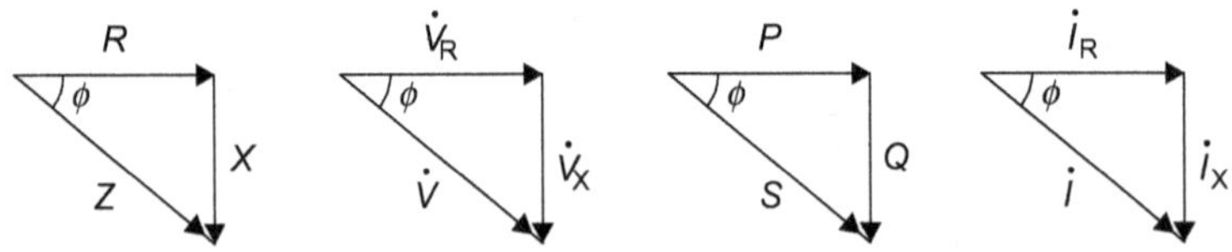

Impedance angle: $\phi = \tan^{-1}\dfrac{X}{R} = \tan^{-1}\dfrac{\dot{V}_X}{\dot{V}_R} = \tan^{-1}\dfrac{\dot{I}_X}{\dot{I}_R} = \tan^{-1}\dfrac{Q}{P}$

- Characteristics of impedance and admittance:
  - The inductive load: $X > 0$ ($X_L > X_C$), $\phi > 0$, $B < 0$ ($B_L > B_C$), $\phi_y < 0$
  - The capacitive load: $X < 0$ ($X_C > X_L$), $\phi < 0$, $B > 0$ ($B_C > B_L$), $\phi_y > 0$
  - The resistive load: $X = 0$ ($X_C = X_L$), $\phi = 0$, $B = 0$ ($B_L = B_C$), $\phi_y = 0$.

Impedances in series and parallel

- Impedances in series: $Z_{eq} = Z_1 + Z_2 + \cdots + Z_n$
- Impedances in parallel: $Z_{eq} = \dfrac{1}{(1/Z_1)+(1/Z_2)+\cdots+(1/Z_n)} = Z_1//Z_2//\cdots//Z_n$

  $$Z_{eq} = \frac{1}{Y_{eq}} \qquad Y_{eq} = Y_1 + Y_2 + \cdots + Y_n$$

  Two impedances in parallel: $Z_{eq} = \dfrac{Z_1 Z_2}{Z_1 + Z_2} = Z_1//Z_2$
- Voltage-divider rule for impedance: $\dot{V}_1 = \dfrac{Z_1}{Z_1 + Z_2}\dot{E} \quad \dot{V}_2 = \dfrac{Z_2}{Z_1 + Z_2}\dot{E}$

  Current-divider rule for impedance: $\dot{I}_1 = \dfrac{Z_2}{Z_1 + Z_2}\dot{I}_T \quad \dot{I}_2 = \dfrac{Z_1}{Z_1 + Z_2}\dot{I}_T$
- The phasor forms of KVL and KCL: $\Sigma\dot{I} = 0 \quad \dot{I}_{in} = \dot{I}_{out}$

  $$\Sigma\dot{V} = 0 \quad \dot{V}_1 + \dot{V}_2 + \cdots + \dot{V}_n = \dot{E}$$

Power of AC circuits

Power	General form	R	L	C
Instantaneous power	$p = ui = VI\cos\phi\,(1 - \cos 2\omega t) + VI\sin\phi\sin 2\omega t$	$p_R = VI - VI\cos 2\omega t$	$p_L = VI\sin 2\omega t$	$p_C = -VI\sin 2\omega t$
Active power	$P = VI\cos\phi$	$P_R = VI = 1/2V_mI_m = I^2R = V^2/R$	$P_L = 0$	$P_C = 0$
Reactive power	$Q = VI\sin\phi$	$Q_R = 0$	$Q_L = VI = I^2X_L = V^2/X_L$	$Q_C = -VI = -I^2X_C = V^2/X_C$
Apparent power		$S = VI = I^2Z = V^2/Z$		

Quantity	Quantity Symbol	Unit	Unit symbol
Instantaneous Power	p	Watt	W
Active Power	P	Watt	W
Reactive power	Q	Volt-amperes reactive	Var
Apparent power	S	Volt-Amperes	VA

- Power: $P = S\cos\phi$, $Q = S\sin\phi$, $S = \sqrt{P^2 + Q^2}$

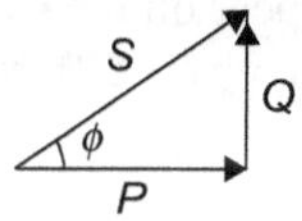

Phasor power: $\dot{S} = R + jQ = \dot{V}\dot{I} = \dot{I}^2Z = \dot{V}^2/Z$
- Power factor
  - Power factor: $\mathrm{PF} = \cos\phi = \dfrac{P}{S}$ $(0 \le \cos\phi \le 1)$
  - Power-factor correction: A capacitor in parallel with the inductive load can increase the power factor (the power factor angle $\phi\downarrow \rightarrow \cos\phi\uparrow$)
    $\cos\phi\uparrow \rightarrow$ line current $\dot{I}\downarrow \left(\dot{I}\downarrow = \dfrac{P}{V\cos\phi\uparrow}\right) \rightarrow$ line power loss $(\dot{I}^2R)\downarrow \rightarrow$ utilize capacity of the source more efficiently.
- Total power
  - Total active power: $P_T = P_1 + P_2 + \cdots + P_n$
  - Total reactive power:
    $Q_T = Q_{LT} - Q_{CT} = (Q_{L1} + Q_{L2} + \cdots) - (Q_{C1} + Q_{C2} + \cdots)$ (Q_{LT} is the total reactive power of inductors, and Q_{CT} is the total reactive power of capacitors.)
  - Total apparent power: $S_T = \sqrt{P_T^2 + Q_T^2}$ $S_T = S_1 + S_2 + \cdots + S_n$
  - Total power factor: $\mathrm{PF}_T = \cos\phi_T = \dfrac{P_T}{S_T}$

- Analysis methods for AC sinusoidal circuits
  All analysis methods that are used to analyse DC circuits with one or two more sources can also be used to analyse AC circuits.

Experiment 9: Sinusoidal AC circuits

Objectives

- To become familiar with the operation of an oscilloscope for measuring the sinusoidal AC voltage.
- To become familiar with the operation of an oscilloscope for measuring the phase difference of two waveforms.
- To verify theoretical calculations of the AC series–parallel circuits through experiment.

Background information

- $X_L = \omega L = 2\pi fL, \quad X_C = \dfrac{1}{\omega C} = \dfrac{1}{2\pi fC}$
- $Z_R = R,\ Z_L = jX_L,\ Z_C = -jX_C$
- Use the oscilloscope to measure the phase difference ϕ (with dual-channel CH I and CH II):

Example L9.1: There are two sinusoidal waveforms A and B with complete cycles of $2\pi(360^\circ)$ as shown in Figure L9.1. If these two waveforms appear on the screen of the oscilloscope and occupy six horizontal grids, determine the phase difference of waveforms A and B in Figure L9.1.

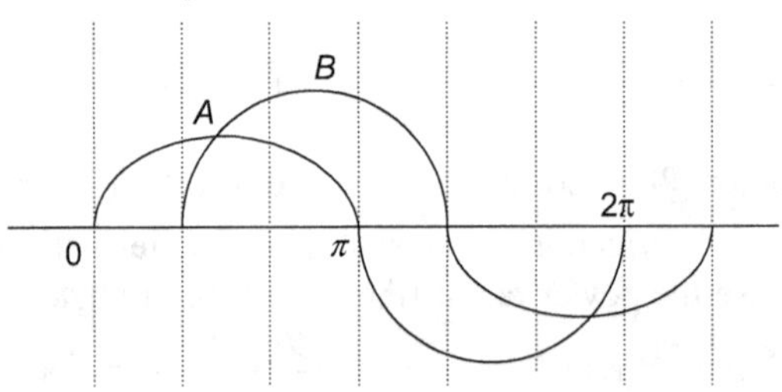

Figure L9.1 Phase difference

Solution: Each grid is 60° ($360^\circ/6\,\text{grids} = 60^\circ/\text{grids}$). Since there is one grid difference between waveform A and B as shown in Figure L9.1, the phase difference of waveforms A and B is 60°. (If the distance between A and B is 0.5 grids, the phase difference will be 30°, i.e. $0.5 \times 60^\circ = 30^\circ$.)

Use the oscilloscope to measure the current in the inductive or capacitive branch (indirect measurement): Connect a small resistor, called a sensing resistor, in series with the inductor or the capacitor. Measure the voltage across the sensing resistor, and then calculate the branch current using Ohm's law. Since the sensing resistance is very small, its impact on the circuit measurement and calculation may be negligible.

Equipment and components

- Multimeter
- Breadboard
- Function generator
- Oscilloscope
- Z meter or LCZ meter
- Switch
- Resistors:15 Ω (two) and 510 Ω
- Inductor: 1.1 mH
- Capacitor: 3 600 pF

Procedure

1. Connect a circuit as shown in Figure L9.2 on the breadboard. Use the multimeter (ohmmeter function) and Z meter or LCZ meter to measure the values of the resistor, inductor and capacitor, and record in Table L9.1.

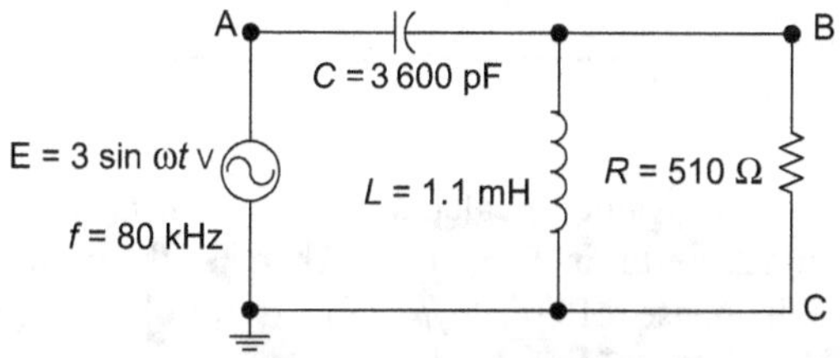

Figure L9.2 Experiment circuit

Table L9.1

	R	R_L	R_C	L	C
Nominal value	510 Ω	15 Ω	15 Ω	1.1 mH	3 600 pF
Measured value					

2. Use the measured R, L and C values to calculate X_C, X_L and Z_{eq} of the circuit shown in Figure L9.2 (at frequency of 80 kHz) and record the results in Table L9.2.

Table L9.2

	X_C	X_L	Z_{eq}*
Formula for calculation			
Calculated value			

*Refer to chapter 9, section 9.2

3. Adjust the frequency of the function generator to 80 kHz. Connect the oscilloscope probe CH I to the point A in the circuit of Figure L9.2, connect the probe ground of the oscilloscope to the ground of the function generator, and then measure the output sinusoidal voltage of the function generator.

Resistor branch

4. Adjust the output voltage of the function generator to 6 V peak–peak value ($E_{p-p}=$ 6 V). Connect the oscilloscope probe CH II to point B in the circuit of Figure L9.2 (choose DUAL channel coupling for the oscilloscope), then measure the voltage across the resistor V_R (peak value) and record in Table L9.3.

Table L9.3

	Resistive branch			Inductive branch			Capacitive branch		
Parameter	V_R	I_R	ϕ_R	V_{RL}	I_L	ϕ_L	V_{RC}	I_C	ϕ_C
Measured value									

5. Use the measured V_R value to calculate the current I_R in the resistive branch (Ohm's law) and record it in Table L9.3. Then use the oscilloscope to observe and determine the phase difference ϕ_R of resistor voltage V_R relative to source voltage E, and record in Table L9.3.

Inductor branch

6. Connect a 15 Ω resistor R_L (sensing resistor) to the inductive branch as shown in Figure L9.3.
7. Connect the oscilloscope probe CH II to the point D in the circuit of Figure L9.3, and measure the peak voltage on resistor R_L and record the result as V_{RL} in Table L9.3.
8. Use the measured V_{RL} to calculate the current I_L in the inductor branch (Ohm's law) and record it in Table L9.3. Then use the oscilloscope to observe and determine the phase difference ω_L of V_{RL} relative to the source voltage E. Record the result in Table L9.3.

Note: The oscilloscope probe CH I is still connected to point A of the circuit in Figure L9.3.

Capacitor branch

9. Connect a 15 ω resistor R_C to the circuit as shown in Figure L9.4.
10. Connect the oscilloscope probe CH II to point E of the circuit in Figure L9.4, measure the peak voltage on resistor R_C and record the measurement as V_{RC} in Table L9.3.

11. Use the measured V_{RC} to calculate the capacitor current I_C (Ohm's law) and record it in Table L9.3. Then use the oscilloscope to observe and determine the phase difference ψ_C of capacitor voltage V_{RC} relative to source voltage E and record in Table L9.3.

Phasor form

12. Calculate the branch currents $\dot{I}_R$, $\dot{I}_L$ and $\dot{I}_C$ in the circuit of Figure L9.2 in phasor form (use peak values) and record in Table L9.4.

Table L9.4

	$\dot{I}_R$	$\dot{I}_L$	$\dot{I}_C$
Formula for calculations			
Calculated value			
Measured value			

13. Convert the measured values I_R, I_L and I_C in Table L9.3 to the phasor form and record in Table L9.4 (as the measured value). Compare the measured values and calculated values. Are there any significant differences? If so, explain the reasons.

Conclusion

Write your conclusions below:

Chapter 10
RLC circuits and resonance

Objectives

After completing this chapter, you will be able to:

- understand concepts and characteristics of series and parallel resonance
- determine the following quantities of series and parallel resonant circuits: resonant frequency, resonant current, resonant voltage, resonant impedance, bandwidth and quality factor
- plot the frequency response curves of current, voltage and impedance for series and parallel resonant circuits
- understand characteristics of the selectivity in series and parallel of resonant circuits
- understand the actual parallel resonant circuits
- understand the applications of the resonant circuits

The resonant phenomena that will be introduced in this chapter have a wide range of applications in electrical and electronic circuits, particularly in communication systems. Resonant circuits are simple combinations of inductors, capacitors, resistors and a power source. However, since the capacitor or inductor voltage/current in a resonant circuit could be much higher than the source voltage or current, a small input signal can produce a large output signal when resonance appears in a circuit. This is why the resonant circuit is one of the most important circuits in electronic communication systems.

Resonance may also damage the circuit elements if it is not used properly. So it is very important to analyse and study resonant phenomena and to know its pros and cons.

10.1 Series resonance

10.1.1 Introduction

Resonance may occur in a series resistor, inductor and capacitor (RLC) circuit, as shown in Figure 10.1, when the capacitor reactance X_C is equal to the inductor reactance X_L.

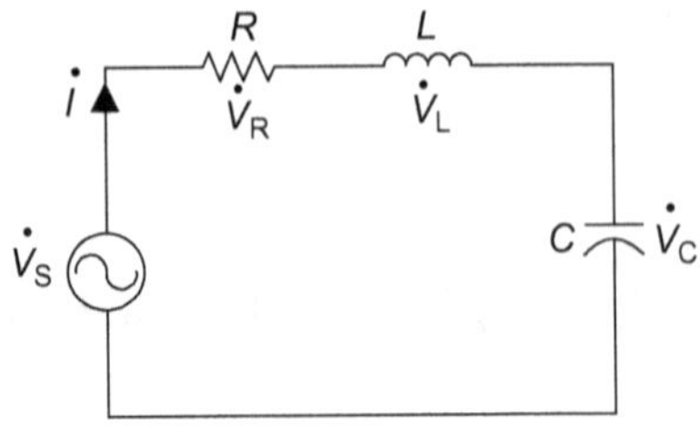

Figure 10.1 An RLC series circuit

When the magnitudes of inductive reactance X_L and capacitive reactance X_C are equal ($X_L = X_C$), or when reactance X is zero ($X = X_L - X_C = 0$), the equivalent or total circuit impedance Z is equal to the resistance R, i.e.

$$\dot{Z} = R + j(X_L - X_C) = R$$

Under the above condition, resonance will occur in the RLC series circuit. That is, when resonance occurs in a series RLC circuit, the energy of the reactive components in the circuit will compensate each other ($X_L = X_C$), and the equivalent impedance of the series RLC circuit will be the lowest ($Z = R$). This is the characteristic of the series resonant circuit.

Series resonance
$X_L = X_C$, $X = 0$, $Z = R$

10.1.2 *Frequency of series resonance*

The angular frequency of the series resonant circuit can be obtained from

$$X_L = X_C \quad \text{or} \quad \omega L = \frac{1}{\omega C}$$

Solving for ω gives $\omega_r = 1/\sqrt{LC}$. (The subnotation 'r' stands for resonance.)

Since $\omega = 2\pi f$, solving for f gives the series resonant frequency as

$$f_r = \frac{1}{2\pi\sqrt{LC}}$$

This is a very important equation for the series resonance. The resonant frequency f_r is dependent on the circuit elements inductor (L) and capacitor (C), meaning that it may produce or remove resonance by adjusting the inductance L or capacitance C in the RLC series circuit.

Frequency of series resonance
Resonant frequency: $f_r = 1/2\pi\sqrt{LC}$
Resonant angular frequency: $\omega_r = 1/\sqrt{LC}$

10.1.3 Impedance of series resonance

As previously mentioned, when series resonance occurs, the circuit's equivalent impedance is at the minimum ($Z = R$). This is illustrated in Figure 10.2, which is the response curve of the impedance Z versus frequency f in the series resonant circuit. When $f = f_r$, the impedance Z is at the lowest point on the curve.

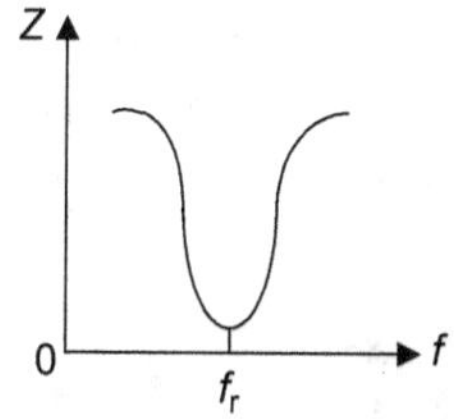

Figure 10.2 The response curve of Z vs. f for series resonance

10.1.4 Current of series resonance

When resonance occurs in a series RLC circuit, the impedance of the circuit is equal to the resistance ($Z = R$), and the resonant current will be

$$\dot{I} = \frac{\dot{V}}{Z} = \frac{\dot{V}}{R}$$

Therefore, when $f = f_r$, $X_L = X_C$, the only opposition to the flow of the current is resistance R, i.e. the impedance is minimum and current is maximum in a series resonant circuit. Figure 10.3 illustrates the response curve of current I versus frequency f in the series resonant circuit, and the current is at the highest point on the curve when $f - f_r$.

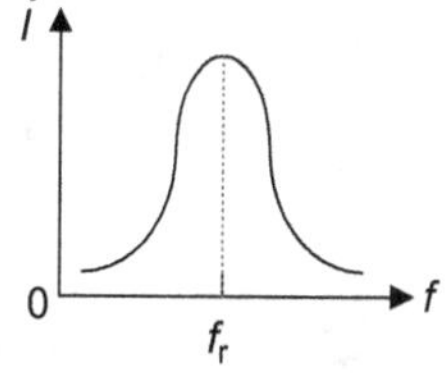

Figure 10.3 The response curve of I vs. f for series resonance

I and Z of series resonance

- Impedance is minimum at series resonance: $\boldsymbol{Z} = R$
- Current is maximum at series resonance: $\dot{I} = \dot{V}/\dot{Z} = \dot{V}/R$

10.1.5 Phasor diagram of series resonance

An RLC series resonant circuit is equivalent to a purely resistive circuit since $Z = R$. The capacitor and inductor voltages in the series resonant circuit are equal in magnitude but are opposite in phase since $X_L = X_C$, $\dot{V}_L = jX_L\dot{I}_L$ and $\dot{V}_C = -jX_C\dot{I}_C$.

The resistor voltage is equal to the source voltage ($\dot{V}_R = \dot{E}$) since $X = 0$ when series resonance occurs. Thus, the current $\dot{I}$ and source voltage $\dot{E}$ are also in phase (since $\dot{V}_R$ and $\dot{I}$ in phase), and the phase difference between $\dot{E}$ and $\dot{I}$ is zero ($\phi = 0$). A phasor diagram of the series resonant circuit is illustrated in Figure 10.4.

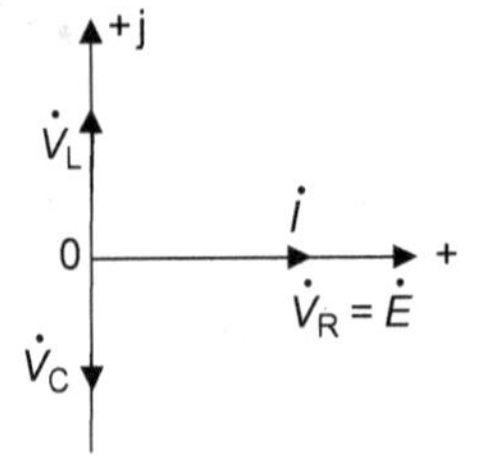

Figure 10.4 Phasor diagram of the series resonant circuit

Phasor relationship of series resonance

- $\dot{V}_L$ and $\dot{V}_C$ are equal in magnitude but opposite in phase.
- $\dot{I}$ and $\dot{E}$ are in phase, and $\phi = 0$.

10.1.6 Response curves of X_L, X_C and Z versus f

The response curves of the inductive reactance X_L, capacitive reactance X_C and impedance Z versus frequency f are illustrated in Figure 10.5.

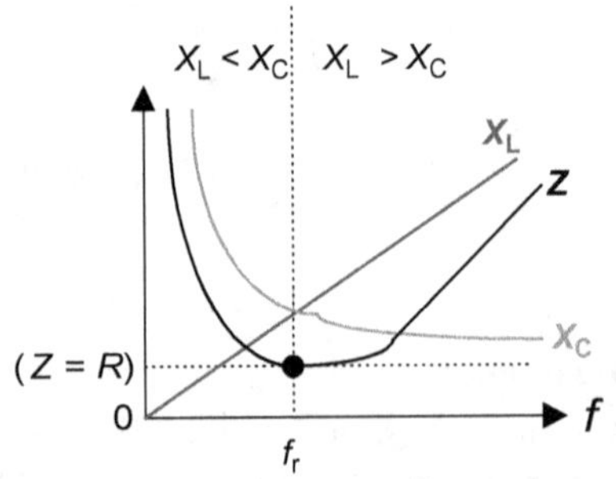

Figure 10.5 Response curves of X_L, X_C and Z vs. f

- X_L and f are directly proportional ($X_L = 2\pi fL$), i.e. as frequency increases, and X_L increases.
- X_C and f are inversely proportional ($X_C = 1/2\pi fC$), i.e. as frequency increases X_C decreases.
- When frequency f is zero in the circuit, $X_L = 0$, X_C and Z approach infinite, $(Z = \sqrt{R^2 + (2\pi fL - (1/2\pi fC))^2} = \sqrt{R^2 + (0 - \infty)^2} = \sqrt{R + \infty^2} \Rightarrow \infty)$.

The response curves of X_L, X_C and Z versus f show that when the circuit frequency is below the resonant frequency f_r, the inductive reactance X_L is lower than the capacitive reactance X_C and the circuit appears capacitive. When the circuit frequency is above the resonant frequency f_r, the inductive reactance X_L is higher than the capacitive reactance X_C, and the circuit appears more inductive. Only when the circuit frequency is equal to the resonant frequency f_r, the resonance occurs in the circuit. Impedance Z is equal to the circuit resistance R and has a minimum value, and the circuit appears purely resistive. These can be summarized as follows:

Characteristics of series resonance

- When $f < f_r$, $X_L < X_C$: the circuit is more capacitive.
- When $f > f_r$, $X_L > X_C$: the circuit is more inductive.
- When $f = f_r$, $X_L = X_C$, $I = I_{max}$, $Z = Z_{min} = R$: the circuit is purely resistive and resonance occurs.

10.1.7 Phase response of series resonance

The phase response of the series resonant circuit can also be obtained from Figure 10.5.

- When the frequency of the circuit is above the resonant frequency f_r, the circuit is more inductive $X_L > X_C$, voltage leads current, and the phase difference is between zero and positive 90° ($0 \le \phi \le 90°$).
- When the frequency of the circuit is below the resonant frequency f_r, the circuit is more capacitive $X_L < X_C$, the voltage lags current, and the phase difference is between zero and negative 90° ($-90° \le \phi \le 0$).
- When the frequency of the circuit is equal to the resonant frequency f_r, $X_L = X_C$, $Z = R$, voltage and current are in phase, and the phase difference is zero ($\phi = 0$).

The phase response of the series resonant circuit can be illustrated in Figure 10.6.

The following characteristics of the series resonant circuit can also be obtained from Figure 10.6.

- When the frequency increases from the resonant frequency f_r to infinite, the phase angle approaches positive 90°.

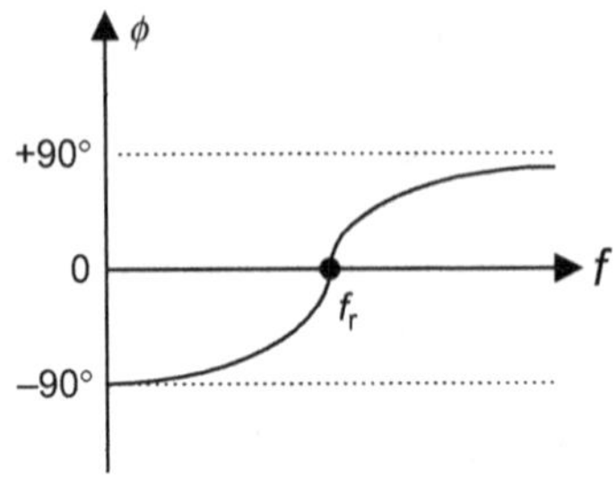

Figure 10.6 Phase response of the series resonant circuit

- When the frequency decreases from the resonant frequency f_r to zero, the phase angle approaches negative 90°.

The expression of the phase angle is

$$\phi = \tan^{-1}\frac{X}{R} = \tan^{-1}\frac{X_L - X_C}{R} = \tan^{-1}\frac{2\pi fL - (1/2\pi fC)}{R}$$

Phase response of series resonance

- When $f \to \infty$, $\phi \to +90°$
- When $f \to 0$, $\phi \to -90°$

10.1.8 Quality factor

There is an important parameter known as quality factor in the resonant circuit, which is denoted as Q. The quality factor is defined as the ratio of stored energy and consumed energy in physics and engineering, so it is the ratio of the reactive power stored by an inductor or a capacitor and average power consumed by a resistor in a resonant circuit, i.e.

$$\text{Quality factor } Q = \text{Reactive power/average power} \tag{10.1}$$

The quality factor can be used to measure the energy that a circuit stores and consumes.

The lower the energy consumption of a resistor (power loss) in a circuit, the higher the quality factor, and the better the quality of the resonant circuit. If substituting the equations of the reactive power and average power into the quality factor equation (10.1), the quality factor of the series resonance will be obtained as follows:

$$Q = \frac{I^2 X_L}{I^2 R} = \frac{X_L}{R} = \frac{\omega L}{R}$$

where R is the total or equivalent resistance in the series circuit.

Similarly, the quality factor Q can also be expressed by the capacitive reactance and the resistance as

$$Q = \frac{X_C}{R} = \frac{1}{\omega CR}$$

The quality factor can be used to judge the quality of an inductor (or coil). A coil always contains a certain amount of winding resistance R_w, which is the resistance of the wire in the winding. The quality factor Q for a coil is defined as the ratio of the inductive reactance and the winding resistance, i.e.

$$Q = \frac{X_L}{R_w}$$

The lower the winding resistance R_w of a coil, the higher the quality of the coil.

Note: Both the quality factor and reactive power are denoted by the letter Q, so be careful not to confuse them. The quality factor is a dimensionless parameter, and the unit of reactive power is Var, which can be used to distinguish between these two quantities.

Quality factor Q

- Quality factor: is the ratio of the reactive power and average power.
- Quality factor of the series resonance: $Q = X_L/R = X_C/R$.
- Quality factor of the coil: $Q = X_L/R_w$.
  (The lower the R_w, the higher the quality of the coil.)

10.1.9 Voltage of series resonant

Multiplying current $\dot{I}$ for both the denominator and nominator of the quality factor equation $Q = X_L/R$, gives

$$Q = \frac{X_L}{R} = \frac{\dot{I}X_L}{\dot{I}R} = \frac{\dot{V}_L}{\dot{E}}$$

Similarly, for $Q = X_C/R$

$$Q = \frac{\dot{I}X_C}{\dot{I}R} = \frac{\dot{V}_C}{\dot{E}}$$

Therefore, when the resonance occurs in an RLC series circuit:

$$\dot{V}_L = \dot{V}_C = \dot{E}Q \tag{10.2}$$

The quality factor Q is always greater than 1, so the inductor or capacitor voltage may greatly exceed the source voltage in a series resonant circuit, as can be seen from the equation (10.2). This means that a lower input voltage may produce a higher output voltage; therefore, the series resonance is also known as the *voltage* resonance. That is one of the reasons that series resonant circuits have a wide range of applications.

When choosing the storage elements L and C for a series resonant circuit, the affordability of their maximum voltage should be taken into account, or else the high resonant voltage may damage circuit components.

The concept of circuit resonance is similar to resonance in physics, which is defined as a system oscillating at maximum amplitude at resonant frequency, so a small input force can produce a large output vibration.

There are many examples of resonance in daily life, such as pushing a child in a playground swing to the resonant frequency, which makes the swing go higher and higher to the maximum amplitude with very little effort. Another example is bouncing a basketball. Once the ball is bounced to the resonant frequency, it will yield a smooth response, and the ball will reach maximum height since a small force produces a large vibration.

Resonance may also cause damage. For example, a legend says that when a team of soldiers walking a uniform pace passed through a bridge, the bridge collapsed since the uniform pace reached resonant frequency that resulted in a small force producing a large vibration.

Voltage of series resonance

- A lower input voltage may produce a higher output voltage.
- Inductor or capacitor voltage may greatly exceed the supply voltage $\dot{V}_L = \dot{V}_C = \dot{E}Q \quad (Q > 1)$

Example 10.1: A series resonant circuit is shown in Figure 10.7. Determine the total equivalent impedance, quality factor, and inductor voltage of this circuit.

Solution:

$$Z = R_T = R + R_w = (2 + 0.5)\Omega = 2.5\angle 0^\circ\ \Omega$$

$$f_r = \frac{1}{2\pi\sqrt{LC}} = \frac{1}{2\pi\sqrt{(2.5\,\text{mH})(0.25\,\mu\text{F})}} \approx 6\,366\,\text{Hz}$$

$$X_L = 2\pi fL = 2\pi(6\,366)(2.5\,\text{mH}) \approx 100\,\Omega$$

$$Q = \frac{X_L}{R_T} = \frac{100\,\Omega}{2.5\,\Omega} = 40$$

$$\dot{V}_L = jX_L\dot{I} = \frac{\dot{E}}{Z}jX_L$$

$$= \frac{2.5\angle 0^\circ\ \text{V}}{2.5\angle 0^\circ\ \Omega} \times 100\angle 90^\circ\ \Omega = 100\angle 90^\circ\ \text{V}$$

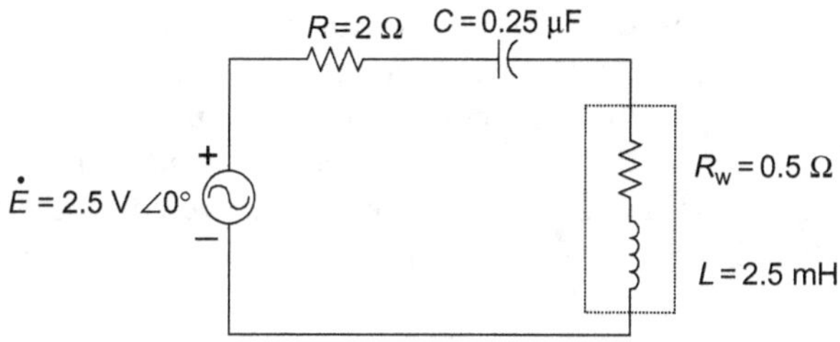

Figure 10.7 Circuit for Example 10.1

This example shows that the inductor voltage of the series resonant circuit is indeed greater than the supply voltage.

$$(\dot{V}_L = 100\angle 90^\circ \text{ V}) > (\dot{E} = 2.5\angle 0^\circ \text{ V})$$

10.2 Bandwidth and selectivity

10.2.1 The bandwidth of series resonance

When an RLC series circuit is in resonance, its impedance will reach the minimum value and the current will reach the maximum value. The curve of the current versus frequency of the series resonant circuit is illustrated in Figure 10.8. As displayed in the diagram, the current reaches the maximum value I_{max} as the frequency closes in on the resonant frequency f_r, which is located at the centrer of the curve.

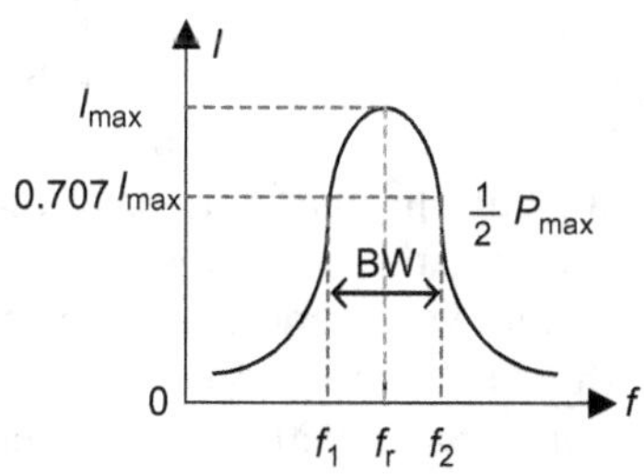

Figure 10.8 Bandwidth of a series resonant circuit

The characteristic of the resonant circuit can be expressed in terms of its bandwidth (BW) or pass-band. The bandwidth of the resonant circuit is the difference between two frequency points f_2 and f_1.

$$\text{BW} = f_2 - f_1$$

where f_2 and f_1 are called critical, cutoff or half-power frequencies.

As shown in Figure 10.8, the bandwidth of the resonant circuit is a frequency range between f_2 and f_1 when current I is equivalent to 0.707 of its maximum value I_{max}, or 70.7 per cent of the maximum value of the curve.

The reason to define the term 'half-power' frequency can be derived from the following mathematical process. The power delivered by the source at the points f_1 and f_2 can be determined from the power formula $P = I^2R$

$$P_{f_1} = I_{f_1}^2 R = (0.707 I_{max})^2 R \approx 0.5 I_{max}^2 R = 0.5 P_{max}$$

and

$$P_{f_2} = I_{f_2}^2 R = (0.707 I_{max})^2 R \approx 0.5 I_{max}^2 R = 0.5 P_{max}$$

Therefore, at both points f_2 and f_1, the circuit power is only one-half of the maximum power that it is produced by the source at resonance frequency f_r, where f_2 is the upper critical frequency, and f_1 is the lower critical frequency.

Bandwidth (pass-band)

- Bandwidth ($BW = f_2 - f_1$) is the range of frequencies at $I = 0.707 I_{max}$.
- f_2 and f_1 are critical, or cutoff or half-power frequencies: $P_{f_{1,2}} = 0.5 P_{max}$.

10.2.2 The selectivity of series resonance

Figure 10.8 shows the frequency range between f_2 and f_1 at which the current is near its maximum value, and the series resonant circuits can select frequencies in this range. The curve in the Figure 10.8 is called the selectivity curve of the series resonant circuit. The selectivity is the capability of a series resonant circuit to choose the maximum current that is closer to the resonant frequency f_r. The steeper the selectivity curve, the faster the signal attenuation (reducing), the higher the maximum current value, and the better the circuit selectivity. For example, in Figure 10.9, the selectivity curve 1 has a bandwidth of BW_1 and a maximum current I_{1max}, which has a better current selectivity than selectivity curve 2 or 3. This means that the series resonant circuit of curve 1 has a higher quality and can be expressed as $Q = f_r/BW$, where Q is the quality factor of the series resonant circuit.

The bandwidth BW is an important characteristic for the resonant circuit. A series resonant circuit with a narrower bandwidth has a better current selectivity. A series resonant circuit with a wider bandwidth is good for passing the signals. Sometimes in order to take into account both aspects, the selectivity curve between narrow and wide curves may be chosen, such as the selectivity curve 2 (BW_2) in Figure 10.9. Therefore, the concepts of bandwidth and selectivity may apply to different circuits with different design choices.

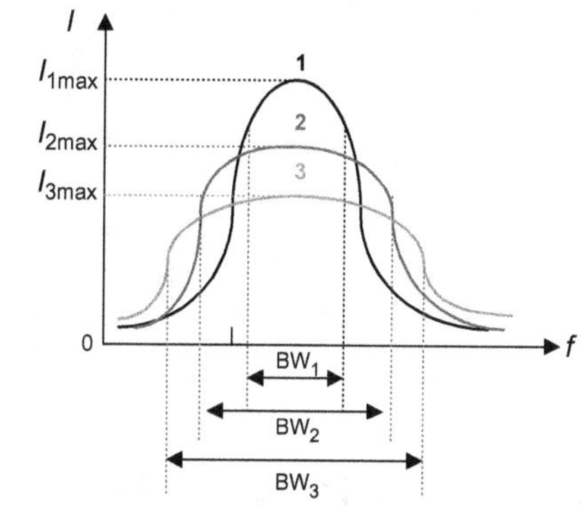

Figure 10.9 Selectivity of a series resonant circuit

Selectivity of the series resonance
The capability of the circuit to choose the maximum current I_{max} closer to the resonant frequency f_r.

10.2.3 The quality factor and selectivity

The quality factor Q in the resonant circuit is a measure of the quality and selectivity of a resonant circuit. The higher the Q value, the narrower the bandwidth (BW $\downarrow = f_r/Q \uparrow$), the higher the maximum current, and the better the current selectivity, which is desirable in many applications. As mentioned earlier, the disadvantage of the narrower BW or higher Q is that the ability for passing signals in the circuit will be reduced. The lower the Q value, the wider the bandwidth (BW $\uparrow = f_r/Q \downarrow$), and the better the ability to pass signals; however, it will have a poor current selectivity. This is the reason that Q is denoted as the 'quality' factor since it represents the quality of a resonant circuit.

Example 10.2: Given a series resonant circuit shown in Figure 10.10(a), determine the bandwidth BW and current $\dot{I}$ (phasor-domain) of this circuit with three resistors that are 50, 100, and 200 Ω, and plot their selectivity curves.

Solution: When $R = 50\ \Omega$:

$$Q = \frac{X_L}{R} = \frac{2\,\text{k}\Omega}{50\,\Omega} = 40, \quad \text{BW}_1 = \frac{f_r}{Q} = \frac{50\,\text{Hz}}{40} = 1.25\,\text{Hz}$$

$$\dot{I} = \frac{\dot{E}}{Z} = \frac{\dot{E}}{R} = \frac{10\angle 0^\circ\,\text{V}}{50\,\Omega} = 0.2\angle 0^\circ\,\text{A}$$

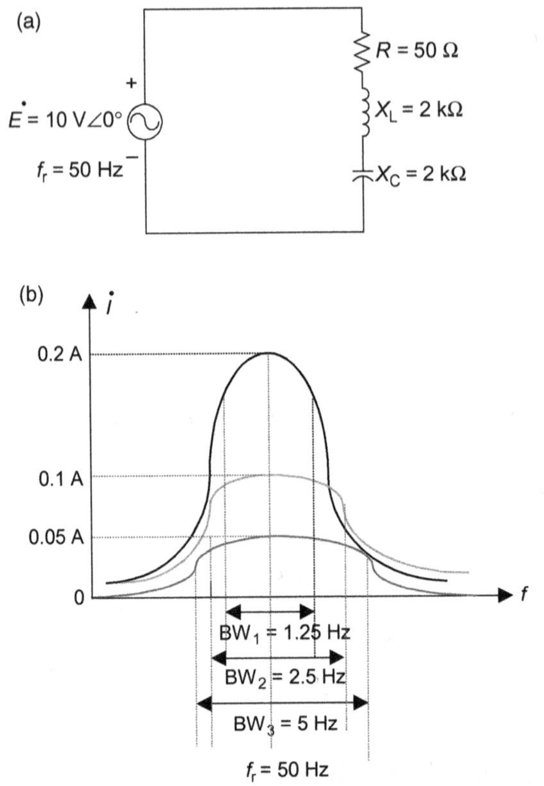

Figure 10.10 (a) Circuit for Example 10.2; (b) Selectivity curve for Example 10.2

When $R = 100\ \Omega$:

$$Q = \frac{X_L}{R} = \frac{2\,\text{k}\Omega}{100\,\Omega} = 20, \quad \text{BW}_2 = \frac{f_r}{Q} = \frac{50\,\text{Hz}}{20} = 2.5\,\text{Hz}$$

$$\dot{I} = \frac{\dot{E}}{Z} = \frac{\dot{E}}{R} = \frac{10\angle 0^\circ\ \text{V}}{100\,\Omega} = 0.1\angle 0^\circ\ \text{A}$$

When $R = 200\ \Omega$:

$$Q = \frac{X_L}{R} = \frac{2\,\text{k}\Omega}{200\,\Omega} = 10, \quad \text{BW}_3 = \frac{f_r}{Q} = \frac{50\,\text{Hz}}{10} = 5\,\text{Hz}$$

$$\dot{I} = \frac{\dot{E}}{Z} = \frac{\dot{E}}{R} = \frac{10\angle 0^\circ\ \text{V}}{200\,\Omega} = 0.05\angle 0^\circ\ \text{A}$$

Example 10.2 shows that the selectivity curve of a resonant circuit depends greatly upon the amount of resistance in the circuit. When resistance R in a series resonant circuit has a smaller value, the selectivity curve of the circuit is steeper, the quality factor Q has a higher value, the current at the resonant frequency f_r has a higher value, and the selectivity is better. However, the pass-band (BW) of the circuit with a smaller R value is narrower, and the ability to pass signal will be poor.

Quality factor and selectivity

- Quality factor: a measure of the quality and selectivity of a resonant circuit $Q = f_r/\mathrm{BW}$.
- $Q\uparrow = X_L/R\downarrow \Rightarrow \mathrm{BW}\downarrow = f_r/Q\uparrow$: the steeper the selectivity curve, the better the current selectivity, but the worse the ability to pass signals.
- $Q\downarrow = X_L/R\uparrow \Rightarrow \mathrm{BW}\uparrow = f_r/Q\downarrow$: the flatter the selectivity curve, the worse the current selectivity, but the better the ability to pass signals.

The analysis method of the series resonant circuit can also be applied to the parallel resonant circuits.

10.2.3.1 Series resonance summary

Characteristics	Series resonance
Condition of resonance	$X_L = X_C$, $X = 0$, $Z = R$
Resonant frequency	$f_r = 1/\left(2\pi\sqrt{LC}\right)$
Impedance	$Z = R$ minimum (admittance Y maximum)
Current	$\dot{I}_T = \dot{V}/R$ (maximum)
Bandwidth	$\mathrm{BW} = f_2 - f_1 = f_r/Q$
Quality factor	$Q = X_L/R = X_C/R$
Relationship of voltage and quality factor	$\dot{V}_L = \dot{V}_C = \dot{E}Q$

10.3 Parallel resonance

10.3.1 Introduction

Resonance may occur in a parallel resistor, inductor and capacitor (RLC) circuit, as shown in Figure 10.11, when the circuit inductive susceptance B_L is equal to the capacitive susceptance B_C.

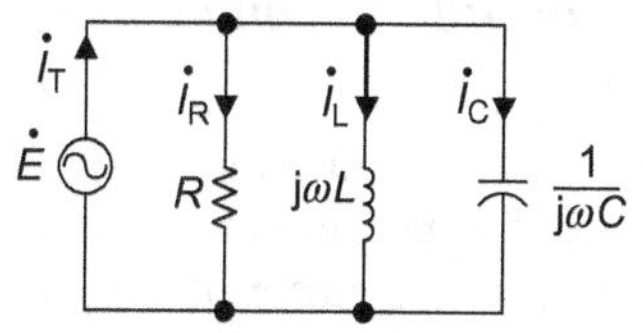

Figure 10.11 A parallel RLC circuit

The analysis method of the parallel resonance is similar to series resonance. When the magnitudes of the capacitive susceptance B_C and the inductive susceptance B_L are equal ($B_C = B_L$), or when the susceptance B is zero ($B = B_C - B_L = 0$), the circuit input equivalent (total) admittance Y is equal to the circuit conductance G, i.e.

$$Y = G + jB = G$$

Under the above condition, resonance will occur in the RLC parallel circuit. That is, when the resonance occurs in an RLC parallel circuit, the energy of the reactive components in the circuit will compensate each other ($B_C = B_L$), and the equivalent admittance of the parallel RLC circuit is at the lowest ($Y = G$). This is the characteristic of the parallel resonant circuit.

Parallel resonance
$B_C = B_L$, $B = 0$, $Y = G$

10.3.2 *Frequency of parallel resonance*

The angular frequency of the parallel resonant circuit can be obtained from

$$Y = G + j(B_C - B_L) = \frac{1}{R} + j\left(\omega C - \frac{1}{\omega L}\right)$$

From $B_C = B_L$ or $\omega C = 1/\omega L$, solving for ω gives $\omega_r = 1/\sqrt{LC}$.

Since $\omega = 2\pi f$, the parallel resonant frequency is $f_r = 1/(2\pi\sqrt{LC})$.

You may have noticed that the parallel resonant angular frequency ω_r and resonant frequency f_r are the same with those in the series resonant circuit. The resonant frequency f_r is dependent on the circuit elements L and C, meaning that if adjusting the inductance L or capacitance C in the RLC parallel circuit, resonance may be produced or removed.

Frequency of parallel resonance

- Resonant frequency: $f_r = 1/(2\pi\sqrt{LC})$
- Resonant angular frequency: $\omega_r = 1/\sqrt{LC}$

10.3.3 *Admittance of parallel resonance*

As previously mentioned, when parallel resonance occurs, the equivalent admittance Y of the circuit is at the minimum ($Y = G$), $B = B_C - B_L = 0$ so the circuit equivalent impedance Z is at a maximum ($Z \uparrow = 1/Y \downarrow$). This is shown in Figure 10.12, which is the response curve of the impedance Z versus the frequency f in the parallel resonant circuit. When $f = f_r$, the impedance Z is at the highest point on the curve and this is opposite to the series resonance.

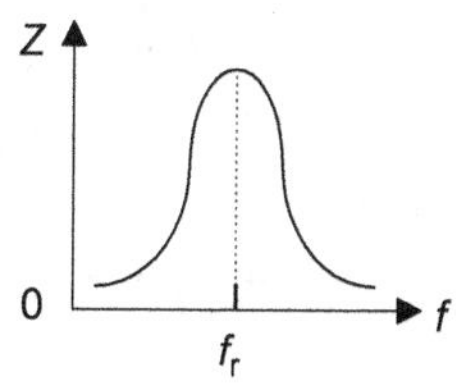

Figure 10.12 The response curve of Z vs. f for parallel resonance

10.3.4 Current of parallel resonance

When resonance appears in a parallel RLC circuit, the impedance of the circuit is equal to the resistance ($Z = R$), and the total current in the circuit will be

$$\dot{I}_T = \frac{\dot{V}}{Z} = \frac{\dot{V}}{R}$$

Therefore, when $f = f_r$, $B_C = B_L$, $Y = G$, the admittance Y is at the minimum $\because\ Y = G + \mathrm{j}\ (B_C - B_L)$, the impedance Z is at the maximum, and the current is at the minimum in the parallel resonant circuit, $\dot{I}_T \downarrow = \dot{V}/Z \uparrow = \dot{V}/R$. Figure 10.13 illustrates the response curve of current I versus frequency f in the parallel resonant circuit and current is at the lowest point on the curve when $f = f_r$. This is also opposite of series resonance.

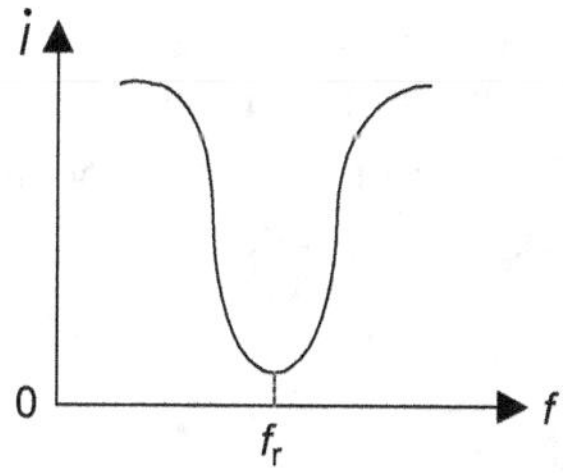

Figure 10.13 The response curve of I vs. f for parallel resonance

I and Z of parallel resonance

- Impedance is maximum at parallel resonance:

$$Z = R\ (B = 0,\ Y = G), \quad \because Y = G + \mathrm{j}(B_C - B_L)$$

- Current is minimum at parallel resonance: $\dot{I} = \dot{V}/\dot{Z} = \dot{V}/R$

10.3.5 Phasor diagram of parallel resonance

An RLC parallel resonant circuit is equivalent to a purely resistive circuit since $Y = G$ and $Z = R$. The capacitor and inductor branch currents in the parallel resonant circuit are equal in magnitude but opposite in phase, since $B_L = B_C$ ($B_L = 1/X_L$; $B_C = 1/X_C$) and

$$\dot{I}_L = \frac{V_L}{jX_L} = -j\frac{V_L}{X_L}, \quad \dot{I}_C = \frac{V_C}{-jX_C} = j\frac{V_C}{X_C} \quad \left(+j = \frac{-1}{j}\right)$$

i.e. $\dot{I}_L = -\dot{I}_C$.

The resistor voltage is equal to the source voltage ($\dot{V}_R = \dot{E}$) in the parallel resonant circuit of Figure 10.13. The total current ($\dot{I}_T$) and the source voltage $\dot{E}$ are in phase (since $\dot{V}_R$ and $\dot{E}$ are in phase) and the phase difference between $\dot{E}$ and $\dot{I}_T$ is zero, i.e. the admittance angle $\phi_y = 0$. A phasor diagram of the parallel resonant circuit is illustrated in Figure 10.14.

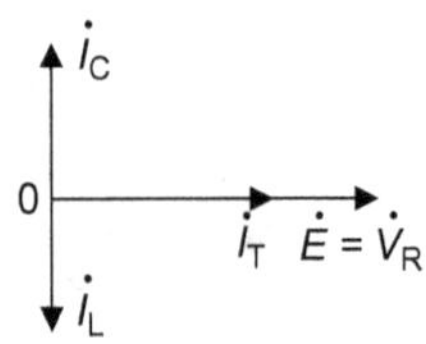

Figure 10.14 Phasor diagram of the parallel resonant circuit

Phasor relationship of parallel resonance

- $\dot{I}_L$ and $\dot{I}_C$ are equal in magnitude but opposite in phase, $\dot{I}_L = -\dot{I}_C$.
- $\dot{I}_T$ and $\dot{E}$ are in phase, and $\phi_y = 0$.

10.3.6 Quality factor

From the previous description, we know that the quality factor is the ratio of the reactive power stored by an inductor or a capacitor and the average power dissipated by a resistor in a circuit, i.e. quality factor Q = reactive power/average power.

If we substitute the expressions of the reactive power and average power in Figure 10.11 into the quality factor equation, the quality factor of a parallel resonance will be obtained as follows:

$$Q = \frac{\dot{E}^2/X_L}{\dot{E}^2/R} = \frac{R}{X_L} \quad (\because \dot{V}_L = \dot{V}_R = \dot{E})$$

Similarly, the quality factor Q can be expressed by the capacitive reactance and the resistance as

$$Q = \frac{R}{X_C}$$

The quality factor of a parallel resonant circuit is inverted with the series resonant circuit. Recall the quality factor of a series resonant circuit:

$$Q = \frac{X_L}{R} = \frac{X_C}{R}$$

Quality factor Q

Quality factor of the parallel resonance:

$$Q = R/X_L = R/X_C$$

10.3.7 *Current of parallel resonance*

Dividing the voltage $\dot{E}$ for both the denominator and the numerator of the quality factor equation

$$Q = \frac{\dot{E}^2/X_L}{\dot{E}^2/R}$$

gives

$$Q = \frac{\dot{E}/X_L}{\dot{E}/R} = \frac{\dot{I}_L}{\dot{I}_T}$$

Similarly

$$\text{For } Q = \frac{\dot{E}^2/X_C}{\dot{E}^2/R}, \quad Q = \frac{\dot{E}/X_C}{\dot{E}/R} = \frac{\dot{I}_C}{\dot{I}_T}$$

Therefore, when resonance occurs in an RLC parallel circuit

$$\dot{I}_L = \dot{I}_C = \dot{I}_T Q \tag{10.3}$$

Usually the quality factor Q is always greater than 1, the inductor or capacitor branch current may greatly exceed the total supply current in a parallel resonant circuit, and this can be seen from the equation (10.3). This means that a lower input current may produce a higher output current, and therefore the parallel resonance is also known as *current* resonance. It is similar to series resonance, and there are benefits and disadvantages to using parallel resonance. When choosing the storage elements L and C for a parallel resonant circuit, the affordability of their maximum current should be taken into account, or else the higher resonant current may damage circuit components.

Current of parallel resonance

- A lower input current may produce a higher output current.
- The inductor or capacitor current may greatly exceed the supply current

$$\dot{I}_L = \dot{I}_C = \dot{I}_T Q \quad (Q > 1)$$

10.3.8 Bandwidth of parallel resonance

The characteristic of the parallel resonant circuit can be expressed in terms of its bandwidth (BW) or pass-band. Recall that $BW = f_2 - f_1$ or $BW = f_r/Q$.

The bandwidth of the parallel resonant circuit is illustrated in Figure 10.15. When the RLC parallel circuit is in resonance, its current reaches the minimum value. The BW of the parallel resonant circuit is a frequency range between the critical or cutoff frequencies f_2 and f_1, when the current is equivalent to 0.707 of its maximum value I_{max}, or 70.7 per cent of the maximum value of the curve.

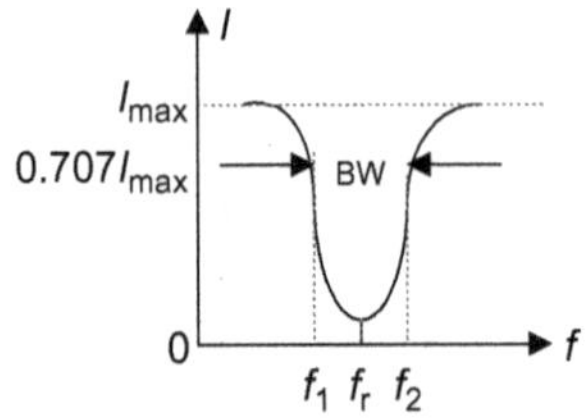

Figure 10.15 The bandwidth of the parallel resonance

10.3.8.1 Parallel resonance summary

Characteristics	**Parallel resonance**
Conditions of resonance	$B_L = B_C$, $B = 0$, $Y = G$
Resonant frequency	$f_r = 1/(2\pi\sqrt{LC})$
Impedance	$Z = R$ maximum (admittance Y minimum)
Current	$\dot{I}_T = \dot{V}/R$ (minimum)
Bandwidth	$BW = f_2 - f_1 = f_r/Q$
Quality factor	$Q = R/X_L = R/X_C$
Relationship of current and quality factor	$\dot{I}_L = \dot{I}_C = \dot{I}_T Q$

10.4 The practical parallel resonant circuit

In practical electrical or electronic system applications, the parallel resonant circuit is usually is formed by an inductor (coil) in parallel with a capacitor. Since a practical coil always has internal resistance (winding resistance), an actual parallel resonant circuit will look like the one illustrated in Figure 10.16.

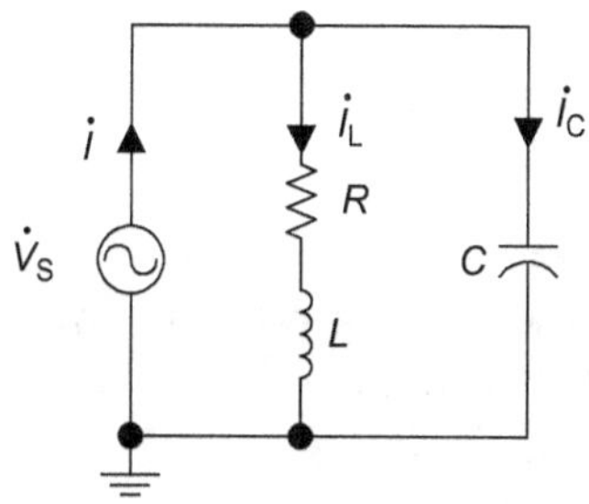

Figure 10.16 A practical parallel circuit

10.4.1 Resonant admittance

The input equivalent admittance of the practical parallel circuit shown in Figure 10.16 is

$$Y = \frac{1}{R + jX_L} + j\frac{1}{X_C}$$

Multiplying $(R - jX_L)$ to the numerator and denominator of the first term in the above expression gives

$$Y = \frac{R}{R^2 + X_L^2} - j\frac{X_L}{R^2 + X_L^2} + j\frac{1}{X_C}$$

or

$$Y = \frac{R}{R^2 + X_L^2} + j\left(\frac{1}{X_C} - \frac{X_L}{R^2 + X_L^2}\right) \tag{10.4}$$

The parallel resonance occurs when the circuit admittance Y is equal to the circuit conductance G $(Y = G)$, so when the resonance occurs for the practical parallel circuit in Figure 10.16, the resonant admittance should be

$$Y = G = \frac{R}{R^2 + X_L^2} \quad (\because Y = G + jB)$$

10.4.2 Resonant frequency

According to the parallel resonant conditions, resonance occurs when the capacitive susceptance B_C is equal to the inductive susceptance B_L, i.e. $B_C = B_L$ Thus, (10.4) gives

$$\frac{X_L}{R^2 + X_L^2} = \frac{1}{X_C}$$

or

$$\frac{\omega L}{R^2 + (\omega L)^2} = \omega C \tag{10.5}$$

The resonance frequency and angular frequency for the circuit in Figure 10.16 can be obtained from (10.5) as follows:

Resonance angular frequency:

$$\omega_r = \sqrt{\frac{L - CR^2}{L^2 C}} = \frac{1}{\sqrt{LC}}\sqrt{1 - \frac{CR^2}{L}}$$

Resonance frequency:

$$f_r = \frac{1}{2\pi\sqrt{LC}}\sqrt{1 - \frac{CR^2}{L}} \quad (\omega = 2\pi f)$$

This indicates that resonance will occur in the circuit of Figure 10.16 only when

$$1 - \frac{CR^2}{L} > 0, \quad 1 > \frac{CR^2}{L}, \quad R^2 < \frac{L}{C}, \quad \text{or} \quad R < \sqrt{\frac{L}{C}}$$

If $1 - (CR^2/L) < 0$, resonance will not occur.

Practical parallel resonance

- Resonant admittance:

$$Y = \frac{R}{R^2 + X_L^2}$$

- Resonant angular frequency:

$$\omega_r = \frac{1}{\sqrt{LC}}\sqrt{1 - \frac{CR^2}{L}}$$

- Resonant frequency:

$$f_r = \frac{1}{2\pi\sqrt{LC}}\sqrt{1 - \frac{CR^2}{L}}$$

- When $1 - (CR^2/L) > 0$, or $R < \sqrt{L/C}$ resonance occurs.

10.4.3 Applications of the resonance

As previously mentioned, resonant circuits are used in a wide range of applications in communication systems, such as filters, tuners, etc. The purpose of resonant circuits are the same – to select a specific frequency (resonant frequency f_r) and reject all others, or select signals over a specific frequency range that is between the cutoff frequencies f_1 and f_2.

The key circuit of a communication system is a tuned amplifier (tuning circuit). Figure 10.17 is a simplified radio tuning circuit for a radio circuit. The combination of a practical parallel resonant circuit and an amplifier can select the appropriate signal to be amplified.

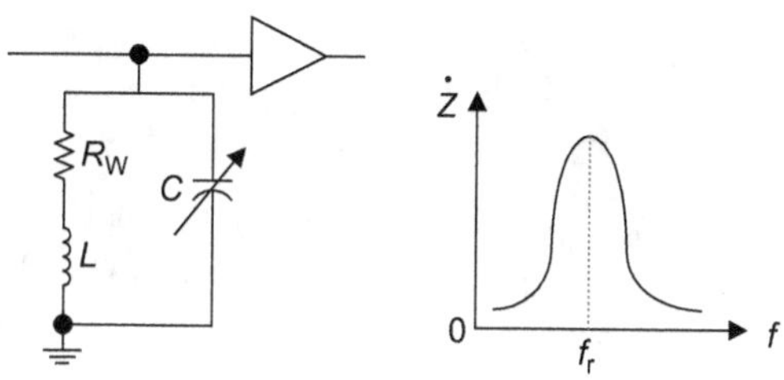

Figure 10.17 A simplified parallel radio tuner

The input signals in the radio tuner circuit have a wide frequency range, because there are many different radio signals from different radio stations. When adjusting the capacitance of the variable capacitor in the practical parallel resonant circuit (i.e. adjusting the switch of the radio channel), the circuit resonant frequency f_r will consequently change. Once f_r matches the desired input signal frequency with the highest input impedance, the desired input signal will be passed, and this is the only signal that will be amplified. After it is amplified by the amplifier in the circuit, this signal of the corresponding station can be clearly heard.

Figure 10.18 is a simplified series radio tuning circuit. It is similar to the parallel tuning circuit. When adjusting the capacitance of the variable capacitor in the series resonant circuit, the circuit resonant frequency f_r will change. Once f_r matches the desired input signal frequency with the highest current, the desired input signal will be passed and amplified.

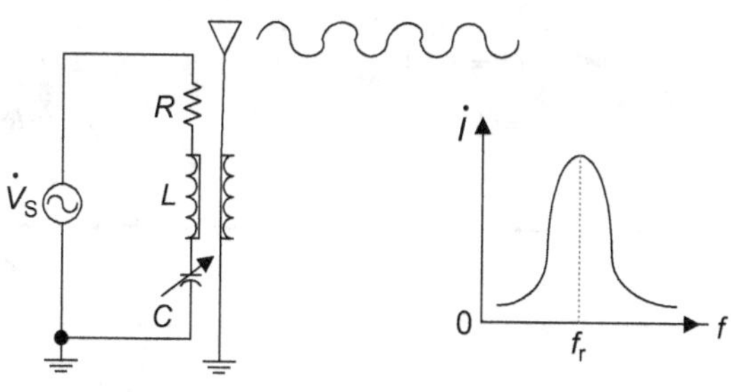

Figure 10.18 A simplified series radio tuner

Summary

Series/parallel resonance

Characteristics	**Series resonance**	**Parallel resonance**
Conditions of resonance	$X_L = X_C$, $X = 0$, $Z = R$	$B_L = B_C$, $B = 0$, $Y = G$
Phasor relationship	• $\dot{V}_L$ and $\dot{V}_C$ are equal in magnitude but opposite in phase. • $\dot{I}$ and $\dot{E}$ in phase $\phi = 0$.	• $\dot{I}_L$ and $\dot{I}_C$ are equal in magnitude but opposite in phase. • $\dot{I}_T$ and $\dot{E}$ in phase $\varphi_y = 0$.
Phasor diagram	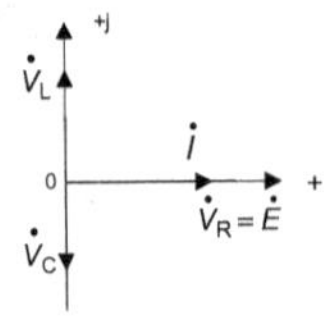	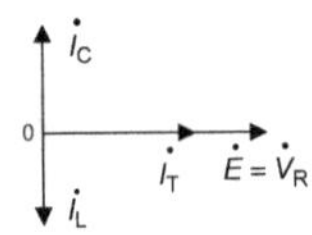
Resonant frequency	$f_r = 1/(2\pi\sqrt{LC})$	$f_r = 1/2\pi\sqrt{LC}$
Impedance	$Z = R$ minimum (admittance Y maximum) 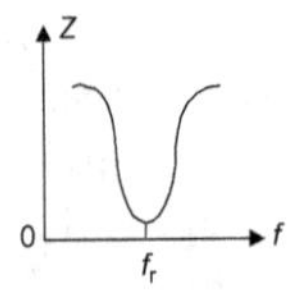	$Z = R$ maximum (admittance Y minimum) 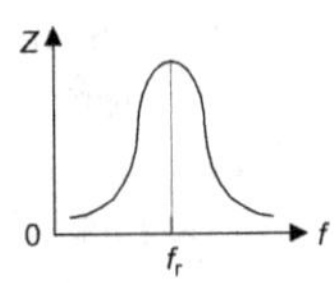
Current	$\dot{I} = \dot{V}/R$ (maximum) 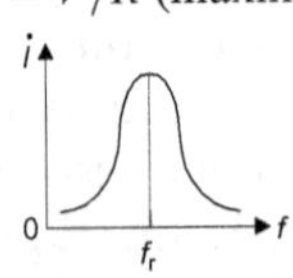	$\dot{I}_T = \dot{V}/R$ (minimum) 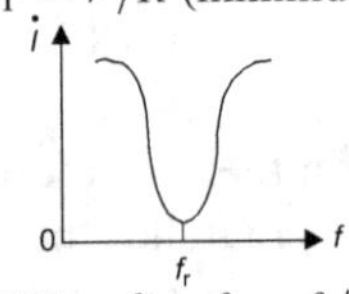
Bandwidth	$\text{BW} = f_2 - f_1 = f_r/Q$	$\text{BW} = f_2 - f_1 = f_r/Q$
Quality factor	$Q = X_L/R = X_C/R$ or $Q = f_r/\text{BW}$	$Q = R/X_L = R/X_C$ or $Q = f_r/\text{BW}$
Relationship of voltage/current and Q	$\dot{V}_L = \dot{V}_C = \dot{E}Q$	$\dot{I}_L = \dot{I}_C = \dot{I}_T Q$

- Bandwidth (pass-band): the frequency range corresponding to $\dot{I} = 0.707\dot{I}_{max}$, and BW $= f_2 - f_1$.
- f_2 and f_1: critical frequencies or cutoff frequencies or half-power point frequencies. $P_{f_{1,2}} = 0.5P_{max}$
- Quality factor: a measure of the quality and selectivity of a resonant circuit.
- Selectivity: the capability of the circuit to choose the maximum current closer to the resonant frequency f_r.
- In a series resonant circuit, V_L or V_C may greatly exceed the supply voltage E, i.e. a lower input voltage may produce a higher output voltage.
- In a parallel resonant circuit, I_L or I_C may greatly exceed the total current I_T, i.e. a lower input current may produce a higher output current.

Experiment 10: Series resonant circuit

Objectives

- Observe and analyse the characteristics of a resonant circuit by using the oscilloscope and function generator.
- Determine the resonant frequency f_r of the series resonant circuit by experiment.
- Observe and analyse the frequency versus current curve and verify that the circuit current reaches the maximum in a RLC series resonant circuit by experiment.

Background information

- Impedance of RLC series circuit: $Z = R + j(X_L - X_C)$
- Resonance: $X_L = X_C$, $V_L = V_C$, $\dot{V}_R = \dot{E}$
- Current reaches maximum at the series resonance: $\dot{I} = \dot{V}_R/R$

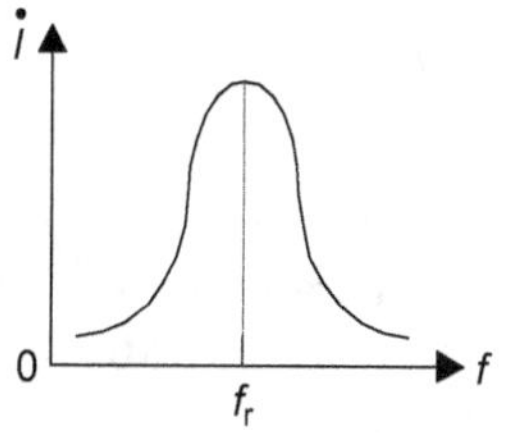

- Resonant frequency: $f_r = 1/2\pi\sqrt{LC}$
- Quality factor: $Q = X_L/R = X_C/R$

Equipment and components

- Multimeter
- Breadboard

- Function generator
- Oscilloscope
- LCZ meter or Z meter
- Resistor: 300 Ω
- Inductor: 12 mH
- Capacitor: 0.0013 μF

Procedure

1. Use multimeter (ohmmeter function) to measure the winding resistance R_w of the 12-mH inductor. Record the value in Table L10.1.

Table L10.1

	R_w	Q	f_r	V_R	V_C	V_L
Value						

2. Calculate the quality factor Q of the series resonant circuit in Figure L10.1 using $Q = X_L/(R + R_w)$ (inductor has a winding resistance R_W). Record the value in Table L10.1.

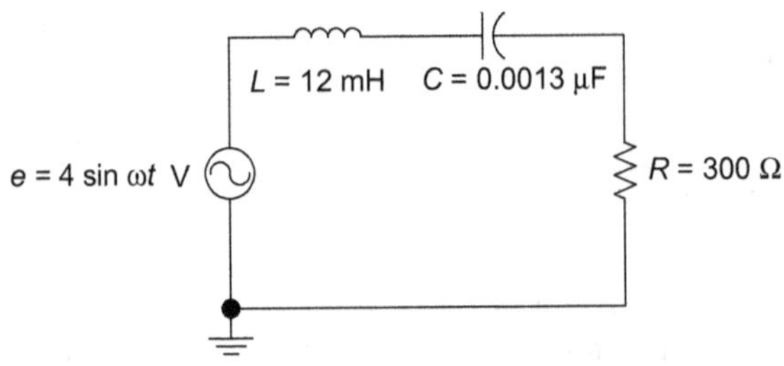

Figure L10.1 Series RLC resonant circuit

3. Compute the resonant frequency f_r and record the value in Table L10.1.
4. Set the sinusoidal output voltage of the function generator to 4 V (peak value), and then adjust the frequency of the function generator to the circuit resonant frequency f_r.
5. Construct a circuit as shown in Figure L10.1 on the breadboard.
6. Use the oscilloscope CH I to measure the peak source voltage E from the function generator in Figure L10.1. Note that the ground of the oscilloscope, function generator and circuit should be connected together.
7. Use the oscilloscope CH II to measure the voltage across resistor V_R (peak value). Record the value in Table L10.1. Is the value of V_R and supply voltage E approximately equal?

8. Exchange the position of the capacitor and resistor in the circuit as shown in Figure L10.2, and so that ground of the oscilloscope, function generator and circuit (capacitor) are connected together. Then measure the voltage across the capacitor (V_C), and record the value in Table L10.1.

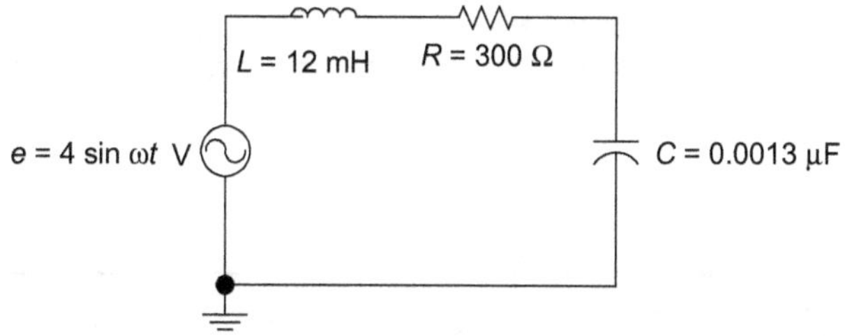

Figure L10.2 The circuit to measure V_C

9. Exchange the position of the capacitor and inductor in the circuit as shown in Figure L10.3, so that the ground of the oscilloscope, function generator and circuit (inductor) are connected together. Then measure the voltage across the inductor (V_L), and record the value in Table L10.1.

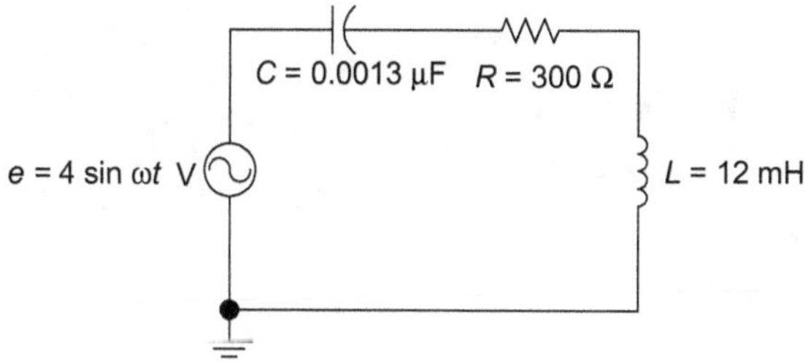

Figure L10.3 The circuit to measure V_L

10. Calculate the currents and voltages of the RLC series resonant circuit at different frequencies given in Table L10.2. Record the values in Table L10.2.

Table L10.2

F	I	V_R	V_C	V_L
$0.3\,f_r =$				
$0.6\,f_r =$				
$f_r =$				
$1.4\,f_r =$				
$1.7\,f_r =$				

11. Adjust the frequency of function generator to 0.3 f_r, 0.6 f_r, f_r, 1.4 f_r and 1.7 f_r, respectively. Repeat steps 7–9 and record the results in Table L10.3.

Table L10.3

f	V_R	V_C	V_L
0.3 f_r =			
0.6 f_r =			
f_r =			
1.4 f_r =			
1.7 f_r =			

12. Does the circuit current reach the maximum at resonant frequency f_r? If not, explain the reason.
13. Based on the data in Table L10.2, plot the current versus frequency curve (with current I in the vertical axis and frequency f in the horizontal axis).

Conclusion

Write your conclusions below:

Chapter 11
Mutual inductance and transformers

Objectives

After completing this chapter, you should be able to:

- understand the concept of mutual inductance
- understand the dot convention concept
- know the basic construction of a transformer
- know different types of transformers
- understand the characteristics of transformers
- determine the turns ratio of an ideal transformer
- calculate the current, voltage, impedance and power of the primary and secondary of a transformer
- understand the concept of impedance matching of a transformer
- know applications of transformers

The concept of self-inductance has been introduced in chapter 6. This chapter will introduce the mutual inductance and transformer. A transformer is a device that is built based on the principle of mutual inductance and can be used to increase or decrease the voltage or current, and transfer electric energy from one circuit to another. It also can be used for impedance matching. Transformers have a very wide range of applications in power systems, telecommunications, radio, instrumentation and many other electrical and electronics fields.

11.1 Mutual inductance

11.1.1 Mutual inductance and coefficient of coupling

As discussed in chapter 6 (section 6.3), when a changing current flows through a coil (inductor), it will produce an electromagnetic field around the coil, and as a result an induced voltage v_L will flow across it. The changing current in a coil that produces the ability to generate an induced voltage is called self-inductance. Mutual inductance is the ability of a coil to produce an induced voltage due to the changing of the current in another coil nearby.

In Figure 11.1, a coil L_1 is placed close to another coil L_2. When AC current i_1 flows through the first coil L_1, the changing of alternating current will produce a changing electromagnetic field and flux ϕ_1, resulting in a self-induced voltage v_1 across the first coil L_1. Since the two coils are very close, there is also a portion of magnetic flux, ϕ_{1-2}, that is produced by changing the electromagnetic field linked to the coil L_2, and consequently produces the induced voltage v_2 across the second coil L_2. The phenomenon of a portion of the flux of a coil linking to another coil is called inductive coupling, and this is the principle of mutual inductance.

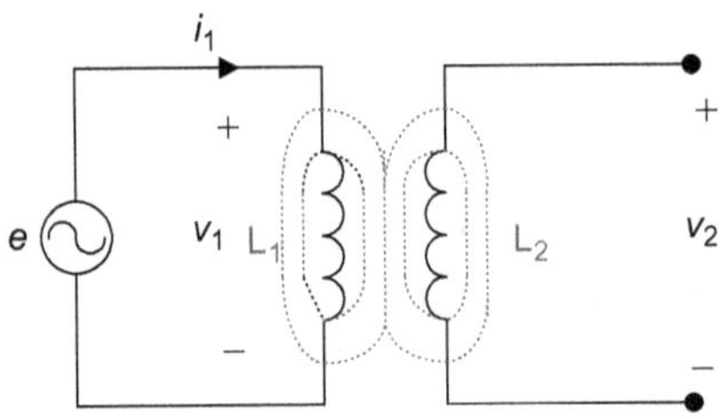

Figure 11.1 Magnetic coupling

Mutual inductance is denoted by L_M and can be expressed mathematically using the following formula: $L_M = k\sqrt{L_1 L_2}$.

Mutual inductance
An inducted voltage in one coil due to a current change in a nearby coil.

$$L_M = k\sqrt{L_1 L_2}$$

There are three factors that affect mutual inductance: inductances of the two coils L_1, L_2 and the coupling coefficient k. The coefficient of coupling k determines the degree of the coupling between the two coils, and it is the ratio of ϕ_{1-2} and ϕ_1:

$$k = \frac{\phi_{1-2}}{\phi_1}$$

ϕ_1 is the magnetic flux generated by the current i_1 in the first coil L_1, and ϕ_{1-2} is the portion of the magnetic flux that is generated by the current i_1 in the first coil L_1 and linked to the second coil as shown in Figure 11.2(a). ϕ_{1-2} is called the crossing link flux.

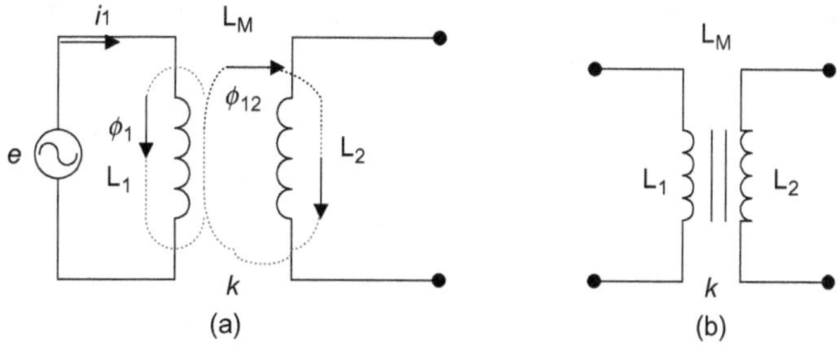

Figure 11.2 Mutual inductance

The induced voltage generated by a changing current (AC) that flows through the self-inductance coil L_1 is given by $v_1 = L(di_1)/dt$ (chapter 6). So when the AC current i_1 flows through the second coil L_2, the induced voltage in the coil L_2 is given by $v_2 = L_M(di_1)/dt$, or $\dot{V}_2 = jL_M\dot{I}_1$ in the phasor form.

In practice, not all of the magnetic flux generated by current i_1 will pass through L_1 and L_2, and the portion of the magnetic flux that does not link with L_1 and L_2 is known as a leakage flux. The closer the two coils are placed (or if the two coils have a common core as shown in Figure 11.3(b)), the higher the cross-linking flux $\phi_{1\text{–}2}$ and the lower the leakage flux.

The full-coupling occurs when $k = (\phi_{1\text{–}2})/(\phi_1) = 1$, i.e. $\phi_{1\text{–}2} = \phi_1$, when all of the flux link coils 1 and 2, and there will be no leakage flux. If the gap between the two coils is large, it will cause the cross-linking flux to decrease, the leakage flux to increase, and the coupling coefficient k to decrease. k is in the range between 0 and 1 ($0 \le k \le 1$).

Coefficient of coupling

- The coefficient of the coupling: $k = (\phi_{1\text{–}2})/\phi_1 \quad (0 \le k \le 1)$.
- ϕ_1: The flux generated by the current i_1 in the first coil L_1.
- $\phi_{1\text{–}2}$: The flux generated by the current i_1 in the coil L_1 cross-linking to coil L_2.

11.1.2 Dot convention

The polarity of the induced voltage across the mutually coupled coils can be determined by the dot convention method. This method can be used to indicate whether the induced voltage in the second coil is in phase or out of phase with the voltage in the first coil.

The dot convention method places two small phase dots (.) or asterisks (*), one on the coil L_1 and the other on the coil L_2, to indicate that polarities of the induced voltage v_1 in the coil L_1 and v_2 in the adjacent coil L_2 are same at these points, as shown in Figure 11.3. This means that the dotted terminals of coils should have the same voltage polarity at all time, and dotted terminals are known as corresponding terminals.

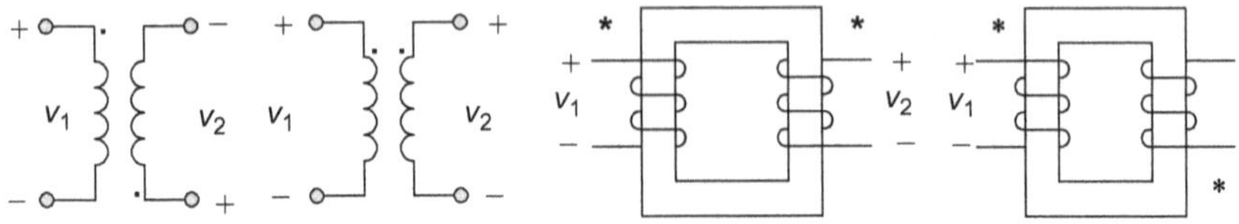

(a) (b)

Figure 11.3 Dot convention

Dot convention
Dotted terminals of coils have the same voltage polarity.

11.2 Basic transformer

11.2.1 Transformer

A transformer is an electrical device formed by two coils that are wound on a common core. You may have seen transformers on top of utility poles. A transformer uses the principle of mutual inductance to convert AC electrical energy from input to output. Recall that mutual inductance is the ability of a coil to produce induced voltage due to the changing of current in another coil nearby.

Figure 11.4 shows two simplified transformer circuits. A changing current from the AC voltage source in the first coil produces a changing magnetic field, inducing a voltage in the second coil. The first coil is called primary winding, and the second coil connected to the load Z_L is called secondary winding.

Structurally, the transformers are categorized as two main types: the air-core and iron-core transformers. The symbols for them are shown in Figure 11.4 (a and b), respectively (inside the dashed lines).

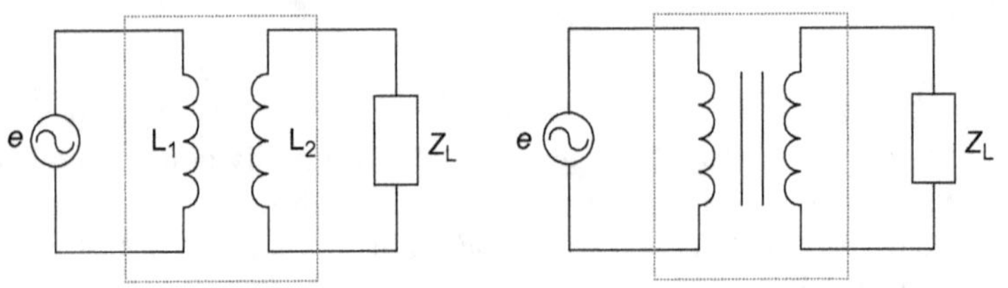

(a) (b)

Figure 11.4 Implified transformer circuits (a) air-core; (b) iron-core

Transformer

A transformer uses the principle of mutual inductance to convert AC electrical energy from input to output.

11.2.2 Air-core transformer

The air-core transformers are usually used in high-frequency circuits, such as in instrumentation, radio and TV circuits. An air-core transformer does not have a physical core, so it can be obtained by placing the two coils L_1 and L_2 close to each other, or by winding both the coils L_1 and L_2 on a hollow cylindrical-shaped core with isolating material as illustrated in Figure 11.5(b). The circuit of an air-core transformer is shown in Figure 11.5(a), there R_1 and R_2 represent the primary and secondary winding resistors of the transformer.

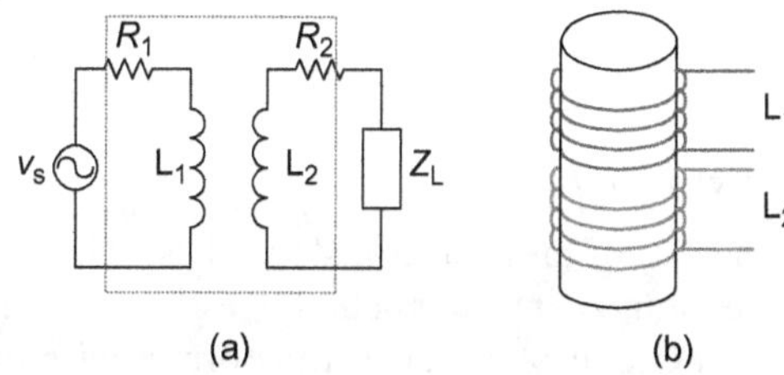

Figure 11.5 Air-core transformer

The air-core transformer is also known as a linear transformer. When the core of the transformer is made by the insulating material with constant permeability, such as air, plastic, wood, etc., it is a liner transformer.

Air-core transformer

A transformer uses the principle of mutual inductance to convert AC electrical energy from input to output.

11.2.3 Iron-core transformer

Iron-core transformers are usually used in audio circuits and power systems. The coils of the iron-core transformer are wound on the ferromagnetic material that are laminated sheets insulated to each other, as illustrated in Figure 11.6.

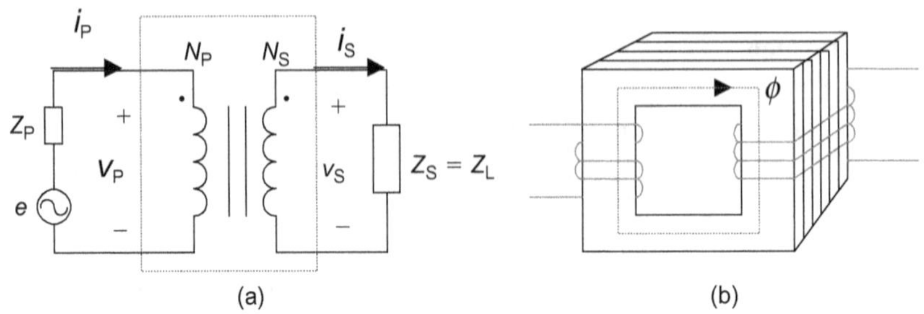

Figure 11.6 Iron-core transformer

When two coils are wound on a common core, it will have higher cross-linking flux and lower leakage flux. The ferromagnetic materials can provide an easy path for the magnetic flux. Furthermore, if two coils are wound on a common core, the flux generated in the coil L_1 will almost all link with the coil L_2. This means that the coupling coefficient k is close to 1, and this is the reason that iron-core transformer is usually considered as the ideal transformer ($k = 1$).

11.2.4 Ideal transformer

The coupling coefficient k of an ideal transformer is 1, i.e. ideal full-coupling, neglecting winding resistance and magnetic losses in the coils of the transformer. Figure 11.6(a) is a circuit of an ideal transformer with the voltage source, and the load, and the portion within the dashed line is the symbol of the ideal transformer.

An iron-core transformer is considered the ideal transformer because it uses ferromagnetic materials with high permeability as its core. Also the primary and secondary windings are wound on a common core, which have near zero leakage flux and can achieve a full-coupling ($k = 1$).

- Transformer parameters: The parameters of an ideal transformer in Figure 11.6(a) are listed in Table 11.1.

Table 11.1

Parameters	Name
v_P	Primary voltage
v_S	Secondary voltage
N_P	Number of turns on the primary coil
N_S	Number of turns on the secondary coil
i_P	Primary current
i_S	Secondary current
Z_P	Primary impedance
$Z_S = Z_L$	Secondary or load impedance

- Turns ratio n: The turns ratio of a transformer is the ratio of the number of turns, i.e. the number of turns on the secondary coil N_S to the number of turns on the primary coil N_P, which can be derived from the voltage ratio of the secondary and primary voltages. From Faraday's law described in chapter 6, $v_L = N\frac{d\phi}{dt}$ we can get:
  - the primary voltage $v_P = N_P \dfrac{d\phi}{dt}$
  - the secondary voltage $v_S = N_S \dfrac{d\phi}{dt}$

  Dividing v_S by v_P gives the transformer's turns ratio n:

$$\frac{v_S}{v_P} = \frac{N_S}{N_P} = n$$

If the transformer is an ideal transformer, i.e. the transformer has no power loss itself, the input power is equal to the output power, i.e. $p_P = p_S$ or $v_P i_P = v_S i_S$, so

$$i_P / i_S = v_S / v_P = n \tag{11.1}$$

$v_P = v_S/n$ can be obtained from (11.1), and also $i_P = n\ i_S$. The primary impedance can be obtained by substituting v_P and i_P into Z_P as follows:

$$Z_P = \frac{v_P}{i_P} = \frac{v_S/n}{n\ i_S} = \frac{1}{n^2} Z_L$$

or

$$n^2 = Z_L/Z_P, \quad n = \sqrt{Z_L/Z_P}$$

where the secondary impedance is the load Z_L

$$Z_L = \frac{v_S}{i_S}$$

Turns ratio

- Instantaneous form: $n = N_S/N_P = v_S/v_P = i_P/i_S = \sqrt{Z_L/Z_P}$, or $Z_L = n^2 Z_P$
- Phasor form: $n = N_S/N_P = \dot{V}_S/\dot{V}_P = \dot{I}_P/\dot{I}_S = \sqrt{Z_L/Z_P}$

Power

- Instantaneous form: $p_S = i_S v_S, \quad p_P = i_P v_P$
- Phasor form: $\dot{P}_S = \dot{I}_S \dot{V}_S, \quad \dot{P}_P = \dot{I}_P \dot{V}_P$

- Conversion of the voltage, current and impedance: The expressions of the transformer's turns ratio indicate that a transformer can be used to convert voltage, current and impedance.

- Voltage conversion:
  - From the primary to the secondary, multiplying by n: $v_S = nv_P$
  - From the secondary to the primary, multiplying by $1/n$: $v_P = (1/n)v_S$
- Current conversion:
  - From the primary to the secondary, multiplying by $1/n$: $i_S = (1/n)(i_P)$
  - From the secondary to the primary, multiplying by n: $i_P = n\, i_S$
- Impedance conversion:
  - From the primary to the secondary, multiplying by $1/n^2$: $Z_P = (1/n^2)(Z_L)$
  - From the secondary to the primary, multiplying by n^2: $Z_L = n^2 Z_P$

(The converted impedance is also called the reflected impedance, meaning the reflection of the primary impedance results in the secondary impedance.)

Transformer parameters conversion

- Voltage conversion: $v_S = nv_P, v_P = (1/n)(v_S)$
- Current conversion: $i_S = (1/n)(i_P), \quad i_P = n\, i_S$
- Impedance conversion: $Z_L = n^2 Z_P, \quad Z_P = (1/n^2)(Z_L)$

Example 11.1: The number of turns on the primary is 40 for an ideal transformer, and the number of turns on the secondary is 100. $\dot{V}_P = 50\,\text{V}$, $\dot{I}_P = 5\,\text{A}$ and $Z_L = 2\ \Omega$. Determine the transformer's turns ratio, secondary voltage, secondary current, primary impedance (reflected from the secondary) and the primary power (the amplitude only).

Solution:

- $n = \dfrac{N_S}{N_P} = \dfrac{100}{40} = 2.5$
- $\dot{V}_S = n\dot{V}_P = (2.5)(50\,\text{V}) = 125\,\text{V}$
- $\dot{I}_S = \dfrac{\dot{I}_P}{n} = \dfrac{5\,\text{A}}{2.5} = 2\,\text{A}$
- $Z_P = \dfrac{Z_L}{n^2} = \dfrac{2\,\Omega}{2.5^2} = 0.32\,\Omega$
- $\dot{P}_S = \dot{I}_S\dot{V}_S = (2\,\text{A})(125\,\text{V}) = 250\,\text{W} = 0.25\,\text{kW}$

11.3 Step-up and step-down transformers

11.3.1 Step-up transformer

A step-up transformer is a transformer that can increase its secondary voltage. Since a step-up transformer always has more secondary winding turns than the primary, the secondary voltage of a step-up transformer (v_S) is always higher than the primary voltage (v_P), i.e. $v_S > v_P$. The value of the secondary voltage depends on the turns ratio (n). The equation of $n = N_S/N_P = v_S/v_P$ indicates that to have a higher secondary voltage, the number of turns on the secondary winding must be greater than that of the primary, i.e. $N_S > N_P$ as illustrated in Figure 11.7(a), meaning the turns ratio $n = (N_S/N_P) > 1$. This is an important characteristic of a step-up transformer.

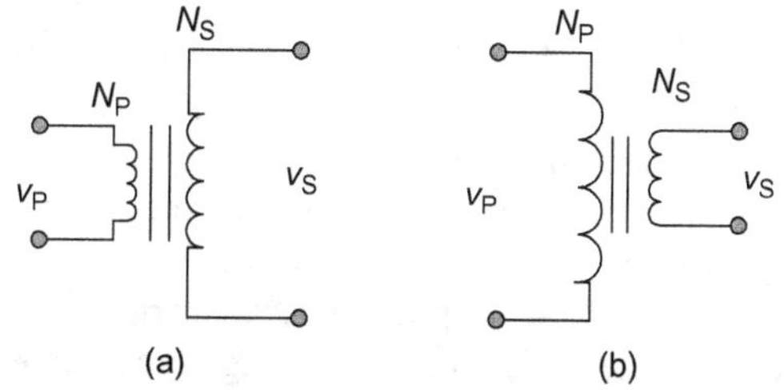

Figure 11.7 (a) Step-up and (b) step-down transformers

Step-up transformer

- $v_S > v_P$
- $N_S > N_P$
- $n > 1$

11.3.2 Step-down transformer

A step-down transformer is a transformer that can decrease its secondary voltage. Since a step-down transformer always has less turns on the secondary winding than the primary, the secondary voltage of a step-down transformer (v_S) is always lower than the primary voltage (v_P), i.e. $v_S < v_P$. The value of the secondary voltage depends on the turns ratio (n). The equation $n = N_S/N_P = v_S/v_P$ indicates that to have a voltage that is lower in secondary than primary, the number of turns on the secondary coil must be lesser than primary, i.e. $N_S < N_P$ as illustrated in Figure 11.7(b), meaning the turns ratio $n = (N_S/N_P) < 1$. This is an important characteristic of a step-down transformer, which is opposite of a step-up transformer.

Step-down transformer

- $v_S < v_P$
- $N_S < N_P$
- $n < 1$

Example 11.2: If a transformer has 125 turns of secondary windings and 250 turns of primary windings, calculate its turns ratio and determine if it is a step-up or a step-down transformer.

Solution: Since $N_S = 125$, $N_P = 250$ and $N_S < N_P$

or $\quad n = N_S/N_P = 125/250 = 1/2 = 0.5 < 1$, i.e. $n < 1$

it is a step-down transformer.

11.3.3 Applications of step-up and step-down transformers

As mentioned in the previous section, transformers can be used to convert voltage, current and impedance. In the power system, the basic usage of transformers is stepping up or stepping down the voltage or current, which will require converting voltage or current from primary to secondary winding. The functions of step-up and step-down transformers are to increase or decrease the voltage of their secondary windings, and have important applications in the power transmission system. A simplified power transmission system is illustrated in Figure 11.8.

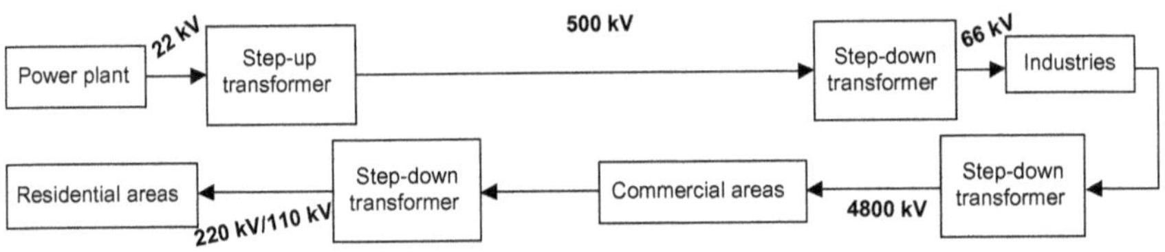

Figure 11.8 Power transmission system

The voltage generated from the generator of a power plant needs to rise to a very high value through the step-up transformer so that it can be delivered through long-distance transmission lines. This can reduce the loss of energy or power created due to the winding resistance in the line ($I^2R_w = P_{Loss}$) for a long-distance-line transmission, and improve the efficiency of the electricity transmission.

Decreasing the current to reduce the power loss on the transmission line may reduce the output power ($P = IV$) of the transmission system. If the voltage is increased through the step-up transformer before the transmission, it can maintain the same output power, but reduce the power loss on the line, i.e.:

$$\vec{P} = (I \downarrow)(V \uparrow) \Rightarrow (I^2 \downarrow)(R) = P_{Loss} \downarrow$$

If a step-up transformer is used to increase the voltage by 100 V, then the current will reduce by 100 A $[v_S \uparrow = (n \uparrow)(v_P), \quad i_S \downarrow = (1/n \uparrow)(i_P)]$, and the loss of the power due to the winding resistance in the line will reduce to 10 000 W, since $I^2R_w = P_{Loss}$ and $I^2 \downarrow R_w \Rightarrow P_{Loss} \downarrow$.

The local distribution stations require step-down transformers to reduce the very high voltage by the long distance transmission and can send it to commercial or residential areas.

11.3.4 Other types of transformers

There are other types of commonly used transformers listed as follows:

- Center-tapped transformer: It has a tap (connecting point) in the middle of the secondary winding, and it can provide two balanced output voltages with the same value, as shown in Figure 11.9(a).

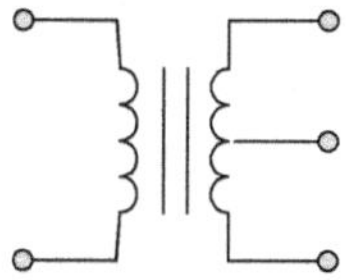

Figure 11.9 (a) Center-tapped transformer

- Multiple-tapped transformer: It has multiple taps in the secondary winding, and it can provide several output voltages with different values, as shown in Figure 11.9(b).

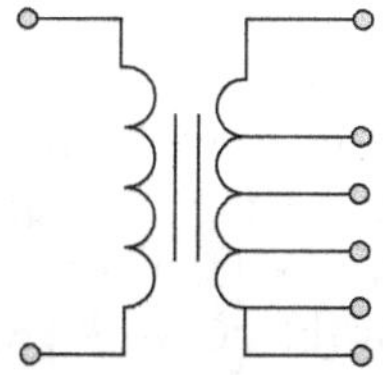

Figure 11.9 (b) Multiple-tapped transformer

- Adjustable (or variable) transformer: The output voltage of adjustable (or variable) transformer across the secondary winding is adjustable. The secondary winding of the adjustable transformer can provide an output voltage that may be variable in a range of zero to the maximum values. An adjustable transformer is shown in Figure 11.9(c).

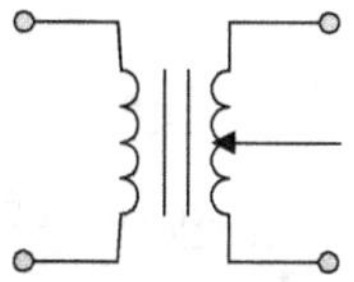

Figure 11.9 (c) Adjustable transformer

- Auto-transformer: It is a transformer with only a single winding, which is a common coil for both the primary and the secondary coils, and a portion of the common coil acts as part of both the primary and secondary coils, as shown in Figure 11.9(d). An auto-transformer can be made smaller and lighter.

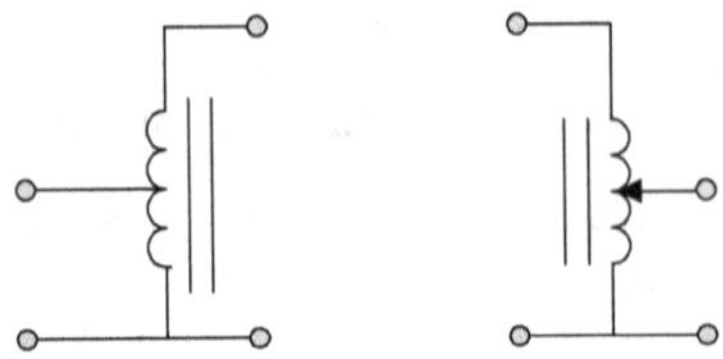

Figure 11.9 (d) Auto-transformer

11.4 Impedance matching

In addition to stepping-up and stepping-down voltages, a transformer has another important application, matching the load and source impedance in a circuit to achieve the maximum power transfer from the source to the load. It is known as impedance matching.

11.4.1 Maximum power transfer

The theory of maximum power transfer in the DC circuits was introduced in chapter 5, i.e. the maximum power delivered from a source to a load in a circuit can be achieved when the load resistance is equal to the internal resistance of the source ($R_L = R_S$), or when the load resistance is equal to the Thevenin/Norton equivalent resistance of the network ($R_L = R_{TH} = R_N$).

This theory can also be applied to an AC circuit by replacing the resistance with impedance. When the load impedance Z_L is equal to the source internal impedance Z_S, the power received by the load from the source reaches the maximum, this is shown in Figure 11.10.

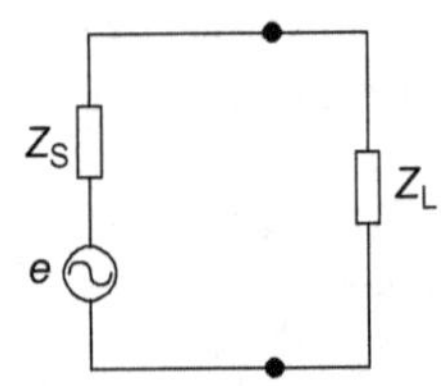

Figure 11.10 Impedance matching

Maximum power transfer
When $Z_L = Z_S$, the power delivered from the source to the load reaches the maximum.

11.4.2 Impedance matching

In the practical circuits (or Thevenin's equivalent circuits), the internal resistance of the source is fixed, usually is not matching with the load impedance, and also not adjustable. In this case, a transformer with an appropriate turns ratio n can be placed between the load and source to make the load impedance and the source internal resistance equal, and to achieve the maximum power transfer, i.e. $n = \sqrt{Z_L/Z_P}$.

Impedance matching
Place a transformer with $n = \sqrt{Z_L/Z_P}$ between the source and the load to achieve maximum power transfer.

Example 11.3: A simplified amplifier circuit is illustrated in Figure 11.11(a). The circuit within dashed lines is Thevenin's equivalent circuit for the amplifier circuit, and its internal resistance is 100 Ω. How do we deliver the maximum power to the speaker if the resistance of the speaker is 4 Ω (so the speaker can have the maximum volume)?

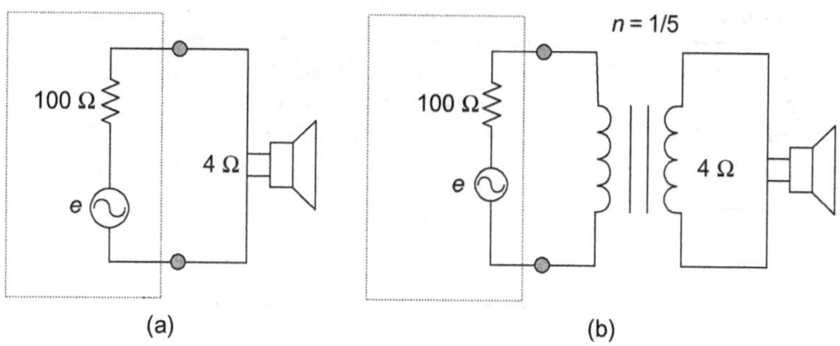

Figure 11.11 Circuits for Example 11.3

Solution:

- Since the load impedance ($Z_L = R_L = 4\ \Omega$) does not match with the source internal impedance ($Z_S = R_S = 100\ \Omega$) currently, the maximum power cannot be delivered to the speaker if the source and load are connected directly.
- Choose an audio transformer with the appropriate turns ratio n, i.e.

$$n = \sqrt{\frac{Z_L}{Z_P}} = \sqrt{\frac{4}{100}} = 0.2 = \frac{1}{5}$$

- Therefore, if placing an impedance matching transformer with the turns ratio of 1/5 between the amplifier and speaker as illustrated in Figure 11.11(b), the speaker will have the maximum volume.

Summary

Mutual inductance

- Mutual inductance: An inducted voltage in one coil due to a change current in the other nearby coil.
- $L_M = k\sqrt{L_1L_2}$.
- Coefficient of coupling: $k = \dfrac{\phi_{1-2}}{\phi_1} \quad (0 \le k \le 1)$.
- Dot conversion: Dotted terminals of coils have the same voltage polarity.

Basic transformer

- Transformer: It uses the principle of mutual inductance to convert AC electrical energy from input to output.
- The parameters of an ideal transformer ($k = 1$):

Parameters	Name
v_P	Primary voltage
v_S	Secondary voltage
N_P	Number of turns on the primary coil
N_S	Number of turns on the secondary coil
i_P	Primary current
i_S	Secondary current
Z_P	Primary impedance
$Z_S = Z_L$	Secondary or load impedance

- Turns ratio: $n = N_S/N_P = v_S/v_P = i_P/i_S = \sqrt{Z_L/Z_P}$ or $Z_L = n^2Z_P$
  In phasor form: $n = N_S/N_P = \dot{V}_S/\dot{V}_P = \dot{I}_P/\dot{I}_S = \sqrt{Z_L/Z_P}$
- Power: $p_S = i_Sv_S, \quad p_P = i_Pv_P$
  $\dot{P}_S = \dot{I}_S\dot{V}_S \quad \dot{P}_P = \dot{I}_P\dot{V}_P$
- Transformer conversion (V, I and Z):
  - Voltage conversion: $v_S = n\,v_P, \quad v_P = (1/n)(v_S)$
  - Current conversion: $i_S = (1/n)(i_P), \quad i_P = n\,i_S$
  - Impedance conversion: $Z_L = n^2Z_P, \quad Z_P = (1/n^2)(Z_L)$
- Step-up transformer:
  - $v_S > v_P$
  - $N_S > N_P$
  - $n > 1$
- Step-down transformer:
  - $v_S < v_P$

- $N_S < N_P$
- $n < 1$

- Impedance matching: Place a transformer with the turns ratio $n = \sqrt{Z_L/Z_P}$ between the source and the load to achieve maximum power transfer from the source to the load.

Experiment 11: Transformer

Objectives

- Experimentally verify the equation for calculating the turns ratio of the transformer.
- Experimentally verify the theory of impedance matching transformer.
- Analyse experimental data, circuit behaviour and performance, and compare them to the theoretical equivalents.

Background information

- The turns ratio of the transformer: $n = N_S/N_P = v_S/v_P = i_P/i_S = \sqrt{Z_L/Z_P}$
- Impedance matching: Place a transformer with the turns ratio $n = \sqrt{Z_L/Z_P}$ between the source and the load (if $Z_{in} \neq Z_L$) to achieve maximum power transfer from the source to the load.
- Formula for the transformer's impedance matching: $Z_L = n^2 Z_P$

Equipment and components

- Multimeter
- Breadboard
- Function generator
- A small transformer with lower secondary voltage (12–25 V)
- A small audio transformer
- Continuously adjustable auto-transformer
- Several resistors
- A small audio speaker

Procedure

Part I: Turns ratio of a transformer

1. Choose a small transformer with lower secondary voltage (12–25 V) in the laboratory, and measure the primary (R_P) and secondary (R_S) resistances of the transformer using a multimeter (ohmmeter function). Record the values in Table L11.1.

Table L11.1

Primary resistance R_P	Secondary resistance R_S	Calculated turns ratio $n = V_S/V_P$	Measured turns ratio n

2. Calculate the turns ratio n of the transformer using the primary and secondary voltages of the transformer (from the nameplate). Record the values in Table L11.1.
3. Plug the primary of the adjustable auto-transformer into the AC power outlet, and connect the secondary adjustable auto-transformer to the primary of the small transformer, as shown in Figure L11.1(a).
   Note: The continuously adjustable auto-transformer is used as a variable AC voltage source. If the primary voltage of the small transformer is not too high, a function generator can be used to replace the adjustable auto-transformer.

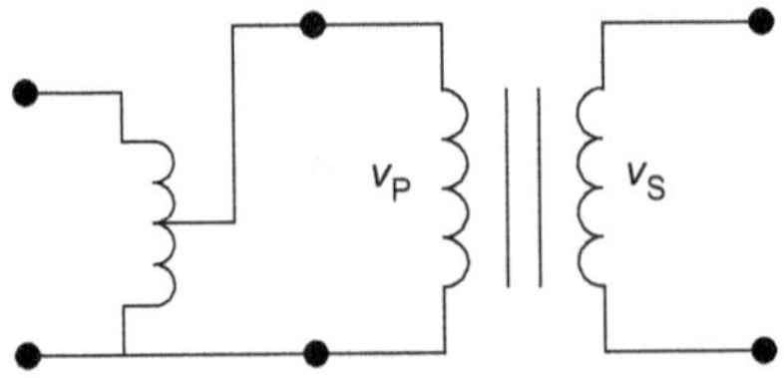

Figure L11.1 (a) Auto-transformer and small transformer

4. Adjust the output voltage of the auto-transformer until it is slightly lower than the rated primary voltage of the small transformer, then measure the primary and secondary voltages (RMS values) of the small transformer using a multimeter (voltmeter function). Calculate the turns ratio n of the small transformer using measured primary and secondary voltages, and record the value in the column 'Measured turns ratio' in Table L11.1.
5. Connect a suitable load resistor R_L to the secondary of the small transformer as shown in Figure L11.1(b). Determine the value of resistor R_L by calculating the secondary current I_S and power P_S, and make sure that I_S and P_S will not exceed the rated current and power of the small transformer after connecting R_L to the secondary. Calculate the primary current I_P ($I_P = P_P/V_P$) and record the value in Table L11.2.

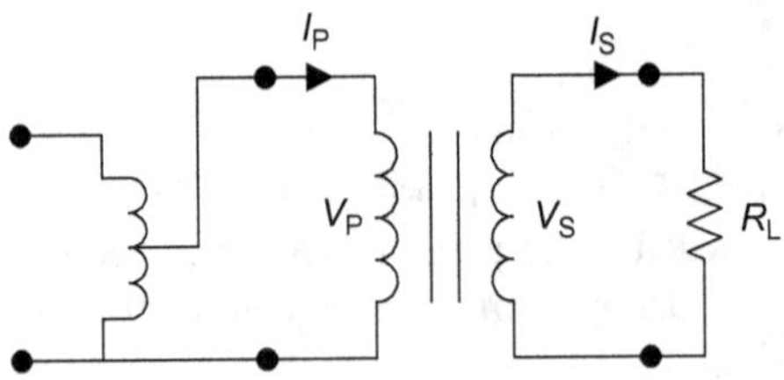

Figure L11.1 (b) The circuit for step 5

Table L11.2

Primary current I_P	Secondary current I_S	Turns ratio: $n = I_P/I_S$

6. Measure the secondary current I_S using either the direct or indirect method. Record the value in Table L11.2.
7. Calculate turns ratio n of the transformer using primary and secondary currents in Table L11.2, and record the value. Compare the turns ratio n of the transformer in Tables L11.1 and L11.2. Are there any significant differences? If so, explain the reasons.

Part II: Impedance matching

1. Set the frequency of the function generator to 2 kHz. Then connect a small audio speaker to the two terminals of the function generator as shown in Figure L11.2(a)

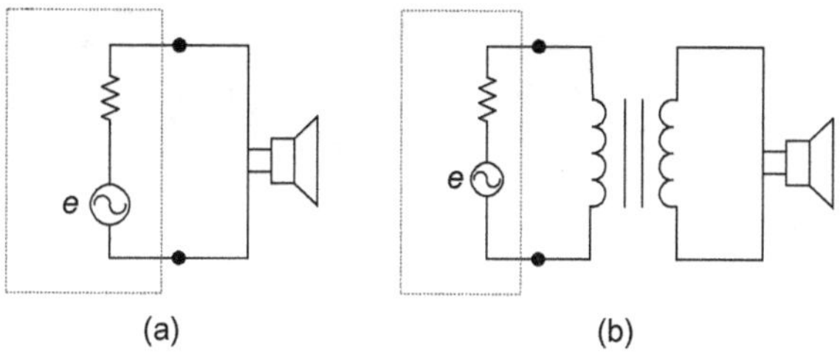

Figure L11.2 Circuit for Part II

2. Adjust the output voltage of the function generator to a suitable value so that the speaker reaches a comfortable listening volume.
3. Measure the voltage across the two terminals of the speaker in Figure L11.2(a). Record the value in Table L11.3 in 'Without transformer' row.

Table L11.3

Voltage across the speaker	Without transformer:
	With transformer:
Transformer turns ratio: $n = \sqrt{Z_L/Z_P}$	

4. Calculate the turns ratio $(n = \sqrt{Z_L/Z_P})$ of a transformer that can be used as the impedance matching. Record the value in Table L11.3.
   - The output impedance of the function generator Z_P usually is about 600 Ω.
   - The impedance of the small speaker Z_L usually is about 8 Ω.
5. Find a small audio transformer that has a turns ratio n closer to the calculated n in step 4, and connect it between the function generator and speaker as shown in Figure L11.2(b).

6. Measure the voltage across the speaker in Figure. L11.2(b). Record the value in Table L11.3 in the row ‘With transformer’.
7. Compare the volume of the speaker *with* and *without* the transformer, and explain the reason in the conclusion.

Conclusion

Write your conclusions below:

Chapter 12
Circuits with dependent sources

Objectives

After completing this chapter, you should be able to:

- understand the concept of the dependent source
- define four types of dependent sources
- know how to convert dependent sources equivalently
- understand the methods for analysing circuits with dependent sources

The DC and AC circuits we have discussed in the previous chapters have independent voltage/current sources (Figure 12.1) and will not be affected by other voltages and currents in the circuit. The circuits we will analyze in this chapter have dependent (or controlled) sources, in which the source voltage or current is a function of other voltage or current in the circuit. Dependent sources are a useful concept in modelling and analysing electronic components, such as transistors, amplifiers, filters, etc.

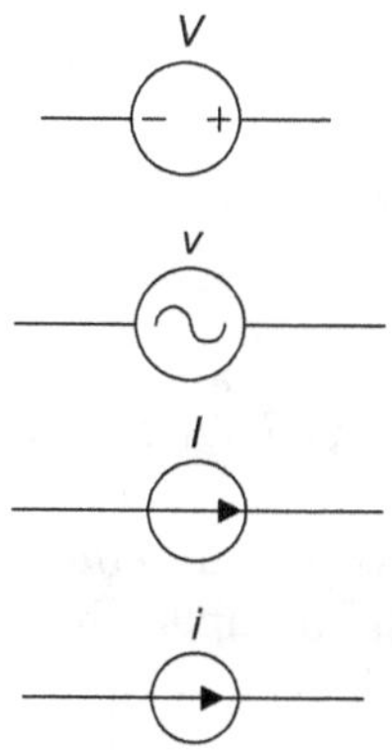

Figure 12.1 Independent sources

12.1 Dependent sources

12.1.1 Dependent (or controlled) sources

As the name suggests, when a source voltage or current is controlled by or dependent on other voltage or current in the circuit, it is called a dependent (or controlled) source. The dependent sources can be categorized into the following four types according to whether it is controlled by a circuit voltage or current, as well as whether the dependent source itself is a voltage source or current source:

- voltage-controlled voltage source (VCVS)
- voltage-controlled current source (VCCS)
- current-controlled voltage source (CCVS)
- current-controlled current source (CCCS)

The above dependent sources can be represented by the symbols in Figure 12.2.

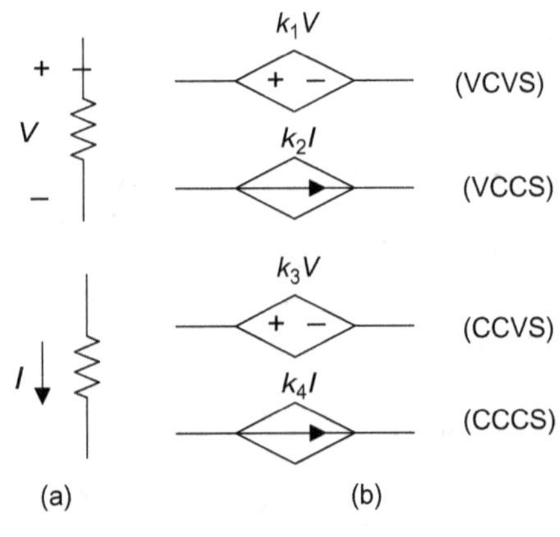

Figure 12.2 Dependent sources: (a) controlling sources and (b) controlled sources

In this figure, k_1, k_2, k_3 and k_4 are called control coefficients or gain parameters.

A voltage-controlled source has a *voltage* across its two terminals that equals to a control coefficient k multiplied by a controlling voltage or current elsewhere in the same circuit. A current-controlled source has a *current* in its branch that equals to a control coefficient k multiplied by a voltage or current elsewhere in the same circuit.

$5\dot{I}$ in the circuit of Figure 12.3(a) represents a CCCS. Its control coefficient k is 5, and the current $\dot{I}$ is a controlling current through the 5 Ω resistor branch in the same circuit. $8\dot{V}$ in the circuit of Figure 12.3(b) is a VCVS. Its control coefficient is 8, and the voltage $\dot{V}$ is a controlling voltage across resistor R_2 in the same circuit.

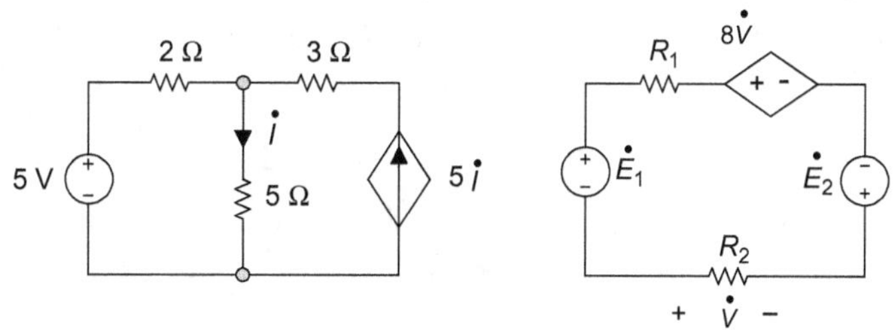

Figure 12.3 Circuits with dependent sources

After you take an analogue electronics course, you will understand that a good example for modelling a CCCS is a transistor circuit. Based on the property of a bipolar transistor, a current amplifier, its large collect current i_c is proportional to the small base current i_b according to the relationship $i_c = \beta i_b$. In this equation, the current gain β is the same as the control coefficient k in the dependent source.

Dependent (controlled) sources

The source voltage or current is a function of the other voltage or current in the same circuit.

12.1.2 Equivalent conversion of dependent sources

Equivalent conversion of dependent sources is the same as the equivalent conversion of independent sources. For instance, the voltage-controlled source in Figure 12.4(a) can be converted equivalently to a current-controlled source as shown in Figure 12.4(b), and vice versa. Internal resistance R_S of the source does not change before and after the conversion, just apply Ohm's law to convert the source.

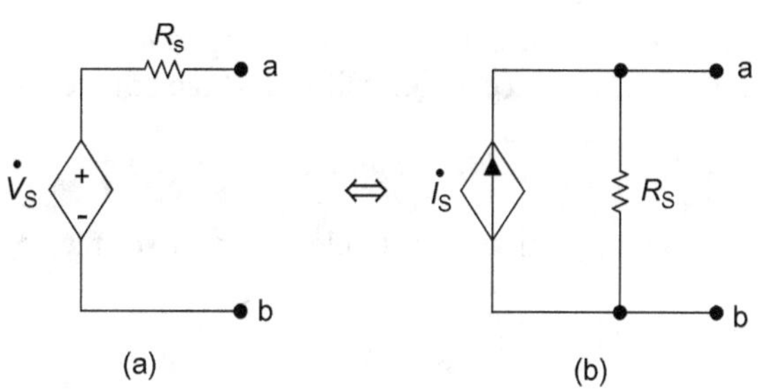

Figure 12.4 Equivalent conversion: (a) $R_S = R_S$, $\dot{V}_S = \dot{I}_S R_S$ and (b) $R_S = R_S$, $\dot{I}_S = \dot{V}_S / R_S$

Example 12.1: The VCVS in Figure 12.5(a) can be converted equivalently to a VCCS as shown in Figure 12.5(b).

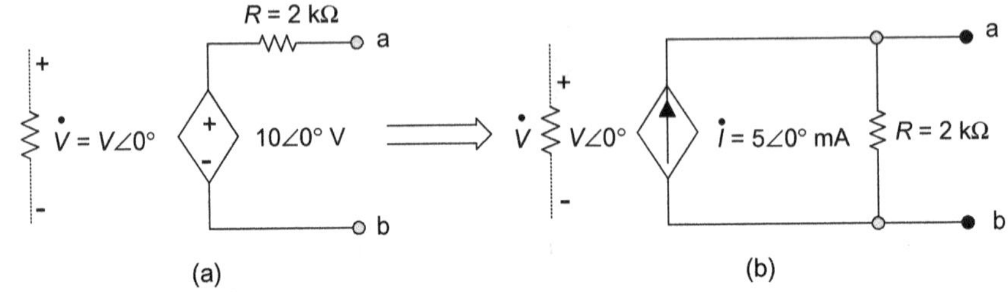

Figure 12.5 Circuit for Example 12.1. (a) Voltage-controlled voltage source (VCVS) and (b) voltage-controlled current source (VCCS), $\dot{I} = \dot{V}/R = 10\ \mathrm{V}\angle 0^\circ / 2\,\mathrm{k}\Omega = 5\angle 0^\circ\ \mathrm{mA}$

Example 12.2: The CCCS in Figure 12.6(a) can be converted equivalently to a CCVS as shown in Figure 12.6(b).

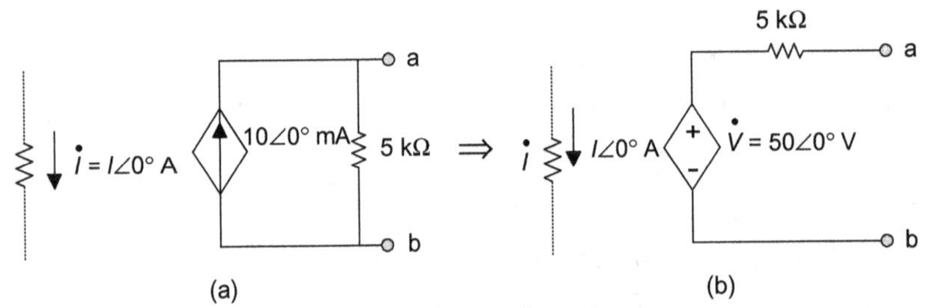

Figure 12.6 Circuit for Example 12.2. (a) Current-controlled current source (CCCS) and (b) current-controlled voltage source (CCVS), $\dot{V} = \dot{I}R_S = (10\angle 0^\circ\ \mathrm{mA})(5\,\mathrm{k}\Omega) = 50\angle 0^\circ\ \mathrm{V}$

Equivalent conversion of dependent sources

The same as the equivalent conversion of independent sources:

- controlled current source→controlled voltage source:

$$R_S = R_S, \quad \dot{V}_S = \dot{I}_S R_S$$

- controlled voltage source→controlled current source:

$$R_S = R_S, \quad \dot{I}_S = \frac{\dot{V}_S}{R_S}$$

12.2 Analysing circuits with dependent sources

The analysing methods for circuits with dependent sources are similar to that for circuits with independent sources. The following examples will describe these methods.

Example 12.3: Determine the current I in the circuit of Figure 12.7(a).

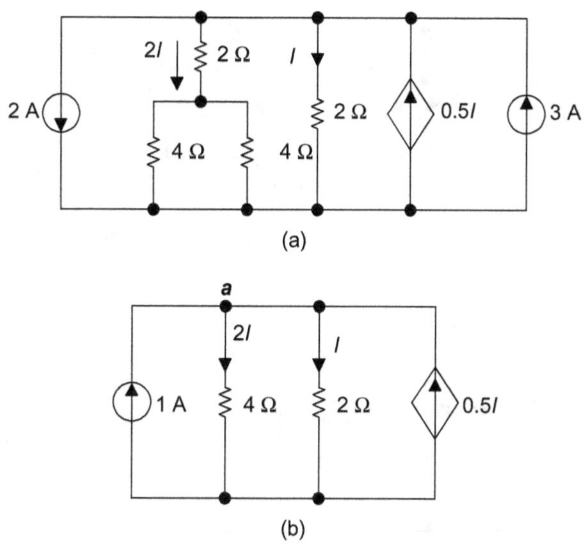

Figure 12.7 Circuits for Example 12.3

Solution: Simplify and convert the circuit of Figure 12.7(a) to that in Figure 12.7(b).

There,

$$3\,\text{A} - 2\,\text{A} = 1\,\text{A}$$

$$2\,\Omega + 4\,\Omega//4\,\Omega = 4\,\Omega$$

Note: This circuit has a CCCS, simplify the circuit without changing the CCCS (both controlling branches and controlled source).

Write the KCL equation for the node a in Figure 12.7(b):

$$\Sigma I = 0: \quad 1\,\text{A} - 2I - I + 0.5I = 0$$

Current I can be solved from the above equation:

$$-2.5I = -1\text{ A}$$
$$\therefore\ I = 0.4\text{ A}$$

Example 12.4: Determine the voltage V in the circuit of Figure 12.8.

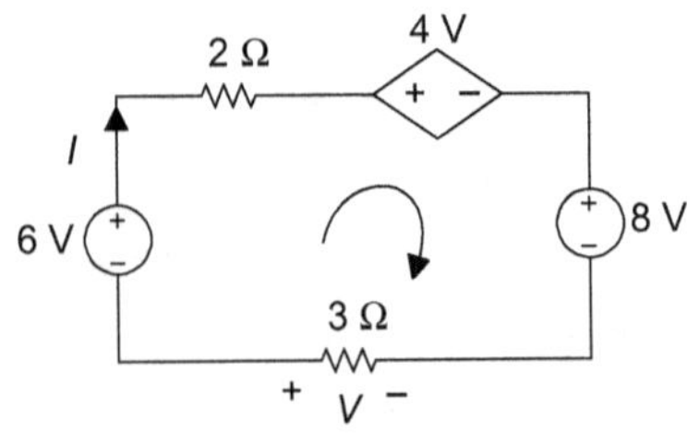

Figure 12.8 Circuit for Example 12.4

Solution:

Applying KVL, $\sum V = 0$:

$$-6 + 2I + 4V + 8 + 3I = 0$$

That is

$$2 + 5I + 4V = 0$$

Substituting $V = -3I$ into the above equation gives:

$$2 + 5I + 4(-3I) = 0$$

$$2 + 5I - 12I = 0$$

Solving for I:

$$2 - 7I = 0, \quad I \approx 0.29\text{ A}$$

Solving for V:

$$V = -3I = (-3\,\Omega)(0.29\text{ A}) = -0.87\text{ V}$$

Example 12.5: Write node voltage equations for the circuit in Figure 12.9 using the node voltage analysis method. (Write KVL by treating the dependent source as an independent source first, and then represent the control quantity as node voltages.)

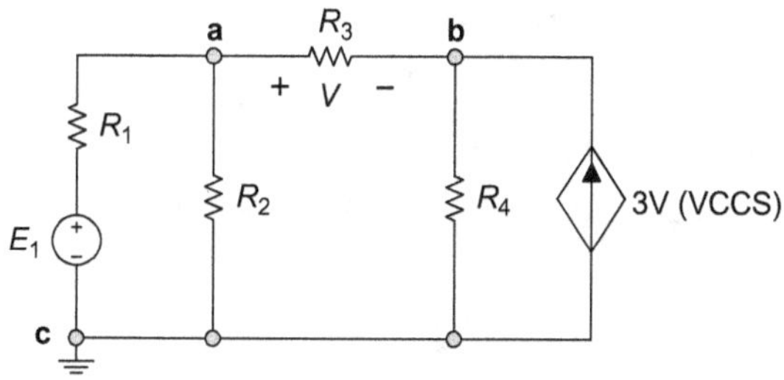

Figure 12.9 Circuit for Example 12.5

Solution: The procedure for applying the node voltage analysis method (chapter 4, section 4) to the above circuit is as follows:

1. Label nodes a, b and c, and choose ground c as the reference node as shown in Figure 12.9.
2. Write KCL equations to $n-1 = 3-1 = 2$ nodes (nodes a and b) by inspection.

Node a:

$$\left(\frac{1}{R_1}+\frac{1}{R_2}+\frac{1}{R_3}\right)V_a-\frac{1}{R_3}V_b=\frac{1}{R_1}E_1 \tag{12.1}$$

Node b:

$$-\frac{1}{R_3}V_a+\left(\frac{1}{R_3}+\frac{1}{R_4}\right)V_b=3V \tag{12.2}$$

Substituting the control voltage $V = V_a - V_b$ to (12.2) gives:

$$-\frac{1}{R_3}V_a+\left(\frac{1}{R_3}+\frac{1}{R_4}\right)V_b=3(V_a-V_b) \tag{12.3}$$

3. Solving (12.1) and (12.3) can determine the node voltage V_a and V_b (if R_1, R_2, R_3 and E_1 are given).

Example 12.6: Use the mesh current analysis method to write mesh equations for the circuit in Figure 12.10. (Write KVL by treating the dependent source as an independent source first, and then represent the controlling quantity as mesh current.)

Solution: The procedure for applying the mesh current analysis method (chapter 4, section 3) to the above circuit is as follows:

1. Label all the reference directions for each mesh current I_1, I_2 and I_2 (clockwise) as shown in Figure 12.10.

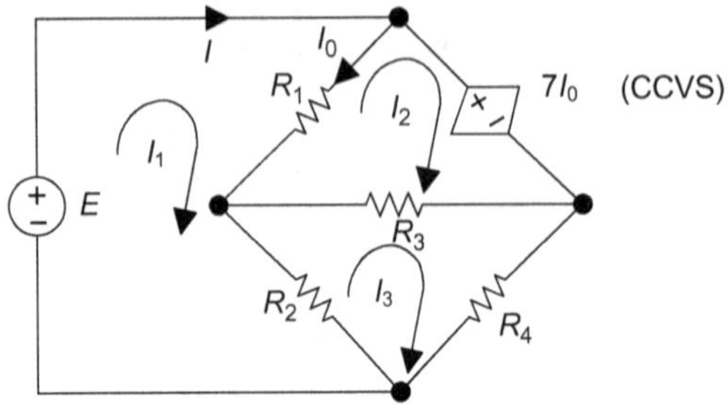

Figure 12.10 Circuit for Example 12.6

2. Apply KVL around each mesh (windowpane), and ensure the number of KVL equations is equal to the number of meshes (there are three meshes in Figure 12.10).
   Mesh 1:

$$(R_1 + R_2)I_1 - R_1 I_2 - R_2 I_3 = E \tag{12.4}$$

   Mesh 2:

$$-R_1 I_1 + (R_1 + R_3)I_2 - R_3 I_3 = -7I_0 \tag{12.5}$$

   Mesh 3:

$$-R_2 I_1 - R_3 I_2 + (R_2 + R_3 + R_4)I_3 = 0$$

Substituting the controlling current $I_0 = I_1 - I_2$ to (12.5) yields

$$-R_2 I_1 + (R_1 + R_3)I_2 - R_3 I_3 = -7(I_1 - I_2) \tag{12.6}$$

3. Solving all the three simultaneous equations, (12.4), (12.5) and (12.6), resulting from step 2 can determine the three mesh currents I_1, I_2 and I_3.

Example 12.7: Determine the branch current I for the circuit in Figure 12.11 by using the superposition theorem. (The dependent source will not act separately

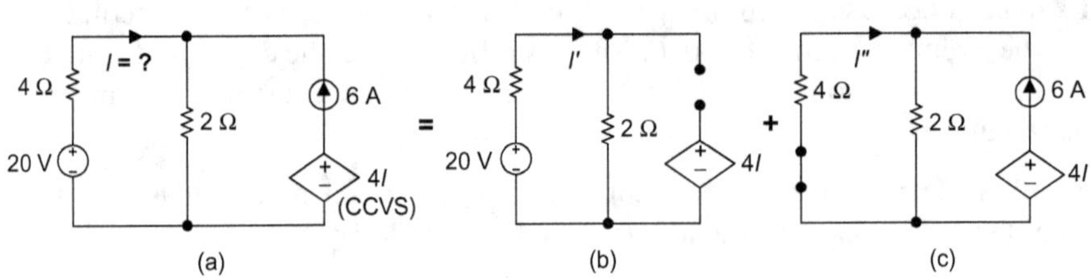

Figure 12.11 Circuits for Example 12.7

in the superposition theorem. Do not change the dependent source in the circuit when another independent source is acting in the circuit.)

Solution: The procedure for using the superposition theorem (chapter 5, section 5.1) to the above circuit is as follows:

1. Choose 20 V voltage source applied to the circuit first, replace the 6 A current source with an open circuit as shown in Figure 12.11(b), and calculate I':

$$I' = \frac{20\ \text{V}}{4\,\Omega + 2\,\Omega} \approx 3.33\ \text{A}$$

2. When a 6 A current source is applied to the circuit, replace the 20 V voltage source with a short circuit as shown in Figurec 12.11(c), and calculate I'':

$$I'' = -6\ \text{A}\frac{2\ \Omega}{2\ \Omega + 4\ \Omega} = -2\ \text{A} \quad \text{(the current} - \text{divider rule)}$$

(The 6 A current is negative due to its assumed direction to be opposite to I''.)

3. Calculate the sum of currents I' and I'':

$$I = I' + I'' = 3.33\ \text{A} + (-2\ \text{A}) = 1.33\ \text{A}$$

Example 12.8: Determine the voltage across the two terminals a and b in Figure 12.12(a) by using Thevenin's theorem.

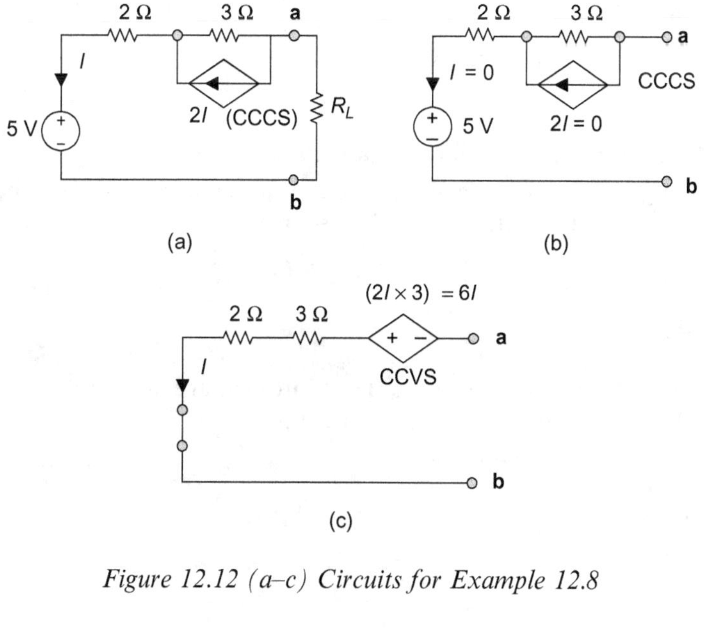

Figure 12.12 (a–c) Circuits for Example 12.8

Solution: The procedure for using Thevenin's theorem (chapter 5, section 5.2) to the above circuit is as follows:

1. Open and remove the load branch resistor R_L, and mark a and b on the terminals of the load branch as shown in Figure 12.12(b).
2. Determine Thevenin's equivalent voltage V_{TH}: Since the branches a and b are open, $I = 0$, and the CCCS is also 0 ($2I = 0$) in the circuit of Figure 12.12(b), so:

$$V_{TH} = V_{ab} = 5\text{ V}$$

3. Determine Thevenin's equivalent resistance R_{TH}: Replace the 5 V voltage source with a short circuit and convert CCCS to CCVS as shown in Figure 12.12(c).

$$R_{TH} = R_{ab} = \frac{V_{ab}}{I} = \frac{6I + (2\,\Omega + 3\,\Omega)I}{I} = 11\,\Omega$$

4. Plot Thevenin's equivalent circuit as shown in Figure 12.12(d).

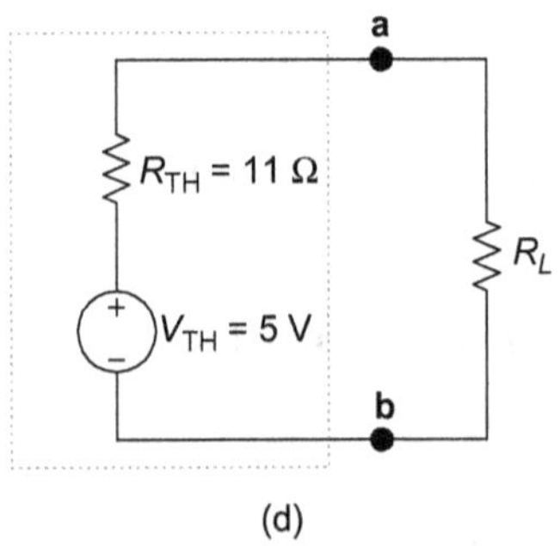

Figure 12.12 (d) Thevenin's equivalent circuit

Analysing circuits with dependent sources

Analysing circuits with dependent sources is similar to the methods of analysis for circuits with independent sources.

Summary

- Dependent (controlled) sources: The source voltage or current is a function of other voltage or current in the circuit.
  - VCVS
  - VCCS
  - CCVS
  - CCCS

- Equivalent conversion of dependent sources is the same as the equivalent conversion of independent sources:
  - controlled current source→controlled voltage source:

$$R_S = R_S, \quad \dot{V}_S = \dot{I}_S R_S$$

  - controlled voltage source→controlled current source:

$$R_S = R_S, \quad \dot{I}_S = \frac{\dot{V}}{R_S}$$

- The method of analysis for circuits with dependent sources is similar to the methods of analysis for circuits with independent sources (ideal sources).

Appendix A
Greek alphabet

Uppercase/lowercase	Letter	Uppercase/lowercase	Letter
Α α	Alpha	Ν ν	Nu
Β β	Beta	Ξ ξ	Xi
Γ γ	Gamma	Ο ο	Omicron
Δ δ	Delta	Π π	Pi
Ε ε	Epsilon	Ρ ρ	Rho
Ζ ζ	Zeta	Σ σ or ς	Sigma
Η η	Eta	Τ τ	Tau
Θ θ	Theta	Υ υ	Upsilon
Ι ι	Iota	Φ φ	Phi
Κ κ	Kappa	Χ χ	Chi
Λ λ	Lambda	Ψ ψ	Psi
Μ μ	Mu	Ω ω	Omega

Appendix B

Differentiation of the phasor

For a sinusoidal function $f(t) = F_m \sin(\omega t + \psi)$, taking the derivative of $f(t)$ with respect to t gives

$$\frac{df(t)}{dt} = F_m\omega \cos(\omega t + \psi)$$

and

$$\frac{df(t)}{dt} = F_m\omega \sin(\omega t + \psi + 90^\circ)$$

$$= J_m\left[\omega F_m e^{j(\omega t+\psi+90^\circ)}\right] = J_m\left(\omega F_m e^{j\omega t} e^{j\psi} e^{j90^\circ}\right) = J_m\left(j\omega \boldsymbol{F} e^{j\omega t}\right)$$

There

$$\boldsymbol{F} = F_m e^{j\psi}$$

and $e^{j90^\circ} = j$ (From Euler's formula, $e^{j90^\circ} = \cos 90^\circ + j\sin 90^\circ = j$)

Therefore, the phasor of $df(t)/dt$ is $j\omega\boldsymbol{F}$ (there $e^{j\omega t}$ is the rotating factor), i.e.

$$\frac{df(t)}{dt} \Leftrightarrow j\omega\boldsymbol{F}$$

Therefore, the derivative of the sinusoidal function with respect to time can be obtained by its phasor $\boldsymbol{F}$ multiplying with $j\omega$; this is equivalent to a phasor that rotates counterclockwise by 90°on the complex plane (since $+j = +90^\circ$).

Bibliography

Robert L. Boylestad. *Introductory Circuit Analysis*. 7th edition. New York, NY: Merrill (an imprint of Macmillan Publishing Company), 1994.

Allan H. Robbins and Wilhelm C. Miller. *Circuit Analysis: Theory and Practice*. Albany, NY: Delmar Publishers, 1995.

Thomas L. Floyd. *Principles of Electric Circuits*. 7th edition. Upper Saddle River, NJ: Prentice Hall, 2003.

Charles K. Alexander and Mathew N.O. Sadiku. *Fundamentals of Electric Circuits*. Boston, MA and London: McGraw Hill, 2000.

David A. Bell. *Electric Circuits: Principles, Applications, and Computer Analysis*. Englewood Cliffs, NJ: Prentice Hall, 1995.

Nigel P. Cook. *Introductory DC/AC Circuits*. 4th edition. Upper Saddle River, NJ: Prentice Hall, 1999.

S.A. Boctor. *Electric Circuit Analysis*. Englewood Cliffs, NJ: Prentice Hall, 1987.

Robert T. Paynter. *Introductory Electric Circuits*. Upper Saddle River, NJ: Prentice Hall, 1999.

Richard J. Fowler. *Electricity: Principle and Application*. 6th edition. New York, NY: McGraw-Hill, 2003.

Electronics Technician Common Core Learning Guider. Burnaby, BC: Ministry of Education, Skills and Training and the Centre for Curriculum, Transfer and Technology, 1997.

David Buchla. *Experiments in Electronics Fundamentals and Electric Circuits Fundamentals*. 6th edition. Upper Saddle River, NJ: Prentice Hall, 2004.

Allan H. Robbins and Wilhelm C. Miller. *Circuit Analysis: Theory and Practice (Laboratory Manual)*. Albany, NY: Delmar Publishers, 1995.

David A. Bell. *Laboratory Manual for Electric Circuits*. Englewood Cliffs, NJ: Prentice Hall, 1995.

Mark Edward Hazen. *Laboratory Manual for Fundamentals of DC and AC Circuits*. Orlando, FL: Saunders College Publishing, 1990.

Lorne MacDonald. *Basic Circuit Analysis for Electronics: Through Experimentation*. 3rd edition. Chico, CA: Technical Education Press, 1998.

Qiu, Guangyuan. *Electric Circuits* (Revised edition). Beijing: People's Education Press. 1983.

Qian, Jianping. *Electric Circuits*. Nanjing: Nanjing University of Science and Technology. 2003.

Web resources

http://www.job-search-engine.com/keyword/electric-electronic-engineering?page=11
http://www.ieeeusa.org/careers/yourcareer.html
http://en.wikipedia.org/wiki
http://www.glenbrook.k12.il.us/gbssci/phys/Class/circuits/u9l3b.html
http://en.wikipedia.org/wiki/Electrical_resistivity#Table_of_resistivities
http://en.wikipedia.org/wiki/Siemens_(unit)
http://www.kpsec.freeuk.com/breadb.htm
http://www.hiviz.com/kits/instructions/breadboard.htm
http://ourworld.compuserve.com/homepages/Bill_Bowden/resistor.htm
http://en.wikipedia.org/wiki/Electronic_color_code
http://www.electronics-radio.com/articles/test-methods/meters/multimeter-resistance-measurement.php
http://www.ee.nmt.edu/~rhb/ee101/labs/2lab/2lab.html
http://www.nanotech-now.com/metric-prefix-table.htm
http://www.wisc-online.com/objects/index_tj.asp?objID=DCE1802
http://www-ferp.ucsd.edu/najmabadi/CLASS/MAE140/NOTES/analysis-2.pdf
http://www.tpub.com/neets/book2/3b.htm
http://www.tpub.com/content/neets/14193/css/14193_138.htm
http://www.allaboutcircuits.com/vol_1/chpt_13/4.html
http://oscilloscope-tutorials.com/Oscilloscope/DisplayControls.asp
http://books.google.ca/books?id=ZzHudUMb7WAC&pg=PA26&lpg=PA26&dq=Dependent+voltage+Source&source=web&ots=78V2Df3xV5&sig=ZKXLpVNMsqZyxiRlYdI858fd84s&hl=en&sa=X&oi=book_result&resnum=2&ct=result#PPA26,M1
http://books.google.ca/books?id=avEjv8zAhQkC&pg=PT615&lpg=PT615&dq=control+coefficients+voltage+controlled+sources&source=web&ots=QjyOKEW-Dv&sig=H_gLC4hWUTo192z9n9B5djI24w8&hl=en&sa=X&oi=book_result&resnum=9&ct=result#PPT615,M1

Index